A HISTORY OF
THE FROZEN MEAT TRADE

THE S.S. *Strathleven.*

(From a Painting in the possession of Messrs. Burrell and Son, of Glasgow.)

Frontispiece.

A HISTORY OF THE
FROZEN MEAT TRADE

AN ACCOUNT OF THE DEVELOPMENT AND PRESENT DAY METHODS OF PREPARATION, TRANSPORT, AND MARKETING OF FROZEN AND CHILLED MEATS

BY

JAMES TROUBRIDGE CRITCHELL

AND

JOSEPH RAYMOND

Foreword by
Theodore A. Raymond, Hon.M.Inst.R.

WITH ILLUSTRATIONS AND DIAGRAMS

1969
DAWSONS OF PALL MALL
London

This reprint authorized by
Constable and Company Ltd.

First published 1912
Reprinted with new foreword 1969

Dawsons of Pall Mall
16 Pall Mall, London, S.W.1.

SBN: 7129 0423 9

Printed in Great Britain
by Photolithography
Unwin Brothers Limited
Woking and London

FOREWORD

SINCE the original edition of "The History of the Frozen Meat Trade" by Critchell and Raymond, published by Constable's in 1912, went out of print in the late thirties, there has continued to be a strong demand for second-hand copies that occasionally come to light in bookshops in and around the Charing Cross Road (and even less frequently elsewhere).

It is most satisfactory to realize that Dawsons of Pall Mall have sensed this need and have acted upon it by publishing a further sizeable edition.

Since meat, let alone frozen meat, cannot be described as a particularly glamorous commodity, some may wonder at the interest this volume has aroused.

Suffice it to say that the work of the great pioneers in mechanical refrigeration in the nineteenth century saved this country from starvation throughout three momentous eras—the "Hungry Forties" and beyond, the outcome of the Industrial Revolution—the First World War when the Kaiser's U-Boats were decimating the merchant carriers of meat to these shores and, finally, the Second World War when Hitler continued the Kaiser's marine strategy. Unquestionably, without the protein derived from imported frozen meat the inhabitants of these isles would have degenerated into a "C3" nation.

So, "The History of the Frozen Meat Trade" *is* a glamorous story. As the son of one of the co-authors of this book, I have pleasure in recommending it, not only as fascinating reading but as an invaluable reference volume to the practices of yesteryear upon which the imported perishable food trades have been built up.

Theodore A. Raymond,
Hon. M.Inst.R.

PREFACE

Much that is extremely interesting, and a little that is romantic, is to be found in the history of the industry to which this book is devoted, and the authors hope that the records which appear in these pages of the rise and progress of the frozen and chilled meat trades will justify the publication of this volume. Many works have appeared describing the beginnings and developments of the great wool industry, but nothing of an historical and exhaustive nature has ever been published in book form (although pamphlets and newspaper articles without number have been issued) about the frozen meat business. To provide food is, at least, as important as to supply clothing, and the one industry lends itself to description as readily as the other.

The *personnel* of the pioneers who worked out the practical and technical problems of the preparation and transport of frozen meat—successfully sometimes, frequently otherwise—included men of a high order of intellect and character. To read Mr. Mort's speech, delivered on September 2, 1875, at the Lithgow Valley Works (see p. 20), is to be thrilled with some sense of the exaltation of spirit which must have inspired his hearers in contemplating the world-wide benefit to follow upon the exportation of Australia's surplus of meat, the hoped-for era so eloquently forecasted by Mort. The engineers engaged upon the freezing formulæ, the shipowners endeavouring to alter their system to grapple with the new position created by the *Strathleven* shipment, the merchants and bankers in London applying themselves to the most important part of the whole trade—marketing the meat—deserve recognition in the printed page before the lapse of time destroys all records. It has been difficult enough, after thirty years from the starting point, to procure data sufficiently reliable to justify the title of this book.

A chapter was written entitled "The Case for Frozen Meat," setting forth specifically the benefit which the new trade has brought to the world at large, but on consideration it was perceived that the whole of the book, including the illustrations, diagrams, and tabular and graphic matter in the appendices, formed a most complete "Case for Frozen Meat," and the authors venture to hope that their volume will assist, to some degree, in popularizing still further the use of frozen and chilled meat in the Northern Hemisphere, inasmuch as it draws attention to the able and distinguished men who pioneered the trade, to the sound quality and drastic inspection of the meat before export, and to the excellent system under which the transport and marketing systems are conducted.

The chapters have been written in a plain style, and technical treatment has been avoided in favour of the general treatise form. Great efforts towards accuracy have been made, and if some errors have crept in, the indulgence of readers is asked for. It has been, of course, the authors' desire to achieve impartiality in discussing the work of the pioneers and in touching on the later developments in which gentlemen and business houses engaged in the trade have been and are concerned.

The part which the journals published in Australia and New Zealand took in helping forward the early efforts of the pioneers by opening their columns to articles and full discussion, deserves a special word of acknowledgment. At a critical time, undoubtedly, this publicity was of considerable assistance to the growing industry.

In order to procure information on trade questions, the authors, wherever possible, have gone to the fountain head, and they owe a heavy debt of gratitude to the many gentlemen and firms who have been appealed to. Hence they desire in the fullest and frankest way to return thanks for the prompt replies to the thousands of letters (over 4,000 in all) written to all parts of the world—only in a few cases has information been withheld. Without such kindly assistance there would not have been any chance of this book being prepared. In particular, acknowledgment is due to some of those whose names appear in the biographical section; to Mr. Gilbert Anderson, Mr. George Goodsir, and Mr. P. B. Proctor, whose statistical reports have been drawn upon; and to the leading firms of importers, agents, and merchants, in Great Britain, who have supplied practical commercial details for certain sections of the work. The special assistance of the following gentlemen has also to be acknowledged: Mr. R. H. Rew, Assistant Secretary of the Board of Agriculture and Fisheries: Mr. A. Scott, Secretary of Lloyd's Register of British and Foreign Shipping; Dr. Sergio Garcia Uriburu, Consul General for Argentina in London; Mr. H. W. G. Millman, Clerk and Superintendent of the London Central Markets; Captain T. R. Mowat, of Messrs. Johnson's Sons and Mowat; Mr. Hal Williams, M.I.Mech.E., M.I.E.E.; Mr. M. T. Brown, B.Sc.; Mr. E. R. Baines, of the Port of London Authority; Mr. T. Douglas Huggett, of the London *Daily Telegraph*, and Mr. Louis H. Furniss, Secretary of the Incorporated Society of Meat Importers.

The plan on which the book has been prepared is, broadly, to touch on the historical part, with a sketch of the conditions which gave birth to the industry, and the personages and leading events figuring in the frozen meat story as refrigeration's aid was found practicable for conveying the surplus meat of the Southern Hemisphere to supply the scarcity of the Northern. The sequence of chapters takes readers to the consideration of the commercial methods under which frozen and chilled meat is transported and handled, and sold in Great Britain. The endeavour mainly has been to describe the various stages through which frozen meat passes from the time it is placed on board the steamer until it reaches its predestined goal, the meat retailer's shop in England or Scotland.

With regard to the pictures appearing in the book, freezing works are more useful than ornamental, but it has been thought that to give illustrations of some of these works would be a fitting accompaniment to the written descriptions of the meat freezing industry. Ships, real argosies of the ocean when food-freighted on account of the workers of the Homeland, need no apology for their presentation in picture form. Many of the pioneers' portraits will be found in the book, and the photographs of some of the leading merchants and importers at present engaged in the trade are also given. It is fitting that in a work of this kind these men, whose business energy and capital have done so much to build up the trade to its present commanding position, should receive this special recognition at the hands of the authors of the History of the Frozen Meat Trade.

The Index and Appendices have been compiled in a very detailed form, and it is hoped that the particulars therein contained, as well as the information recorded in the twenty-nine chapters, will make this volume a useful and necessary work of reference.

The authors acknowledge, with thanks, the courtesy of the proprietors of the journals named below, in giving permission for the use of photographs:—*Engineering*, for the two views of the La Negra Works on p. 82; the *Cold Storage and Produce Review*, for the portrait of M. Charles Tellier; *Syren and Shipping* for the illustration of the *El Zarate* on p. 344; *Ice and Cold Storage* for the picture of the insulated van on p. 344; and the *Canterbury Times*, New Zealand, for the page view of the Islington Works of the Christchurch Meat Company on p. 66.

LONDON,
Easter, 1912.

CONTENTS

CHAPTER XXIX.

APPENDICES.

Appendices—*continued.*

ILLUSTRATIONS

A HISTORY OF THE FROZEN MEAT TRADE

CHAPTER I

BRITAIN'S CALL FOR OVERSEAS SUPPLIES: SHORTAGE AND SURPLUS

It must be rather difficult for those who are engaged in the various branches of the great overseas frozen meat trade to realize that only thirty years separate them from the time of the foundation of what is now recognized as one of the most important of the world's industries. A commerce that unites continents, and is an essential factor in the progress of human civilization to-day, might well be believed to be more than a generation old, but it does not need a student of history to recall that this is not the case.

The industry that hangs on the slender piston-rod of a refrigerating machine, yet feeds nations with a regularity that defies famine, had its birth within the recollection of many who are not yet old. The great changes it has wrought in the world during the first three decades of its existence make all the more interesting some preliminary inquiry into the circumstances which led to its establishment.

Before coming to the conditions which directly gave rise to and immediately preceded the refrigerated meat industry, it may be well to deal somewhat fully with two subjects not completely bearing upon the title of this book, though not by any means foreign to it, (1) the dwindling flocks and herds in

Great Britain for some decades prior to the frozen meat era; and (2) the canned meat industry in Australasia. This trade represented the first attempts of the colonists to ship beef and mutton to the old country, and was the evolutionary form of the great meat industry, the real and lasting development of which was to be the preservation and export of meat under conditions of refrigeration.

A Nation's Need.

It was, of course, the demand for greater meat supplies that called for the discovery of some means for the safe delivery of those supplies. The uneasiness that was felt respecting the food supply in England became very marked by the early fifties, and the gradual growth of the manufacturing industries made it clear that Great Britain must be an increasingly important meat consumer. Manufacturing demanded energetic, flesh-fed men, but meat supplies and their prices were on a most unsatisfactory basis for a generation before the establishment of the great overseas dead meat trade.

Official statistics of the flocks and herds of the United Kingdom at the middle of the nineteenth century do not exist; in fact, no official enumeration of live stock was made in the United Kingdom before 1867. But the estimates of Mulhall and McCulloch were, no doubt, fairly accurate. According to Mulhall, in the ten years 1851—1860, the decade in which meat imports were first brought into the United Kingdom, the average production of meat—beef, mutton, and pork—in England, Scotland, and Ireland was 910,000 tons, which gave 72 lbs. per head per annum; this was supplemented by an average import of live cattle furnishing 44,000 tons, making in all 75 lbs. of meat per head of population yearly. In the 1861—1870 decade the average home production of meat had increased to 1,036,000 tons, and imported meat to 131,000 tons. It is significant that the home production of these meats showed no increase from that point up to the days of the beginnings of frozen meat imports. In the year 1882 the home production was 1,090,000 tons, and the Continental and overseas supplies had grown to 654,000 tons; these quantities yielded a per capita annual supply of meat equal to 110 lbs., of which the imported supply claimed 43 lbs. The home-produced meat in 1882 consisted of 690,000 tons of beef, 305,000 tons of mutton, and 95,000 tons of pork, and the imported meat was still mainly in the form of live cattle and sheep. The population of the United Kingdom in 1851—1860 averaged 28,265,000, and had grown by 1882 to 35,606,000.

The following totals of the food animals of the United Kingdom—cattle, sheep, and pigs together—may be useful as a record and as showing the decline in home stocks between 1867 and 1880, the arrest of that fall, and restoration to the figures of 1867, brought about by the importation of live and dead meat in the period of thirty years up to 1910:—

1851—1855	40,676,000	(Mulhall)
1867	46,770,524	(official)
1880	42,974,261	,,
1910	46,491,521	,,

So, prior to the introduction of frozen meat, supplies of home stock were being overtaken by consumption, and it was plain that the inhabitants of England would have to be content with less meat or pay fancy prices for it, or arrange for largely increased supplies of dead meat to be brought across the seas to be sold at a moderate price.

Mulhall helps us again in showing how, with stagnant home supplies of meat, the price advanced. In the ten years 1851—1860, the "average" of the wholesale "prices" of first quality meat—beef, mutton, and pork together—was $6\frac{1}{2}$*d.* per lb., during the next decade it had risen to 7*d.* per lb., and in 1882 the "price" was $8\frac{7}{8}$*d.* per lb. A parliamentary return issued in 1911 gives the following average prices for beef of first quality:—1851, $4\frac{3}{4}$*d.* per lb.; 1861, $6\frac{3}{4}$*d.* per lb.; 1871, 8*d.* per lb.; and 1881, $8\frac{1}{4}$*d.* per lb.

It is interesting to examine the meat import movement into the United Kingdom of the last fifty years. For the quinquennial period 1861—1865 the average quantity of fresh beef, mutton, and pork imported amounted to only 0·1 lb. per head of the population. The coming of frozen meat in 1880 sent up the average imports for 1881—1885 to 3·5 lbs. For the five years 1891—1895 the average imports were 12·4 lbs., and for 1906—1910 each unit of the population was provided for to the extent of 28 lbs. of fresh meat imported from British possessions and foreign countries.

Harking back to the days of 1850–1860, the difficulty that lay ahead had not escaped official notice, for about 1860 the Privy Council discussed the question of the national food supply, and numbers of societies and institutions followed the lead thus given. In 1863 the Privy Council laid down a rule "that, to avoid starvation diseases, the weekly food of an average adult must contain 28,600 grains of carbon and 1,300 grains of nitrogen." Dr. Brown, in "The Food of the People," published in 1865, wrote: "The plague spot, the skeleton in the closet of England, is that her people are underfed." This condition of things was accompanied by the abuse which one would expect, terrible and shameless adulteration, and the poor were further defrauded by traders giving short weight.

The most practical step in the direction of providing a more ample food supply was the formation of a committee of the Society of Arts, which first met on December 21, 1866. Amongst those present were Messrs. H. C. E. Childers, M.P., Harry Chester, W. Ewart, M.P., Benjamin Shaw, and Lord Robert Montagu, M.P. As early as 1853 Mr. Chester, in delivering the centenary address of the Society, asked why Australia should be content with exporting wool and tallow, "and not the mutton itself to the hungry masses of this country?" The proceedings at the committee's meetings make most interesting reading. The committee subdivided itself into four sections: meat, milk, fish, and cooking. That was the time when canned meats were on their trial. Mr. C. G. Tindal, Mr. Robert Tooth, Mr. McCall, and others, were examined, and explained their processes. Dr. Bancroft's "pemmican" and Mr. Alexander's powdered beef were tested. (Both these gentlemen were Queenslanders.) The committee found that, weight for weight, the dried beef was four times more nutritious than ordinary beef. It was stated that 200 patents had been registered for preserved meat processes, so widely recognized was the scarcity of meat in England. All sorts of ideas were expounded to the committee. Medlock and Bailey averred that by dipping meat in their bisulphide of lime solution "anything of animal origin, from a beefsteak to a bullock, from a whitebait to a whale, can be preserved sweet for months." Possibly the bullock and the whale might have objected! C. Nielson proposed to fix blood in the form of sausages, puddings, cakes, and so on. The Rev. M. J. Berkeley delivered a stirring address on fungi, but somehow the mushroom palliative failed to impress the committee as a substitute for the roast beef of Old England. De la Peyrouse's idea was to pack meat in barrels, and to pour in fat at a temperature of 300° F. all round the stored viands. Professor Gamgee loomed large, and his method, though revealing a touch of Max Adeler, certainly possessed genius. He suggested that cattle should be happily dispatched by being made to inhale carbonic oxide gas, at a cost of 2*s*. to 3*s*. per animal. The flesh of oxen so slain was declared to retain its fresh and bright appearance, and the committee reluctantly and warily tasted chops from a sheep killed in this way, reporting, doubtless to the chagrin of the Professor, that the meat was "slightly flat." A tin of meat forty-one years old, from the stores of H.M.S. *Blonde,* was tested and found sound. Professor Redwood advocated raw meat preserved in paraffin.

Scores of different processes for tinning meat were tested. Dr. Hassall's "Flour of Meat," Australian "mutton hams," meat dried by sulphurous acid, and many other inventions were put before this committee, evidence which contained the germs of many of the modern methods of preserving and handling animal substances for food. The committee's records are packed with good things. For instance, we hear of the stimulus given to the Australian tinned meat trade by the Franco-German war. In October, 1875, two huge tin-lined cases of meat arrived from Melbourne, 30 lbs. of meat, in joints, in each case. The meat was wrapped in prepared calico, and the whole packed in charcoal. Alas! on opening, the

Melbourne meat "was found to be in an advanced state of decomposition." As far back as 1843 the Society of Arts gave its medal for attempts made in Australia to render down the lean of meat by means of the water bath and the introduction of the extract in a solid form. Samples of this extract were sent on a voyage to Buenos Aires and back, and failure resulted. This, the authors believe, was the earliest attempt made in Australia to preserve meat for export.

It was a matter for regret that this committee of the Society of Arts, after a vigorous and most useful campaign of fifteen years, came to a sudden stop in 1881. In that year the committee delivered a gloomy report, and found itself unable to award the £100 prize which Sir Walter Trevelyan had presented for the best means of preserving fresh meat. This £100 was disposed of by being divided into five sums of £20 and granted to food and cooking exhibits at the 1884 Health Exhibition. Without doubt, the introduction of frozen meat in 1880 settled the whole difficulty which the Society of Arts' committee had spent so many years in trying to solve, and it could only have been blindness to facts—the success of the *Strathleven's* trial was common knowledge in 1881—that made the committee in its report neglect its obvious duty of stating that the introduction of frozen meat removed all its difficulties. Emphatically, Sir Walter Trevelyan's £100 prize should have been awarded to Thomas Sutcliffe Mort, of Sydney, as Chapter II. of this volume will show.

It may be remarked at this point, where attention has been drawn to the necessity of Britain, already fully stocked with cattle and sheep, looking abroad for her further needful food supplies, that, apparently, France, Germany, Austria, Italy, and Switzerland are reaching, or have now reached, the same stage in their economic development. In the early part of the twentieth century the peoples of these Continental countries are making the same investigations as to the nature and source of their future imported meat, and entering upon the same struggles in their initial efforts to secure supplies from the lands of the south, as did the inhabitants of England in the sixties and seventies.

Surplus of Live Stock in the Southern Hemisphere.

So much for a nation's need and the experimental and unpractical efforts made to relieve it. Next may be considered the circumstances in the countries in the Southern Hemisphere, Australia, New Zealand, and Argentina, from which, to a limited extent by means of meat preservation, and eventually more thoroughly through refrigeration's aid, the real relief was to come.

In Australia and New Zealand in the seventies sheep had increased so rapidly in number that it was apparent that, as a wool producer only, the full value of the animal was not being realized. The flocks grew, the population remained small, and there was no means of adequately dealing with the surplus stock. It was in the eventful period 1868 to 1879 that the frozen meat trade had its genesis, when Harrison and Mort in Australia, and Tellier and Carré in France, were experimenting, and when the successful voyage of the *Strathleven* set all doubts at rest as to whether the surplus meat of the New World could be brought in a perfectly fresh and sound state to supply the shortage in the Old.

Tinned meat export had been instituted in Australia prior to the establishment of the frozen meat trade, and of this collateral industry some brief particulars as to its pioneering may be given here.

Apropos of meat tinning or canning, it is interesting to note that the first person to preserve meat in closed jars by employing heat was a Frenchman, Appert. Earliest mention of his process occurred in 1809. At first glass bottles were used, and soon afterwards iron tins (English patent 3,310, 1810, Heine). The soldered tins made of tinplate were introduced through the French brothers Pellier about 1850. The great German chemist, Liebig, gave a lead to the Australian pioneers of the canning industry by applying chemistry to the invention of extract of beef. Replying to an inquiry as to whether Liebig ever studied the problem of applying cold to meat storage and transport, Dr. L. Geret, of the Liebig Co.'s scientific department, Antwerp, reports that Liebig never considered refrigeration in this connection. In Liebig's "Chemical Letters" (Vol. XI., 32nd letter, p. 139, 4th edit.) he alludes to the surplus cattle in Australia and the River Plate as "merely export material for tallow and hides production." Liebig's first experiments on a commercial scale in the manufacture of extract were carried out in the River Plate about 1863 or 1864. Mr. G. C. Giebert was the first general manager in South America of the Liebig's Extract of Meat Co.

The first men to preserve meat by practical tinning methods in Australia were the late Henry Dangar, of the Hunter River, New South Wales, and his brother, the late William Dangar, of Turanville, Scone. Discontented with the wretchedly low values of cattle—they had sold a mob of splendid bullocks in Sydney for £2 12*s.* 6*d.* a head in 1846—Messrs. Dangar set about starting tinning works. Towards the end of 1847 they began operations at Honeysuckle Point, near Newcastle, New South Wales. Mr. Charles Gedye was manager; the meat was packed in 4 lb. and 6 lb. tins, and hermetically sealed, the tins being painted and labelled in London. The meat (beef, mutton, tongues, and soup and bouilli) sold readily in London, and the Admiralty took quantities of it. The factory was carried on with success—meat at first cost only ¾*d.* per lb.—but the gold discoveries in 1851 raised the price of cattle to such a prohibitive figure that the works were closed early in the fifties, and now Honeysuckle Point is covered with dwelling-houses. Messrs. H. E. and M. Moses, of New South Wales, were also canning meat about that time.

In 1850 there were 110 boiling down establishments in Australia, and the production of tallow was enormous. The sheep slaughtered for this purpose numbered 800,000, cattle 73,000: the tallow exported in that year was close on 11,000 tons, valued at £301,000. At one works (Russell's, Hunter River) 12,000 sheep were boiled down in four weeks. In 1851 the scale of boiling down enterprise greatly increased, and probably about 10 per cent. of all the sheep in Australia fell victims to this wasteful process in that year. Dr. Lang, one of Australia's ablest pioneers in the problems of public life, religion, and education, inveighed against "this wholesale and enormous destruction of valuable animal food going on in New South Wales for eight years. . . . Viewed in connection with the fact that there are millions 'at home' on the brink of starvation, this destruction is discreditable to Great Britain and her rulers, and cannot but be peculiarly offensive in the sight of Heaven."

The tinning trade was certainly a useful outlet for the surplus sheep and cattle in the early days. A few years before the time when the millions of visitors to the Great Exhibition of 1851 were confronted with Australian tinned mutton, cattle in the grazing regions of Australia were worth only £2 to £4 per head, but the gold discoveries of 1851 proved a time of blessing to cattle owners, prices then going up to £8 to £10 per head. But the increasing flocks were ever a problem to the Australian sheep farmer. Boiling down for tallow, the earliest method of supplementing the pastoralist's returns from his wool, was a crude system, with strict limitations. The following table gives some idea of how the pastoralists and graziers of Australia stood with regard to their local market, and why they were forced to find markets abroad :—

—	Population.	Cattle.	Sheep.
1851 . . .	403,889	1,894,834	15,993,954
1861 . . .	1,153,973	3,846,554	20,135,286
1871 . . .	1,668,377	4,277,228	40,072,955
1881 . . .	2,252,617	8,010,991	65,078,341

After Messrs. Dangar's meat tinning enterprise of the late forties, not much appears to have been done in Australia till 1865 and 1866. In the former year Mr. Robert Tooth began making extract at Yengaree, New South Wales, and Tooth's Extract of Meat Co. (London Bridge) states that Messrs. Allen and Hanbury were the consignees of the first parcel to be imported, on July 24, 1866. About 1875 Mr. Tooth retired from the business, and devoted himself to sugar growing in Manila. Mr. C. G. Tindal, a pioneer in meat preservation, studied Liebig's works when he was a young man, and came across a chemist at Clapham named Deane, who was making splendid extract and selling it at 35*s*. per lb.; Brady, of Newcastle, and Reynolds, of Leeds, were making a cheaper extract. Mr. Tindal worked with Deane for some time, and then began making Liebig's extract of meat at Ramornie, New South Wales, on September 13, 1866. From that year manufacture and export have been continuous. Mr. Alban Gee, the present manager of the Sydney Meat Preserving Co., went out from England to Ramornie in 1866, and the late Mr. Thomas Cordingley went to the same place in 1872. In 1875 Mr. Cordingley started preserving mutton in Botany Bay, and later, supported by Mr. Tindal, he formed the North Queensland Meat Export Co. at Alligator Creek, North Queensland, a successful concern now working under the management of Mr. Harold Cordingley. Mr. C. G. Tindal, in answer to an inquiry, writes as follows: "I hold an autograph letter from Baron Liebig on the subject of making his extract, which has been made use of in two trials at law, and which established the right of myself and other makers to call the extract we make 'Liebig's Extract' throughout Great Britain. But the Law Courts on the Continent decided against us."

The Melbourne Meat Preserving Co. was formed in 1868, Mr. S. S. Ritchie, who had been a partner with Mr. John McCall, of London, being mainly instrumental in its establishment. Messrs. J. McCall and Co. were intimately associated with the building up of the tinned meat trade.

The Sydney Meat Preserving Co., Ltd., of Sydney, which was a concern established in 1869 for the purpose of preventing the fluctuation of the price of fat stock brought to the Sydney market, has a paid-up capital of £20,315, and reserves amounting to £67,684, but no dividends are paid upon the capital. The mode of operation is for persons who send stock to Flemington, Sydney, to allow the Sydney Meat Preserving Co. a rebate of 2½ per cent. on its purchases; this, in ordinary years, means an eighth of a penny per lb. The payment of this rebate is not a compulsory charge, and is not allowed by all the persons who sell stock, but all the larger stockowners allow the rebate to the company. The purchases of stock vary, of course, from year to year, in accordance with the condition of the market; if the markets are good, very little stock is purchased. The account up to June 30, 1910, shows that during the previous half-year the company bought 498,509 sheep and 1,472 cattle. The stock purchased by the company are killed at its yards. None of the meat is sold, but the whole of the carcasses are canned or turned into tallow, so that the purchases of the company do not come into competition with those of the ordinary butchers.

Australian canned meat began to be known to the British public by the year 1867. Before that time preserved meat had only been used by the services, by explorers, and on sailing ships, but it appealed to the public very soon after the first imports came along—the Midlands took a great fancy to it, and clamoured for it. Shipments were easily absorbed. No American meat of any kind was imported during the sixties, and Australia was the pioneer of the tinned as she was of the frozen meat trade. South American tinned meat was first imported in 1871. Boiled mutton was the principal article turned out when the business began, and corned beef was also shipped. The growth of this trade was remarkable; in 1867 the United Kingdom's imports from Australia were 286,526 lbs., in 1868 no less than 878,444 lbs., while in 1869 they advanced to 2,000,000 lbs. In 1880 Great Britain imported 16,000,000 lbs. of canned meat.

The position in New Zealand was much the same as in Australia, there being a large surplus of sheep which the small population was quite unable to deal with profitably. After shearing it was not an uncommon thing for the old and inferior animals to meet the fate of the Gadarene swine. Boiling down works were the first means introduced to deal with this surplus, and on many of the stations in New Zealand are still to be seen the primitive plants erected for that purpose. There was, of course, enormous waste; the sheep were kept during the flush of grass in the summer and then boiled down for tallow—wool and tallow were the only products. All sorts of plans were tried; legs of mutton packed in tallow were shipped to England, and although the meat arrived in good condition, the enterprise failed.

The next step was canning. About 1869 a large company, the New Zealand Meat Preserving Co., established works in various districts—Templeton, Styx, Kakanui, Washdyke, Green Island, and Woodlands, in the South Island, and in a few North Island centres, though at that time the North Island carried but few sheep. At these works the best joints were preserved, and the rendering of tallow from the rest of the carcass was also carried on. Preserving was rough and ready, as there was no chilling process available to hold the meat for any time. All the offal and the skin were wasted. Owing to unreliability of quality, the canning business did not pay, and all the works were ultimately closed down. The sheep industry in New Zealand at this period was unprofitable ; the surplus animals often went for 6*d.* or 1*s.* per head, and, as a rule, the measure of value was the skin on their backs. Various attempts were made to preserve meat by chemical and other means, but were not successful, and from 1865 to 1882 run-holders in New Zealand had a very bad time. It is a matter of great regret that very few of them were able to stand against the adverse conditions till the better day brought by the frozen meat export trade had dawned.

The population of New Zealand at the start of the frozen meat trade was about 500,000, and the statistics of New Zealand's herds and flocks from 1851 to 1881 are as follow :—

—	Population.	Cattle.	Sheep.
1851 . . .	26,707	68,000*	233,043
1861 . . .	99,021	193,285	2,761,383
1871 . . .	256,393	436,592	9,700,629
1881 . . .	489,933	698,637	12,985,085

* Approximate.

The Real Genesis of Meat Export.

Australia and New Zealand were not, of course, the only countries striving to " realize " on their surplus live stock, and the United States of America was the first country to inaugurate a meat trade dependent on artificially cooled storage during transport. That this was a welcome industry to America goes without saying, seeing that in 1874, when beef was first exported to Great Britain, cattle on farms in the United States of America numbered 27,000,000—the population was well under 50,000,000. Undoubtedly, the real genesis of the meat export trade under conditions of refrigeration is to be found in the shipments of chilled beef from the United States of America in the seventies; by the end of 1880, when only 400 carcasses of mutton had reached home from Australia, Great Britain had imported from North America 120,000 tons of fresh beef. (See pp. 190, 191, for other references to this trade.) But as the general conditions and lines of development of the North American chilled beef trade were so widely different from those associated with the Australasian and South American frozen meat trades, extended references to it do no come within the scope of this book.

Argentina: The Problem of the Pampas.

One can go back a long way in tracing the introduction and history of the cattle and sheep which roam the pampas of the Argentine Republic. The progress of this great industry, the backbone of Argentina's prosperity, must have been interwoven with adventure and romance; of this we get an occasional glimpse in reading the scanty literature which may be consulted by anyone wishing to acquaint himself with the intermediate steps between the fine freezing stock now entering the frigorificos and the first stock introduced into the Republic, which introduction took place in the sixteenth century.

One Captain Nuflo Chaves brought the first sheep to Argentina in 1550. As to horned stock, Juan de Salozer y Espinosa introduced in 1552 seven cows and one bull, which are said to have been the foundation of the mighty herds that are scattered over the Campo to-day. Spanish colonists soon settled in the Plate district, and Juan Torre de Vega y Aragon, recognizing the suitability of the pampas for stock breeding, distributed 4,000 cows and bulls and 4,000 sheep amongst the colonists. Soon the herds and flocks went beyond the needs of the small population, and even in those early days the question of export was mooted to relieve the congestion caused by overstocking. But the ideas of the cattle-owners did not soar beyond hides, to export which to Spain and Brazil the "Governor and Captain General of the province of the River Plate" issued licences. This was the position in 1616, when the ship *Our Lady of Refuge* left Buenos Aires with 1,281 hides, valued at 10,248 reales—about £117. So, three hundred years ago, pastoral products were of little account in the River Plate. From the first the herds and flocks multiplied enormously, and in the seventeenth century it was recorded that "all the wealth of these inhabitants consists in their animals, which multiply so prodigiously that the plains are covered with them, particularly with bulls, cows, sheep, horses, mares, mules, asses, pigs, deer, and other sorts, in such numbers that, were it not for the dogs that devour the calves and other tender animals, they would devastate the country" —a sort of internecine strife between the animals that preceded the latter day organized onslaught of the frigorifico.

Mr. Herbert Gibson, in his book on "The Sheep Breeding Industry in the Argentine Republic," states that "sheep were neglected and despised. They were almost classed with wild beasts and fowl, looked upon as public property, and allowed to roam at will, and increase or die off as the years were clement or severe." They were of two classes, the pampa sheep, descended from the mountain long-wools imported from Spain, and the criollo, the much degenerated descendants of the Spanish merinos. It is curious to note that the two great sheep countries of the world, Australia and the River Plate, introduced the improved Spanish merino at the same date. In 1794 Don Manuel José de Labarden exported ten rams and twenty ewes from Spain to the Banda Orientál (the old romantic name for Uruguay), which at that time was one of the provinces of the River Plate Viceroyalty. In 1813 Mr. Henry Lloyd Halsay imported 100 improved Spanish merino sheep and founded the first fine-wooled merino flock in the province of Buenos Aires. The first introduction of English sheep took place in 1825, with the purchase of thirty Southdowns, and the first Lincolns were imported in the forties. Particulars in detail of British pedigree stock imported into Argentina will be found on p. 98.

Harking back to the general position of live stock, the estancieros had now to consider how to find an outlet for the enormous herds of cattle and flocks of sheep that were running almost wild over the River Plate plains. Some bold spirits about 1717 started a large salting works at Buenos Aires, and the beef was exported. In 1794 the live stock breeders of Buenos Aires and Monte Video presented a petition to the Minister, Don Diego Cardogin, urging the free exportation of tallow and jerked beef, the trade to be assisted by the introduction of "eight or nine hundred Irishmen, bachelors and Roman Catholics." In the early eighties so great was the congestion of the sheep that in one case a flock was driven to the coast and a portion were precipitated over the cliffs into the sea. By 1822 the export trade in hides, tallow, and wool, had grown to a total for the year valued at $3,300,000. We find news of increasing quantities of salted meat passing through the Buenos Aires customs house for export from 1862 (357,860 quintals) to 1866 (430,781 quintals).

The importance of this trade to-day is seen when it is stated that nearly one-half of the cattle of Argentina depend on the up-river saladeros, where the number slaughtered exceeds the total number handled at the frigorificos—such is the importance of the Argentine salted and jerked beef business. The first Shorthorns were imported in 1865 by Don Juan N. Fernandez—a historic event. The foundation of the Argentine Rural Society quickly followed, and marked the systematizing of the laudable efforts of cattle farmers who had already commenced to convert the primitive cattle raising business into a well-organized and intelligent pastoral industry. It was from 1850 to 1860 that the importation of pure-bred stock was started on a commercial scale for the purpose of improving the herds of Argentina. In 1868 the Government of Argentina offered $8,000 for the discovery of a practical means of preserving fresh meat, and in 1877 the export duties on fresh meat were suspended for five years. In 1882 these duties were abolished. It may be interesting to give here the figures of the four official stock censuses that have been taken in Argentina :—

—	Cattle.	Sheep.
1875 . . .	13,337,862	57,501,261
1888 . . .	21,963,930	66,701,097
1895 . . .	21,701,526	74,379,562
1908 . . .	29,116,625	67,211,754

At this stage of the great Argentine Republic's live stock industry, the estancieros perceived that the means hitherto employed of dealing with the ever-increasing surplus of animals were both unscientific and obsolete. Boiling down, the last resort of the stock breeder, was found to be as unprofitable as it was wasteful, and in the seventies the sheep of Argentina had increased to such an extent that even this desperate remedy for the accumulating flocks failed to dispose of the surplus. The saladeros' profitable consumption of cattle was also found to be limited. The ideal before the stock breeders was (1) to use their fat cattle and sheep to good purpose financially ; (2) to handle the surplus in such a way as would lead to a trade at once permanent and increasing ; and (3) to see to it that the new outlets should involve a steady improvement in the standard of quality of the herds and flocks. It was perceived that only one avenue promised the realization of these conditions—the export of the Republic's stock to supply the needs of European countries, where fat sheep and cattle were few and men were many. So in 1874 the export of live cattle and sheep began, and by the end of 1879 close on 1,000,000 head of cattle and 165,000 sheep had been shipped away. On p. 75 appears a full statement of the rise and fall of this industry.

The essay in 1876 of the steamer *Frigorifique*, with Charles Tellier at the helm, was possibly the stimulus which helped to set going the Argentine frigorificos on their successful career. (But the success of the *Strathleven* venture had more to do with it.) The steamer *Paraguay*, two years later, with frozen meat on board for the Old World, kept the ball rolling, and prepared the country for the real start in 1883 of the great Argentine export trade in dead meat. When the River Plate Fresh Meat Co., at Campana, and Messrs. S. G. Sansinena and Co., at Barracas, built their meat works, the sheep of the country were by no means pretentious. "Woollies" were purchasable at $2 to $3 a head. The improvement of the Argentine sheep by the introduction of the pure-bred English Lincoln and other breeds was yet to come; the average weight of the frozen carcass was 35 lbs. This was confirmed by the fact that the Smithfield salesmen were wont to speak of the Argentine mutton carcasses when they first arrived as "rats." The advantage of exporting their stock in the form of dead meat, as compared with the live stock trade, was quickly appreciated by the cattle and sheep breeders of Argentina, who found, as the industry took root and expanded, that in the frozen meat trade they had found a way of dealing with their stock at once profitable, economic, and scientific.

As the improvement in quality, both of sheep and cattle, became general, the frozen meat from the Argentine commanded a better price in the home markets, and enabled the freezing companies to outbid the live-stock exporters even before the outbreak of foot-and-mouth disease in 1900 brought the trade of the latter to a termination. In subsequent years the average prices paid for wethers and steers have exceeded the highest ever obtained in the period when the live-stock exporter competed with the freezer, and the Argentine breeder is now persuaded that the dead meat trade is his most profitable market. The economy it represents is too obvious to merit discussion; the labour employed in the factories, the hides, tallow, and offal that remain at the Argentine end, the economy in space and in freight, all combine to secure for the exporting country the maximum quantity of the total value of the animal. In the improved methods of handling and carrying the dead meat, its collection in cold storage in the Argentine and its distribution in the European markets, the River Plate has achieved the most scientific application of its commerce.

CHAPTER II

THE WORK OF THE PIONEERS

WITH the light of less than four decades shining on the brilliant achievements of those who played the part of pioneers in the frozen meat industry, the task of according each name concerned the exact importance it bears in relation to succeeding progress is not an easy one. For instance, while the date on which one inventor patented a certain freezing process may be prior to the launching of a scheme by another, pioneering pride of place may belong to a third whose foresight of the ultimate situation was clear, and whose early work, therefore, was more material in setting the industry on its legs.

The work of the French chemist and engineer Carré must always be regarded for its early date ; James Harrison, whose record of early struggle, achievement, and failure, is tersely recorded in a Geelong cemetery, can never be forgotten; while the efforts of Mort in Australia, and the Americans who established the earliest refrigerated trade across the Atlantic, must have a pre-eminence of their own.

Thomas Sutcliffe Mort.

The figure of Thomas Sutcliffe Mort stands out boldly amongst all the pioneers and experimenters in Australia and elsewhere whose efforts laid the foundations of the frozen meat trade. Mr. Mort cheerfully spent a large fortune in experimental enterprises in practical meat freezing, and his conception of the future that awaited the industry was prophet-like, so sympathetic and keen was his grasp of the subject—as will be seen in reading the sentences extracted from his speech, which are given below.

Mr. Mort was born at Bolton, Lancashire, on December 23, 1816, emigrating to Australia in 1838, and later founding the

THOMAS SUTCLIFFE MORT, AND THE STATUE ERECTED TO HIS MEMORY AT SYDNEY.

To face p. 18.

great financial and wool-broking firm of Mort and Co. This firm afterwards amalgamated with that of R. Goldsbrough and Co., Ltd., under the name of Goldsbrough, Mort and Co., Ltd. As early as 1843 Mr. Mort turned his attention to meat matters, and was later introduced by Mr. Augustus Morris to the French engineer Nicolle. The pair took up the subject of freezing meat for export, and experiments were conducted, Mort supplying the capital and Nicolle the engineering skill. Partial freezing, "chilling," Tellier's plan, was tried and rejected, as it was soon realized that thorough congealing was far preferable for the proper preservation of meat. **Mr. Mort in 1861 established at Darling Harbour, Sydney, the first freezing works in the world.** Thirteen years later Mr. Mort's company became the New South Wales Fresh Food and Ice Co. The original freezing process at these works was applied in two large apartments, each about 75 feet square and 9 feet 9 inches high, and enclosed by brick walls 4 feet 6 inches thick. The freezing room below was used for the treatment of meat for export. In 1875 the collateral enterprise, the slaughtering works at Lithgow Valley, Blue Mountains, was completed: the two establishments were intended to supply the Sydney market. Ammonia compression refrigerating machinery was used at these works.

At an inaugural lunch on September 2, 1875, at which 300 persons attended, including Sir John Hay, the Hon. J. Docker, Sir Saul Samuel, the Hon. (afterwards Sir) John Robertson, and Mr. (afterwards Sir) Henry Parkes, Mr. Mort made his famous speech, the peroration of which stands out as a white stone in the annals of the Australian meat trade, clearly showing him to have been a man of imagination, noble aims, and high character. Mr. Mort in this speech said that Mr. Morris first suggested the "diabolical idea" of freezing meat to send to England. "I can tell you that not once but a thousand times have I wished that Mr. Morris, Mr. Nicolle, and myself had never been born." Mr. Mort mentioned that the Sydney Chamber of Commerce about 1867 had put up a sum of money for him to provide meat for distribution in England, and to overcome the English prejudice against "frozen" meat. (This reads curiously, for in 1867 not an ounce of (mechanically) frozen meat had reached England !) The meat upon which Mr. Mort feasted his 300 was, of course, all frozen, and he stated that some of it had been kept since June, 1874. He told his guests that Australia was destined to become the great feeder of Europe.

"Before long France and England will look to us almost entirely for their supply of food." Mr. Mort suggested the breeding of Highland cattle, as this breed secured top prices in English markets. Mr. Mort's classic peroration must be given verbatim.

"I feel, as I have always felt, that there is no work on the world's carpet greater than this in which I have been engaged. Yes, gentlemen, I now say that the time has arrived—at all events, is not far distant—when the various portions of the earth will each give forth their products for the use of each and of all ; that the over-abundance of one country will make up for the deficiency of another ; the superabundance of the year of plenty serving for the scant harvest of its successor ; for cold arrests all change. Science has drawn aside the veil, and the plan stands revealed. Faraday's magic hand gave the keynote, and invention has done the rest. Climate, seasons, plenty, scarcity, distance, will all shake hands, and out of the commingling will come enough for all, for 'the earth is the Lord's and the fulness thereof,' and it certainly lies within the compass of man to ensure that all His people shall be partakers of that fulness. God provides enough and to spare for every creature He sends into the world ; but the conditions are often not in accord. Where the food is, the people are not ; and where the people are, the food is not. It is, however, as I have just stated, within the power of man to adjust these things, and I hope you will all join with me in believing that the first grand step towards the accomplishment of this great deed is in that of which you yourselves have this day been partakers and witnesses."

Mr. Mort is supposed to have spent £80,000 in connection with his freezing experiments, and £20,000 besides was put up by Australian squatters for a trial shipment to England about 1876, for which the sailing ship *Northam* was chartered. Mr. Andrew McIlwraith, who happened to be in Sydney at the time the vessel was being fitted with an ammonia installation similar to that used on land, supplies for these pages the following special account of this historic effort:—

"We found that a considerable space in the square of the main hatch had been bulkheaded off as a meat chamber. The insulation consisted of a 15-inch space between two bulkheads run in with tallow, and inside the chamber the cold brine pipes were fixed. It occurred to us as we went round that there was a danger of destruction to the meat in having the pipes in the chamber, should the movement of the ship strain the pipes and a leakage occur, and, as a matter of fact, that is what took place, but fortunately before the vessel left the harbour."

The meat had to be discharged, thus making an end of the experiment before it had really begun. This failure was a terrible blow to Mr. Mort, and hastened his death. He died at Bodalla, New South Wales, in 1878, and to his memory a monument was erected by public subscription. Had Mort spent less time in research work on the mechanical side of refrigeration, his actual achievement in starting the new trade might have been greater. There were many engineers working out the scientific problems in Europe at the time, and had Mort depended on their labours rather than spending his time experimenting along with Nicolle, it is possible that the commercial beginning of the trade might have been in 1876 instead of 1879.

Before passing to another pioneering record a word may be said about Mr. Mort's associate, Eugene Dominique Nicolle. Mr. Nicolle was born at Rouen in 1824, visited Australia in 1853, and became manager for the well-known Sydney house, P. N. Russell and Son. In the days when Mort and Nicolle were actively engaged in overcoming the difficulties of meat and provision export (1860 to 1877), ice was imported from America, and Mr. Nicolle did interesting work in running a factory at Darlinghurst for the manufacture of ice by chemical process. As a preliminary to the placing on board ship of a freezing machine to be used for the frozen meat trade contemplated by Mr. Mort and himself, Mr. Nicolle erected a special chamber at the back of the Royal Hotel, George Street, Sydney, where the process was tested for fifteen months. Negotiations were first opened up with the owner of the well-known trader *Whampoa*, but fell through when it was heard that liquefied ammonia was to be the freezing agent. In the same year as Mr. Mort died, Mr. Nicolle retired from business and settled on Lake Illawarra, where he died a comparatively short time ago.

James Harrison.

James Harrison, another Briton who made Australia the scene of pioneer work in meat freezing and also in the manufacture of ice, was born in Glasgow in 1816, the same year as Mort first saw the light. That Harrison deserves place as one of the pioneers, if not *the* pioneer, of the frozen meat trade admits of no shadow of doubt. In an article in the *Melbourne Age* (of which important journal Harrison was editor for some time), written in 1893, at the time of his death, on September 3, 1893, aged 77, at Geelong, occurs this passage: "It is a striking proof of his insight that he was the first to see the enormous source of wealth that lies still undeveloped in the export of meat from the Australian pastures. The very industry which Mr. Russell tells us has pulled New Zealand out of the shoals into calm water was receiving Mr. Harrison's strenuous advocacy thirty years ago." Long before the year indicated by the *Age* (1863) Harrison was not only "advocating" but busily experimenting with his ice-making machinery.

He emigrated to Sydney about 1837. In 1840 he settled at Geelong, "Australia Felix,"—as Victoria was styled in the early days—taking up journalism, and in 1850, having some leisure, devoted himself to the working out of an ice-making scheme. He acquired land at Rodey Point, on the Barwon, and there erected his first ice factory at a cost of £1,000. In 1851 the brewing firm of Glasgow and Co., Bendigo, installed a refrigerator of the Harrison type. This was the world's pioneer of such machines. Perceiving that the works were too small

JAMES HARRISON.

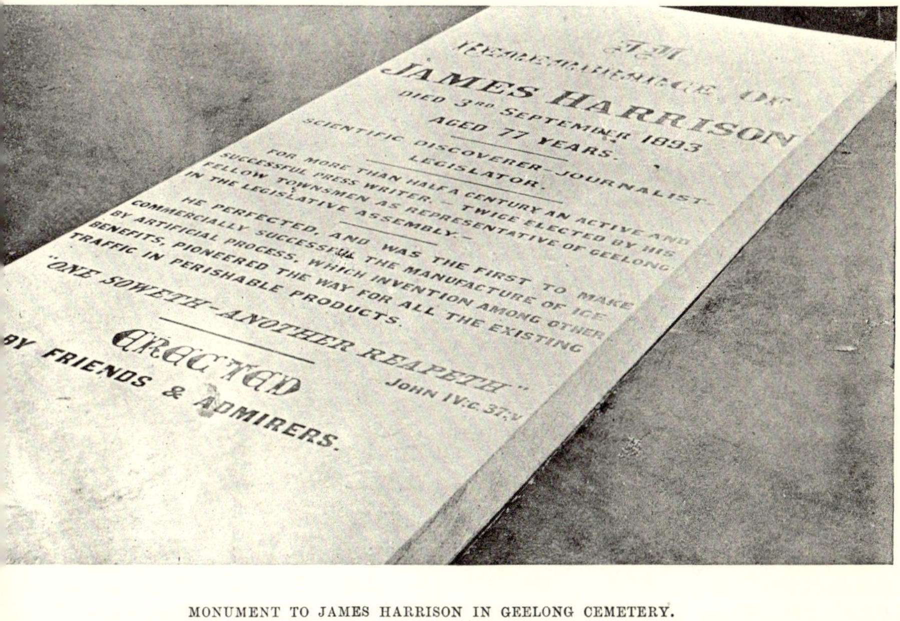

MONUMENT TO JAMES HARRISON IN GEELONG CEMETERY.

To face p. 22.

to make a commercial profit, Harrison travelled to London in 1857. His two first English patents (see Appendix) are dated March 28, 1856, and September 10, 1857; two years previously the machine had been patented in Australia. He had corresponded with Faraday and Tyndall, and discussed freezing problems with those distinguished men. He also got into touch with Siebe Brothers, who had a jobbing foundry in Red Lion Court, Holborn, and Mr. Siebe made a large machine for him, which was taken to Hobson's Bay in the ship *Tricolor* on Harrison's return to Australia. It may be mentioned that Mr. H. J. West, an early inventor of refrigerating machinery and founder of the firm of refrigerating engineers which bore his name, was manager of, and later a partner in, the firm of Siebe Brothers, and knew Mr. Harrison well. Mr. West died in 1910.

On his return to Australia Harrison devoted himself specially to meat freezing, and before attempting export he experimented in the preservation of meat for lengthy periods on land. At Melbourne, about 1873, he publicly exhibited his cold-producing machine, and by its means several carcasses of sheep, sides of beef, poultry, fish, etc., were frozen, and six months afterwards these viands were consumed at a public banquet.

Like Mort, three years later, James Harrison failed when he put his process of meat preservation to the test of sea voyage, and the failure ruined him. All the profits of his newspaper were eaten up in his experiments and by the disaster which befel the shipment when the 20 tons of mutton and beef placed on board the s.s. *Norfolk* went bad on the journey to London. The vessel sailed in July, 1873, from Sandridge Railway Pier, Melbourne; the meat had been frozen on board in "two tanks," ice and salt freezing mixture being used to effect the refrigeration of the cargo. The tanks leaked, and when the vessel arrived at London in October the meat was unsaleable. This terrible blow, no doubt, crushed the inventor's spirit, as it ruined his fortunes, for he soon sought retirement in London, where he spent some years in scientific study.

The refrigerating machine of Harrison's which was put to work in a paraffin factory in England in 1860 was probably the first refrigerating plant ever applied to a manufacturing process, though the Americans state that Professor Twining, about the same time, had an ether machine at work at Cleveland, Ohio. The Patent Office is probably the best guide to settle priority in inventions, and, according to this, Harrison, excepting Perkins, whose 1834 patent never seems to have got beyond Chancery Lane, was years ahead of all rivals; Carré was four years after him, Tellier eight years, Mort eleven, Little twelve, Pictet thirteen, and Postle seventeen years later, according to London Patent Office records. The accompanying reproduction of a photograph of the inscription on James Harrison's tomb at Geelong tells briefly the story of the man's great work.

Mr. J. D. Postle must be numbered among Australian pioneers as an experimenter, about 1869, at Melbourne with the chilling of meat by means of an air compression machine. Evidence of the early spread of enterprise in the new industry is afforded by the fact that the Melbourne *Australasian* from 1868 to 1880 contained a great mass of correspondence dealing with the earliest days of meat freezing and the merits of the various systems of refrigeration.

The Bell-Coleman Machine.

The name "Bell-Coleman" must, where meat freezing is concerncd, ever remain an honoured one in the two hemispheres, for it was through the agency of a Bell-Coleman refrigerating machine that there was landed in London early in 1880 the first shipment of fresh meat ever successfully carried from Australia. In the career of Mr. (now Sir) Henry Bell is contained the history of this and other pioneering efforts in meat refrigeration, and some account of his early work has proper place here.

Mr. Bell's first connection with meat refrigeration was when, early in 1877, he took up the Glasgow agency of the dressed beef shipping business from New York established by Mr. T. C. Eastman. To Mr. Eastman, by the way, must be given the credit of having first introduced chilled meat into Great Britain (*vide* a personal note below). The refrigerating process then used consisted in allotting 25 per cent. of the whole of the space occupied to an ice-container and filling the latter with blocks of natural ice, circulating through the ice a current of air by means of a fan. A modification, or rather elaboration, of this plan was patented by a Dr. Craven, and adopted for cooling the shipments of a Mr. Gillette, who sent meat across the Atlantic by some other lines of steamships. Dr. Craven's method consisted in using ice and salt, by means of which a lower temperature could be obtained than by ice alone, and this freezing mixture was used to cool brine circulated in pipes in the vessels' holds.

Seeing the costly nature of iced transport, in which a quarter of the space paid for was not available for the meat cargo, to say nothing of the boats sometimes arriving with all their ice melted, Mr. Henry Bell and his brother James (now Sir James) Bell set themselves to study all the available literature on mechanical refrigeration, and decided in their own minds that refrigeration by means of cold dry air was the method most suitable for use on board ship. They approached Sir William Thomson (afterwards Lord Kelvin), who was then a professor at Glasgow University, and asked his advice as to whether it was practicable to use that method for the transport of meat. The professor said he thought it was, but that it would prove too costly, promising, however, to look into the subject if they would supply him with data as to the quantities to be carried. When he found that the figures were large, he agreed that there was a future in the business, but he added that he was too busy to take it up. This incident probably indicates how near Lord Kelvin's name was to becoming a household word in commercial refrigeration. The Glasgow professor said, further, that he would introduce the brothers Bell to a Mr. J. J. Coleman, and with the latter gentleman Messrs. Bell formed a partnership under the name of the Bell-Coleman Mechanical Refrigeration Co., and took out patents in 1877, the first being dated June of that year.

The first ship on which Messrs. Bell and Coleman fitted their machine was the s.s. *Circassia*, of the Anchor Line, in the American trade, in the spring of 1879, but in order to satisfy the world that they could bring meat across the ocean they fitted up in the engineering works of Messrs. D. and W. Henderson and Co., of Glasgow, a space similar to the 'tween decks of an ordinary steamer, and therein put meat and kept it chilled for ninety days, subsequently bringing it up to Smithfield market and selling it there. That consignment was kept at a temperature of 30° F.; the cattle were slaughtered close to the chilling chamber, and the meat was not moved after it was put into the chamber, never being actually frozen. All through 1878 and 1879 the inventors were receiving meat from the United States on Mr. Eastman's account, cooled by freezing mixture.

Queen Victoria's Approval.

Although it is only an incident, it is worth mentioning here that as early as 1875 American chilled beef met with Royal approval. The following information received from Mr. F. Link, of the Central Markets, forms an interesting and a Royal reminiscence:—

"Mr. T. C. Eastman was the shipper, from New York in October, 1875, of the first lot of American chilled beef, and to him must be given the credit and honour of this innovation, undertaken at considerable risk and expense to himself. A baron of that beef was sent to Queen Victoria at Windsor Castle, and Eastmans, Ltd., have the Royal Seal in connection with that transaction. The Queen pronounced the meat 'very good.' The shippers of this early American chilled beef included Toffee Brothers, Gillette (Jersey City), Martin Fuller (Philadelphia), and Sherman (Philadelphia), and Mr. J. D. Link acted as agent for Mr. Eastman up to the time when Messrs. John Bell and Sons took over the agency."

Charles Tellier.

Having recounted the early efforts of Mort and Harrison in Australia, one is tempted to proceed at once to a description of the first frozen meat shipment successfully finding its way from those shores to the Mother Country. It must not be forgotten, however, that refrigerating experiments were being

CHARLES TELLIER AND THE S.S. *Frigorifique.*

To face p. 26

carried on elsewhere, and it is well to review the part played by such a scientist and engineer as the Frenchman, Charles Tellier, who was responsible for a shipment of meat brought at a chilling temperature from Buenos Aires to Rouen as early as 1877, this being in fact the first meat cargo shipped through the tropics under refrigeration. Tellier was the inventor of an ammonia-absorption refrigerating machine as early as 1859, and in 1867 he produced an ammonia-compression refrigerating plant. M. Tellier's first essay at shipping meat under refrigeration was in 1868. Financially supported by Mr. Francisco Lecocq, of Monte Video, he put an ammonia compression machine into the *City of Rio de Janeiro.* He shipped 300 kilos of beef from London to that city as a test, the intention being on the homeward journey to import meat from Uruguay into France. But twenty-three days out an irreparable accident occurred to the apparatus, and the meat was eaten on board. The temperature was 32° F. Next we come to the *Frigorifique,* a slow steamer, previously named the *Eboe,* 210 feet in length, and with a meat hold 85 feet long, 25 feet wide, and 13 feet high. This was bought in Liverpool in 1875 for 210,000 francs (£8,400) by a company formed in France to import fresh meat from La Plata, Texas, or Madagascar. Three of Tellier's refrigerating engines were installed at the stern, and all the room forward of the engines was given up to cooled space for meat. The insulating material was powdered cork and chaff. The vessel sailed from Rouen on September 19, 1876, and arrived at Buenos Aires on December 25, carrying some meat from France. "Dark spots" were reported on some of this meat when inspected by the deputy president of the Argentine Rural Society, who also said that "at table they gave us small dishes prepared from the meat, the flavour of the most part of it was rather unpleasant." (Turf, brought as ballast, was the alleged cause of this.) After considerable difficulty in "assembling" a return cargo of meat, the ship *Frigorifique* sailed for Rouen, where she arrived on August 14, 1877, after a voyage of 104 days: some of the meat when landed there had been preserved for 110 days. *Le Rappel de Paris* of December 2, 1877, stated that a certain portion of the meat did not arrive in good condition, and a "rather careful selection had to be made." But, it adds, "the problem was solved." No particulars are to hand as to the realization of the cargo, but the French people were indifferent, and even in those early days it was in England where there were the best chances of exploiting Tellier's process, for it seems that one of Mr. Tallerman's companies, the London Meat Importation and Storage Co., arranged to buy the *Frigorifique* and send her out again to South America. The negotiations, however, fell through, and the vessel experienced many vicissitudes, almost knocking down one of the Seine bridges on one occasion, and being put up for sale in May, 1879. A Mr. Robert MacAndrew was introduced to the *Frigorifique* company to finance it. Actually ten tons of the desiccated beef brought over by the ship reached London. "Not an atom of mould was on it," Mr. Tallerman says, and he adds: "the meat was like leather, and had lost 30 per cent. of weight, which was regained in cooking!" This loss of weight may seem incredible. But in Harrap and Douglas's "Public Abattoirs and Cattle Markets" it is stated that a piece of beef 116 ounces in weight lost 50 ounces in 67 days (43 per cent.) by "air cooling." M. Tellier patented his process in all the countries of Europe, and in Victoria, Australia, from 1874 to 1878.

The Shipment by the *Paraguay* in 1877.

Although the shipment by no means marked the start of the refrigerated meat export trade from South America, the successful carriage of a cargo of frozen meat on the s.s. *Paraguay* from Buenos Aires to Marseilles in 1877 must be chronicled as actually the first entirely successful frozen meat shipment in the world's history. About 1860 an ammonia-compression machine called the "engine Carré" was constructed, an invention with which M. Charles Tellier was in some way associated. M. Carré's name must stand out as that of the pioneer in Europe of the *frozen* meat trade; for Tellier never brought his meat below freezing point, and the Argentine meat brought by the *Frigorifique* was a thing apart—nothing has ever since been seen like it. Records of the solitary attempt made by the "Society Jullien Company for the transport and preservation of fresh meat by means of cold (Carré-Jullien system), Boulevard National, 386, Marseilles," are fairly clear. Messrs. Jullien were shipowners, running vessels engaged in the Mediterranean fish trade. They fitted a Carré ammonia machine in the s.s. *Paraguay,* 1,120 tons. Captain Lefèvre was in charge, the engineer was M. Lescornet, and the refrigerator was constructed by Messrs. Hanthonille, of Marseilles, from M. Carré's designs. The engineers were so determined to make a new departure and congeal their meat that they kept the temperature during the homeward passage at about $-17°$ F.! The chronicler says the meat was "petrified, as hard as a stone." Extraordinary precautions were taken with some meat sent out from France, as a test. The Argentine Vice-Consul at Marseilles sealed up the refrigeration chamber and the four quarters of beef and ten sheep therein. The vessel sailed on August 13, 1877, and arrived at Buenos Aires on September 29. The *Paraguay* commenced to take in her meat from San Nicolas for freezing on board on October 7, and she did not arrive at Havre till about May 7, 1878, owing to having been compelled to put into St. Vincent for repairs after a collision. There she stayed four months, yet she arrived at the French port with the meat in tip-top condition—a marvellous performance for 1878! Her cargo consisted of 5,500 carcasses of mutton. The reporter at Havre was enthusiastic. "The congealing completely destroys the germ of putrefaction," he wrote, and the good people at Havre received the consignment with joy. The 80 tons were "used to the last morsel." The garrison troops feasted on it and, *mirabile dictu*, the Grand Hotel in Paris used the meat for a whole week. As the French were so ready to welcome frozen meat, and partook of it so freely, one concludes that the regulations now built up against its import must be artificial or engineered. The Jullien Co. prepared the *Paraguay* for a second essay on a grander scale (800 tons of meat), but, as Dr. Pierre Bergés has recorded, "as it happened, the project was never realized, and this new industry of the freezing of meat was abandoned by the French."

Dr. Bergés says, a little bitterly, in another place: "As has often happened in the history of industries, it has been the French who have made the discoveries, and the English who have turned them to account to their profit. The refrigerating industry belongs to this number." A parcel of this meat was sent to London. Messrs. John Schmidt and Co., 69, Mark Lane, E.C., were the consignees, and the mutton was reported to be of "extraordinarily good flavour, but very small." Messrs. Jullien wished to build a fleet of steamers, 2,000 tons each, to convey to France Argentine frozen meat and remounts for the army. Had the French capitalists come forward, the whole course of the commercial genesis of the meat trade would have been altered.

The *Strathleven.*

The *Strathleven* shipment was the outcome of an inquiry instituted early in 1878 by Queensland squatters, who, having heard of the *Paraguay* voyage from South America to Havre, cabled to London, and Mr. Andrew McIlwraith and Mr. Beardmore Buchanan went to Havre to inspect the steamer. They reported adversely to the application to the Australian trade of the refrigeration system employed, and shortly afterwards Mr. McIlwraith got into touch with Messrs. Bell and Coleman in Glasgow. Experiments were made with dry air at 30° F., which resulted in the conviction that Mr. Mort's freezing theories were correct. Negotiations were opened with Messrs. Bell and Coleman for a machine designed and arranged for shipboard, and Mr. Andrew McIlwraith chartered the s.s. *Strathleven* (gross tonnage 2,436) from Messrs. Burrell and Son, and installed the Bell-Coleman machine on board. The vessel sailed from Plymouth in 1879 under the control of Mr. James Campbell, a civil engineer, and Mr. Matthew Taylor Brown, B.Sc., went out as representative of the Bell-Coleman Co., and returned with the vessel on its epoch-making voyage. The captain was Mr. C. W. Pearson. Mr. (afterwards Sir) Thomas McIlwraith, Mr. B. D. Morehead, and Messrs. William and Robert Collins were amongst the Queensland gentlemen who took a leading part in helping forward the movement, and

Mr. Hastings Cuningham and Mr. George Fairbairn in Victoria rendered assistance, but Mr. McIlwraith's firm, McIlwraith, McEacharn and Co., bore the risk of the venture.

The late Sir Malcolm McEacharn left London for Sydney to make all arrangements there for the *Strathleven's* loading. The vessel sailed thence on November 29, 1879; Melbourne was made a second port of call, and the vessel left that port on December 6, 1879, for London, where she arrived on February 2, 1880, with 40 tons of beef and mutton. The meat had been frozen on board. One of the first persons to inspect the new commodity, destined to revolutionize the world's meat trade, was the produce representative of the New Zealand Loan and Mercantile Agency Co., and it was recorded in the market report of that company that "On inspection of the meat while the vessel lay in dock, it was found to be in a perfectly sound state, frozen quite hard and covered with an artificial rime." The representative was, however, doubtful about several things, including the possible early decomposition upon thawing, deficient flavour, and partial destruction of the natural juices. But, on the whole, he took a sanguine view, and his comments form the pioneer report upon a trade which has attracted the most exhaustive attention from commercial writers. The meat, which had cost from 1½*d.* to 2*d.* per lb. in Australia, was placed in the hands of Mr. J. D. Link, of Smithfield Market, and realized 4½*d.* to 5½*d.* per lb. for the beef and 5½*d.* to 6*d.* per lb. for the mutton. A lunch to celebrate the success of the venture took place on the *Strathleven* on February 6, 1880. Messrs. Andrew and Thomas McIlwraith, Sir R. R. Torrens, Colonel Taylor, Messrs. W. Westgarth, W. Jordan, and E. Alford Wallace, amongst others, were present. A carcass of lamb was sent to Queen Victoria and a sheep to the Prince of Wales. A joint was given to the Travellers' Club, and the late Lord Hatherton happened to be lunching at the Club when the said joint was brought to table. He found it so good that he asked for another helping, and was surprised when he learned that the meat which he had enjoyed so much represented a hidden attack on his, or, rather, his tenants', industry.

The newspapers and the trade gave a most friendly reception to this pioneer shipment of frozen meat; the *Times*, the *Daily Telegraph*, the *Morning Advertiser*, the *Mark Lane Express*, the *Globe*, etc., were enthusiastic. The *Daily Telegraph*, in referring to the meat, wrote: "It has been tested by the ordinary methods of cooking, and found to be in such good condition that neither by its appearance in the butchers' shops, nor by any peculiarity of flavour when cooked for the table, could it be distinguished from freshly killed English meat." The following interesting credential emanated from the London Central Markets:

This is to certify that we, the undersigned meat salesmen, doing business at the Central Meat Market, London, inspected the meat imported from Australia ex *Strathleven*. We found it in perfectly sound marketable condition, and it readily fetched prices averaging $5\frac{1}{2}$*d*. per lb. Both beef and mutton were excellent in quality, for those of us who tasted the meat when cooked pronounced it tender, and its flavour very good. From the success which has attended this shipment, we are of opinion that similar supplies from Australia will find a large and ready sale in this market.

(Signed)	J. D. Link.	Wm. Bowyer.
	Hannah Ward and Co.	Archer and Malthouse.
	Charles Mathew and Son.	H. S. Fitter.
	B. W. Frost and Co.	H. Killby and Sons.
	B. Venables and Sons.	H. Hicks and Son.

The *Strathleven*, after its historic voyage, was stripped of its refrigerating machinery and insulation, and was sold by Messrs. Burrell and Son in 1899 to Messrs. Abram and Addie, of Glasgow, who kept her running in the American trade. In January, 1901, she was lost in the Atlantic during a heavy gale while on passage from the Gulf of Mexico to the Mediterranean.

The Australian Frozen Meat Export Company.

It was felt by the leading pastoralists of the Riverina district of New South Wales and of Victoria that this highly successful opening of the new trade should be followed up in a business way. Both the squatters and the mercantile community entered enthusiastically into the consideration of how to proceed in the most practical manner to exploit this industry. Meetings were held in Melbourne at the end of 1879 and in 1880 at Scotts' Hotel, and Sir James McCulloch, Sir Samuel Wilson, and Messrs. George Fairbairn, J. H. Douglas, J. L. Currie, Lloyd Jones, James Blackwood (Dalgety, Blackwood and Co.), Hastings Cuningham, and C. M. Officer, were prominent amongst the gentlemen who supported the movement. The Australian Frozen Meat Export Co., Ltd., was formed with a subscribed capital of £100,000, in £100 shares, and some £80,000 was spent in developing the trade. The gentlemen above named were the directors of the company, Sir James McCulloch being Chairman—all these gentlemen have passed away. Mr. F. W. Armytage joined the board later. Mr. John Hotson was appointed secretary and manager.

At that time there were neither freezing works nor fitted steamers, but during the early months of 1880 temporary works were put up at Maribyrnong, and subsequently substantial freezing and storing works were erected at Newport, near Melbourne. Then the company chartered the s.s. *Protos*, and fitted up the vessel with refrigerating machinery; the insulation of the chamber consisted of nine inches of wool. Refrigerating machines were manufactured in Melbourne by Messrs. Robison Brothers, engineers. The machine was a duplicate of the Giffard refrigerator which the company had imported. Toward the end of 1880 the *Protos* was loaded and despatched to London with about 4,600 sheep and lamb carcasses, and 100 tons of butter, all of which produce was landed in London in excellent condition on January 17th, 1881, at a moment when the city was cut off from supplies owing to heavy snowstorms. The meat sold from 5½*d*. to 7*d*. per lb., and the butter fetched 1*s*. 3*d*. per lb. These prices, of course, represented a substantial profit. After the success of the *Protos* shipment, Mr. Thomas Brydone, of New Zealand, visited the Melbourne works, and no doubt the information gathered by him there was of great help in preparing the *Dunedin* shipment—to be referred to later on in this chapter. Another steamer was chartered and fitted in Melbourne, the ss. *Europa*, the insulation being charcoal—the *Protos* wool insulation had been unfitted in London and sold in good condition. A larger and more powerful refrigerator was made for the *Europa* by Robison Brothers.

Some 9,000 carcasses of mutton and lamb, and some quantity of butter, were despatched to London, where this shipment also was landed in excellent condition. The prices realized were about 3½*d.* per lb. for the meat, and 1*s.* 1*d.* per lb. for the butter.

After these ventures it occurred to the directors of the company that one of the regular lines of steamers trading from Australia to London should take up the trade. The Orient Line was approached in 1881, and readily entered into the suggestion. The s.s. *Cuzco* and two other steamers of the line were fitted up by the Orient Co. with Haslam machines. The *Cuzco* was the first vessel to sail with meat; she took 4,000 frozen carcasses, the freight paid at that time was 2¼*d.* per lb. The *Orient* and *Garonne* followed. Difficulties were many at this stage of development, and the Australian Frozen Meat Export Co. had to struggle with low prices for their produce as well as high rates of freight, and soon the capital of the company was seriously reduced. Drought set in, and exports from Melbourne had to be suspended for a time.

The company was wound up in 1886. In 1887 the Newport freezing works were purchased by the Victorian Government, which used them in connection with the starting of the export butter business. In 1893 the firm of John Hotson and Co. leased the Newport works from the Government for a period of years, and in 1896 the works passed into the hands of the present owners, Messrs. John Cooke and Co., who in 1899 reorganized and reconstructed them with new plant. On a calm survey of the events just chronicled, it is clear that the enterprise of the directors and shareholders of the Australian Frozen Meat Export Co. at Melbourne, coming just after the inauguration of the frozen meat trade by the *Strathleven* shipment, was very helpful in giving the industry a good start.

Queensland's First Freezing Enterprise.

The first freezing enterprise in Queensland was that of the Central Queensland Meat Export Co. at Lake's Creek, near Rockhampton. This company had its origin in a boiling-down works at Laurel Bank, on the Fitzroy River, opened in 1868 by Messrs. Berkelman and Lambert. Messrs. Whitehead and Co.

subsequently acquired the factory. The Lake's Creek works were started in 1871 by the Central Queensland Meat Export Co. This company had but a few years of precarious existence. Difficulties occurred as time went on; cattle became scarce and dear, and in 1874 the works were closed. The preserving works had absorbed the surplus cattle and raised the stock in the district to prohibitive prices. The works remained closed until 1877, at which date Whitehead and Co. purchased the property from the liquidators, and transferred the plant at the Laurel Bank factory to the Lake's Creek works. About 1880 No. 2 Central Queensland Meat Export Co. was formed, with Mr. Bertram as manager, and about 1883 the company added a freezing plant to its establishment. The freezing chambers were full of meat ready for shipment in September, 1883, by the s.s. *Fiado.* The vessel was a fortnight late in arrival, and on September 13 what would have been Queensland's pioneer enterprise in the shipment of frozen meat was wrecked by a disastrous fire at the works, which caused a loss of £30,000. (It is curious that in 1884 another of Nature's wrecking forces, a hurricane, frustrated the Poole Island pioneers in their initial effort to ship frozen meat by the same vessel: see p. 36.) Not until August, 1884, did the works restart. In 1885 the company went into liquidation, and in 1886 the property was purchased by a Melbourne syndicate, including in its members Messrs. Andrew Rowan, George Fairbairn, and John Living. Mr. M. C. Thomson was managing director of the new company formed, which retained the former style, and Mr. Bertram was general manager until 1890, when he was succeeded by Mr. Alexander Paterson. Mr. Paterson held office for nine years, and during his period of management excellent results were obtained from the works. Mr. W. S. Lambe (who has of late years been associated with the management of South American frigorificos) in 1899 became works manager. An average of the annual outturn of the Lake's Creek works for five years struck at this stage of the company's history showed 50,000 cattle treated and 273,000 sheep; meat frozen 11,966,000 lbs., and meat preserved 7,750,000 lbs. In 1901 the company and business were taken over by a syndicate of London capitalists, including Sir Montague Nelson, Mr. George Mackenzie, and Mr. James Caird. The Central Queensland Meat Export Co. was registered in London, where the office is at 14, Dowgate Hill, E.C. Mr. G. H. Hopper was installed as the general manager at Lake's Creek, a position he still holds. The property is one of the largest meat preserving and freezing establishments in Australia, possessing modern machinery and plant. The land held by the company is about 30,000 acres, of which 18,000 are freehold. When the works are in full operation 700 to 1,000 hands are employed. Quite a small town is built round the factory for the employees; a dining hall to seat 200 men, a mission hall, school, school of arts, rowing, cricket, and football clubs, are amongst the social features of the Lake's Creek meat works.

Absolutely the first shipment of frozen meat actually despatched from Queensland left Moreton Bay, Brisbane, on May 20, 1884. Concerning this shipment, the *Brisbane Courier* of May 21, 1884, said: "The British India Co.'s steamer *Dorunda*, with the first shipment of frozen meat for London from this colony, left the Bay yesterday afternoon. The frozen cargo consisted of 3,594 sheep and 100 quarters of beef, and the shipment may be attributed to the enterprise of the Queensland Freezing and Food Export Co. The cattle were the property of Mr. Collins, one of the earliest and most consistent of the supporters of the meat export trade. The shipment was not a success, but the company afterwards demonstrated the vessel's capacity for carrying meat."

Poole Island.

An early and gallant attempt to follow up the success achieved by the *Strathleven* with its epoch-making shipment was made in Queensland in 1881, when the Australian Co., Ltd., was registered in London on April 29 of that year. Mr. Robert Christison, owner of large flocks and herds on his Lammermoor and other stations in North Queensland, conceived the idea of establishing meat freezing works in that part of Australia and of forming a London company to work the enterprise. At that time there was no market in the North of Queensland to speak of for fat cattle. There was just the boiling-down works and nothing more. Mr. Christison went to London, and was successful in meeting with some leading City men. The Australian Co., Ltd., was formed in London, Sir Richmond Cotton, M.P., afterwards Lord Mayor of London (then the senior partner in the well-known London firm of produce brokers, Culverwell, Brooks and Cotton), becoming chairman. The other directors of this concern, the first purely frozen meat company formed in England, were Messrs. B. T. Bosanquet, R. Campbell, jun., J. Jackson, Samson Lloyd, Laidley Mort, T. Salt, and A. Van de Velde, the secretary being Mr. R. M. Stephenson. The subscribed capital was £61,613.

Before referring to Poole Island and what happened there, it may be mentioned that the Australian Co. was wound up in 1888, but prior to that stage its consignment business was passed over to the New Zealand and Colonial Consignment Co., Ltd., which was registered in London in November, 1885, with a capital of £50,000, the directors comprising Messrs. R. M. Stephenson (managing director), G. M. Mackenzie, B. T. Bosanquet, and A. Van de Velde. This company ran a shop at Hounslow as a part of its operations. As Messrs. Nelson Brothers, Ltd., were desirous of receiving consignments, they purchased the goodwill and business of the New Zealand and Colonial Consignment Co. from June, 1886, and took over Mr. Stephenson, giving him the appointment of country manager.

The Australian Company, Ltd., was granted a lease by the Queensland Government of Poole Island, near Bowen, North Queensland, and the works—the second freezing establishment erected in Australia—were put up. Bowen has a fine natural harbour, and Messrs. Houlder's steamers now take delivery of the frozen meat shipped thence by the Merinda works of Bergl, Australia, Ltd. After many delays the factories were completed. A stumbling block to the company's progress was that the underwriters would not entertain a policy on its frozen meat shipments to London, no matter how high the premium offered; nevertheless in 1884 the first cargo of meat from the works was put on board the s.s. *Fiado*, which was fitted with Bell-Coleman refrigerating machinery. The vessel was bound for Batavia, and a remunerative price for the meat was anticipated. Here, unfortunately, has to be recorded a very aggravated instance of the ill-luck which so often dogs the footsteps of pioneers and wrecks their enterprises. On January 30, 1884, just as the *Fiado* was ready to sail, occurred one of the terrible cyclones that occasionally visit the North Queensland coasts. When the storm abated, it was found that the *Fiado* lay high and dry on the mainland. The company lost heavily, and after mature consideration it was decided to wind up affairs. Ultimately the buildings and effects were sold to the British India Steam Navigation Co., which concern, employing steamers of their own, and those of the Ducal Line, owned and managed by Messrs. J. B. Westray and Co., began in 1885 the lifting of the meat from the Poole Island works. The first shipment of 1,000 tons was made by the s.s. *Duke of Westminster* in the year named, running on the British India-Queensland mail line. In the freezing of this meat various difficulties presented themselves, and at one point so critical was the position that the zinc-lined piano case of Mr. Stevens, the manager, was impounded and turned to use as an extra condenser. One or two shipments were made subsequently on the joint account of growers and shipowners, but the works lacked that efficiency and perfection of detail equipment since attained elsewhere. The new owners of the works found the enterprise was attended with such lack of success that in 1886 it was discontinued. It must be remembered that the works had been erected on an island, the only approach to which was at low tide, when the cattle had to cross, sometimes up to their girths in water. Apart from this inconvenience, the early management had to contend with the difficulties usually experienced by pioneers, lack of trade organization and markets, besides the prejudice against frozen meat, which in those days was very marked. The British India Co. still own the site and what remains of the works, but the machinery and boilers have been sold, the latter to the Alligator Creek Works, North Queensland.

New Zealand enters the Field.

It will be convenient at this point to turn to New Zealand and consider the beginnings of the export trade from that Colony in frozen mutton. In 1851 there were 233,043 sheep in New Zealand, and by 1880 the flocks had grown to 11,530,623, owned by 6,857 farmers.

A notable pioneer of the New Zealand meat export trade was the New Zealand and Australian Land Co., the head-quarters of which concern are at Edinburgh. Mr. W. S. Davidson, who became general manager of that company in 1879, provides an exceedingly interesting and detailed account of the circumstances in which New Zealand's pioneer frozen shipment was made on the *Dunedin*. Realizing the great need of an outlet for the large surplus of sheep, and perceiving the possibilities of the frozen export trade already indicated by the first attempts from Australia, Mr. Davidson investigated the whole business, his directors agreeing that a trial shipment from New Zealand might be arranged for on the part of the New Zealand and Australian Land Co., at the same time authorizing a preliminary expenditure of £1,000. In February, 1880, Mr. Davidson communicated with the late Mr. James Galbraith, a director of the Albion Shipping Co., and with him had a first interview with Messrs. Bell and Coleman in Glasgow. It was under Mr. Coleman's able supervision that the sailing ship *Dunedin* was fitted up, his early assurance to the promoters being that if the carcasses were hard frozen they would suffer no deterioration in the long voyage of 100 days or more in a sailing ship; that they could be frozen on board the ship itself without any refrigerating works on shore to assist; and, moreover, that with thorough insulation of the meat chambers in the ship, and with a proper system for the circulation of the cold air, the carcasses when frozen solid might be stored for the voyage as closely as they could be packed without risk of their being crushed. This was, indeed, a marvellous piece of prescience. Careful investigation having led Mr. Davidson to the conclusion that a Bell-Coleman cold air compression refrigerating machine was the best to adopt, a contract was finally concluded whereby the Albion Shipping Co. agreed to fit up one of its best and fastest sailing ships with insulated meat chambers, boilers, and Bell-Coleman refrigerating machinery, the Land Co. undertaking to find a cargo of meat (up to 7,500 sheep, if necessary) to fill the chambers, and to pay a freight of $2\frac{1}{4}$*d.* per lb., taking all the risk of the cargo arriving in a marketable condition.

In support of this enterprise, Messrs. William Ewing and Co., insurance brokers, of Glasgow, were plucky enough to accept what was a totally unknown risk by covering all contingencies attached to the carriage of the meat at the moderate premium of five guineas per cent. The ship *Dunedin*, of about 1,200 tons, commanded by Captain Whitson, was selected for the venture, and Mr. Davidson sent his instructions to Mr. Thomas Brydone, the Land Co.'s superintendent in New Zealand, to erect a killing shed in which to slaughter the sheep, to secure first-rate butchers, and in every way to prepare for the provision of a cargo of the most attractive classes of sheep. Under Mr. Brydone's able direction this work was carried out, and the slaughterhouse was erected on the company's Totara Estate. Looking back now, too much praise can hardly be given to the extraordinary wisdom and pluck shown by these *Dunedin* men. Failure in their enterprise might well have thrown back the industry for years.

The Voyage of the *Dunedin*.

It was decided to freeze on board, and the work was entered upon in a 'tween decks chamber on the *Dunedin* at Port Chalmers on December 7, 1881, when Mr. Davidson and Mr. Brydone personally stowed the first frozen sheep ever loaded in New Zealand, the question with them being as to whether the carcasses, after they had been frozen on board the ship, should be placed "thwart ship" or "fore and aft" in the chambers. All went well until December 11, when a fracture of the engine's crank-shaft owing to a flaw in

THE SHAW, SAVILL SAILING SHIP *Dunedin* WITH PORTRAIT OF CAPTAIN WHITSON.

To face p. 40.

the casting stopped the work and compelled the sale of the 641 sheep then in the chambers and of the 360 killed and in transit. Thus New Zealanders themselves were the first consumers of their own frozen meat. A repair was made, and the loading was completed on February 11, 1882, the ship sailing on the 15th of that month, and arriving in London Docks on May 24, after a long passage of ninety-eight days. During that period the refrigerating machine had worked steadily; sometimes, in cool weather, it was only run two or three hours in the twenty-four.

The anxiety as to the fate of the meat is aptly described by Mr. Davidson, who had then returned to London. "Captain Whitson," he says, "came on to London ahead of his ship in a pilot boat, looking very strained and careworn as he entered the shipping company's office. He was not quite sure about the condition of the cargo, but thought that most of it was sound. The vicissitudes of his experimental voyage were related, the captain's anxieties about the cargo having been aggravated by his dread that his masts would be burnt, as the sparks from the funnel set fire to the sails on several occasions. Then, in the tropics the ship was for a long time on one tack, and owing to its steadiness the cold air was not sufficiently diffused amongst the carcasses, and, in fact, the temperature in the upper chamber remained so high that the engineer was almost in despair." At last Captain Whitson had determined to alter the circulation of the air, which was evidently defective, and to do this he had to crawl down the main trunk, and in the process of cutting fresh openings for the better escape of the cold air he became so benumbed by the frost that he was only rescued from his perilous position by the mate crawling in behind him and attaching a rope to his legs by which means he was pulled out of the air trunk!

It was found at London Docks that the cargo had arrived in a sound condition, and its sale at Smithfield was at once arranged for, the cargo embracing 3,521 sheep and 449 lambs belonging to the Land Co., and 939 sheep supplied by "outsiders."

The results of the marketing of this pioneer shipment were

officially communicated to New Zealand, and the following is a brief extract of the particulars then forwarded :—

" The discharging of the cargo commenced three days after arrival, and the whole shipment was sold within a fortnight, the meat being taken out at night and conveyed to Smithfield market, so that the sheep were hard frozen when butchers went to buy them in the morning. There are no auction sales of dead meat in London, and the carcasses were sold at Smithfield in the usual way, by placing so many in the hands of some half-dozen salesmen who made as good prices for them as they could. At first the salesmen were rather doubtful about the venture being a success, especially as it was the first trial from New Zealand, but when they saw the fine big sheep, which, though many of them had been frozen over four months, were as clean and bright as newly-killed mutton, they quickly changed their opinion and pronounced the meat to be 'as perfect as meat could be.' New Zealanders will be pleased to learn that the shipment was mentioned even in the House of Lords. Excepting the very fat coarse sheep weighing over 100 lbs. each—several weighed over 150 lbs., and one 182 lbs.—the mutton was quite suitable for the English market. Out of the whole cargo only one sheep was condemned. Including some eight or ten sheep and lambs given away, which are entered at the average price of those sold, the following is an exact statement of the actual results of the shipment :—

	No.	Weight.		Per sheep. £ s. d.	Per lb. d.	£ s. d.
Sheep :	3,136	244,073	lbs. sold in London	2 2 7	6·56	6,675 9 8
	373	29,415	,, ,, ,, Glasgow	2 3 0	6·54	801 13 6
	8	477	,, ,, captain	1 9 9	6	11 18 6
	3		to order of manager	2 2 7		6 7 9
	1		condemned			
	3,521					
Lambs :	425	16,846	lbs. sold in London	1 1 4	6·45	453 0 11
	24	950	,, ,, ,, Glasgow	1 5 0	7·60	30 1 0
	449					£7,978 11 4
Pigs :	22	1,164	,,			31 2 11
						£8,009 14 3

CHARGES.

	£	s.	d.	£	s	d.
Brought forward				£8,009	14	3
Calico for bags	91	8	9			
Keeping meat frozen in *Dunedin* after her arrival in London	43	11	9			
Freight on 296,477 lbs.	2,779	9	5			
Insurance on £7,500	414	13	9			
Supervision during discharge, telegrams, etc. . .	29	13	6			
Dock Company's account for discharging . . .	78	5	7			
Carriage to Smithfield	65	10	0			
Carriage 400 sheep London to Glasgow . . .	51	4	3			
Sale, commission, bank charges, etc.	239	5	4			
				3,793	2	4
Net proceeds				£4,216	11	11

The net return per sheep in Port Chalmers is £1 0*s*. 11¾*d*. or 3·23*d*. per lb.
" " " " lamb " " " " 10*s*. 9*d*. " 3·25*d*. "

"I calculate there will be a net return of fully 9*d*. per sheep in New Zealand from sale of skins and tallow, after paying cost of killing and putting on board ship, so that the company has netted £1 1*s*. 8¾*d*. for its sheep, averaging rather under 81 lbs.

"The loss in selling weight, as compared with shipping weight, amounts to only a little over 1 lb. on each carcass. The charges between Port Chalmers and London, including insurance and freight, amounted to 2·73*d*. per lb., and, after the ship arrived in London, 0·41*d*. per lb. The sheep sent home would only have netted some 11*s*. or 12*s*. per head in the Dunedin market at the time of shipping, so that their value was about doubled. The 939 sheep sent by other shippers sold at about the same prices as the company's sheep, all but some very heavy ewes, which did not fetch so much per lb., but nevertheless brought full value considering the quality of the meat."

Mr. Davidson states that "After the success of the *Dunedin* shipment the Land Co. engaged the *Marlborough*, another fast sailer belonging to the Albion Shipping Co., and these two vessels each brought home very successfully about half a dozen cargoes of mutton, the former ship carrying about 9,000 carcasses and the latter about 13,000. Tragedy, however, awaited them, as, while still carrying frozen meat for the Land Co., both vessels left New Zealand within six weeks of one another, and neither was ever heard of again, the supposition being that they ran on icebergs off the Horn, these being very numerous about that time."

Mr. Brydone's Forecast.

Mr. Brydone took the deepest interest in the trade, for the successful inauguration of which he was largely responsible. In 1881 he read a paper at the annual meeting of the Otago Agricultural and Pastoral Association, giving an account of his investigations in Australia and elsewhere. His faith never wavered, for even as late as 1892, when many people in New Zealand thought that the export then reached of 2,000,000 carcasses was about the "top notch," he stated: "I should say there is every prospect of New Zealand being able to export 4,000,000 sheep ten years hence as easily as we do 2,000,000 now." Mr. Brydone's forecast was correct!

Canterbury and Otago erected a building in the Agricultural Hall, Christchurch, to his memory, called the Brydone Hall, and the farmers of the Oamaru district have erected a handsome memorial cairn on the Totara Estate, where the first sheep for the frozen meat trade were killed and prepared for export in the ship *Dunedin*. In reviewing the circumstances that led up to the successful pioneer shipment from New Zealand, one can have no doubt of the important part played by Thomas Brydone. It is clear that he had grasped some of the possibilities of the new trade and what they promised for the sheep farmer of New Zealand. Deservedly may he take front rank amongst the pioneers.

Supplementing Mr. Davidson's statement as to the outturn of the meat ex *Dunedin*, the market circular of the National Mortgage and Agency Co. of New Zealand, dated June 2, 1882, declared the shipment to have arrived in first-class condition. The carcasses were pronounced excellent in quality, exception, however, being taken to the very fat and heavy sheep. The consignment, it was stated, was selling at prices ranging from $5\frac{1}{2}d.$ to $6\frac{1}{2}d.$ per lb., lambs at $6\frac{1}{2}d.$ to $7\frac{1}{2}d.$ per lb., and it was added that the experiment was considered a great success. The other consignors of meat by the *Dunedin* referred to

THOMAS BRYDONE.

To face p. 44.

above were Messrs. James Elder (Maheno), J. H. Smith (Invercargill), Murray, Roberts and Co., and James Shand.

The *Times* had a leading article on the shipment, and called the venture a " prodigious fact," which it indeed was. The English butcher, with the fine New Zealand wether sheep before him, equal in quality to home-bred mutton, realized that a new era was opened up in the retail meat trade. He was well justified in his realization, for since May, 1882, some 78,000,000 New Zealand frozen carcasses have been handled by retailers in Great Britain, much to their profit, and more to the well-being of the British public who have consumed them.

CHAPTER III

THE FREEZING WORKS OF AUSTRALIA

THE development of freezing in Australia has not been on such an extensive scale nor so continuous as in New Zealand. This has been partly due to the periodic droughts from which Australia has suffered and the distances separating the sheep-breeding districts from the shipping ports. Moreover, in New Zealand the freezing industry arrived in the nick of time to place farming in that country on a paying basis, and the Colony found it absolutely necessary to follow it up with regularity. On the other hand, Australia's merino wool production—one of the grandest industries in the world—except in periods of depressed prices for the golden fleece, has always put the freezing of sheep into a secondary position: only the surplus sheep have been exported. Again, in New Zealand freezing has proved fairly profitable, and the business has been organized on systematic lines, while in Australia the freezing works as a whole have gone through many vicissitudes. Large schemes were propounded for covering portions of the continent with meat-freezing works, but before these schemes could be properly carried out the 1895—1903 drought intervened and prevented the proper development of the enterprises.

In a considerable number of the Australian works, especially those in New South Wales, stock are killed at various centres and then either railed or carted to the freezing works. After being frozen, the carcasses are again either railed or carted to the steamer, in most instances in insulated vans. The best works, however, are conducted on the same principle as in New Zealand and the Argentine Republic, the animals being slaughtered alongside the freezing chambers.

Queensland.

In the State of Queensland, where the general freezing scheme has been well designed, there have been big fluctuations in meat exports, as the few following figures will show:—

—	Cattle in the State.	Frozen Beef Exports to the United Kingdom.
		Cwts.
1893 . . .	6,693,200	207,000
1899 . . .	5,053,836	513,000
1903 . . .	2,481,717	77,000
1910 . . .	5,131,699	800,000

Most of the meat works are situated at the point of shipment.

Central Queensland Meat Export Co.—The first freezing enterprise in Queensland was that of the Central Queensland Meat Export Co. at Lake's Creek, near Rockhampton. As a meat works Lake's Creek was opened in 1871, and a freezing plant was added in 1883 (see p. 34 for a sketch of the early history of this concern).

Queensland Meat Export and Agency Co.—The Queensland Meat Export and Agency Co., with works at Eagle Farm, on the Brisbane River, and at Ross Creek, Townsville, was formed in 1890. Sir Thomas McIlwraith was a prime mover in floating the company with £1,000,000 capital, and Mr. John Cooke, who had newly come from New Zealand, assisted in its establishment. Of this nominal capital less than £100,000 was subscribed. Messrs. W. Weddel and Co. were appointed agents in London, and a five years' freight contract was made with Houlder Brothers and Co. for the conveyance of 1,200 tons of meat per month at $\frac{13}{16}$*d.* per lb. for mutton and $\frac{3}{4}$*d.* per lb. for beef. The Queensland Meat Export Co., which was formed to give graziers a market in the Old World for their surplus cattle, has had a varied career, bad luck attending the early shipments of beef, which suffered terribly from "bone taint." The quality of the cattle handled was excellent, and after the fault referred to had been overcome Queensland beef met a good demand at Smithfield. But shipments soon overtaxed the English market, and prices fell so low that the Continent of Europe was tried as a market; in some cases the Queensland squatters not only drew no cash for their consignments, but had to pay a reclamation charge. Then came the drought, with the consequent gradual reduction of shipments. The company, supported in the first place by squatters' consignments, had to buy on its own account, and of late years has depended for its business on shipments of frozen and preserved meats on its own account to British and Eastern markets. Mr. C. Ross is the manager of the Brisbane works, and Mr. Robert Stewart is in charge of the Ross Creek works. Some little time ago Messrs. G. S. Yuill and Co. purchased a controlling holding in the company, the London office of which is at 120, Fenchurch Street, with Mr. Warrington Laing as its London representative. The Queensland Meat Co.'s works form a good example of sound engineering equipment, and the refitting of the Townsville works, recently undertaken, brings them into line with the most up-to-date freezing works in existence in either hemisphere. The works now have the most modern machinery and appliances for handling produce, etc., that can possibly be procured, the original freezing machinery and other plant being discarded, and new freezing and electric plant being erected in its place. The whole of the new steam engines, both for freezing and electric service, are triple expansion with a superheated steam supply at 220 lbs. pressure, supplied by mechanically fired boilers, and the entire coal supply is worked by the latest coal-handling machinery. The whole of the freezing rooms, stores, etc., have been piped with direct expansion coils, in lieu of the air circulating batteries which were formerly in use. When these works were designed about twenty years ago, the old dry air system was installed but

QUEENSLAND MEAT EXPORT AND AGENCY COMPANY, LIMITED. ROSS RIVER WORKS, TOWNSVILLE.

QUEENSLAND MEAT EXPORT AND AGENCY COMPANY, LIMITED. EAGLE FARM WORKS, PINKENBA, NEAR BRISBANE.

To face p. 48.

a few years later the machines were replaced by ammonia compression plant, and the freezing rooms were operated by air circulating batteries. About two years ago it was decided to fit half of the total freezing capacity with direct expansion piping, and this, after working for a season, gave such excellent results and so thoroughly satisfied the directors of the company that they decided to pipe the remainder of the freezing block at large expenditure. In Townsville coal is expensive, the water supply limited, and drainage difficult; nevertheless, the improvements now made render the works second to none for economy and convenience in working.

Gladstone Meat Works.—The Gladstone Meat Works of Queensland had its inception in 1893, when a meeting was called by Messrs. J. H. Geddes and Co. to consider the question of establishing freezing works at Gladstone. The company was formed in 1894, and shipments began in 1896. Mr. W. B. Shaw was the prime mover in the formation of this company. The works can treat 150 bullocks and 2,000 sheep a day. The freezing chambers are substantially built of cemented brickwork, and the risk of fire is minimized by the use of pumice insulation. The refrigerating power is supplied by one of Haslam's 80-ton compound ammonia machines and one 100-ton compound Linde ammonia machine. The slaughterhouse is provided with hydraulic plant for handling the carcasses during dressing. The company has its own pier, at which ocean-going vessels berth. Mr. N. W. Kingdon is the manager.

Meat Works at Bowen.—The Merinda Meat Works, near Bowen, were established in 1895, Bergl, Australia, Ltd., being the proprietors, and Mr. Frank H. Houlder and Captain Thomas Hutton (Australia) its directors. The original concern, taken over by the Bergl Co., was the Bowen Meat Export and Agency Co. The works are a few miles up the railway, and the meat is trucked to the Bowen jetty and put on board the Houlder liners. The Merinda works have been kept thoroughly up-to-date, and many additions and improvements have been made, including the establishment of a canning plant. The works can treat 150 head of cattle for freezing and 50 head for canning daily, with a storage capacity of 1,500 tons.

Birt and Co.'s Works.—The important interests held in Australia in the way of freezing works by Messrs. Birt and Co., Ltd., Sydney—represented in London by Messrs. Birt, Potter and Hughes, Ltd.—may be set forth as follow. In 1895 Messrs. Birt and Co. leased the Government meat markets at Sydney, and later in the same year they also leased the Musgrave Wharf, Brisbane, and built freezing works on the property adjoining. From that time onwards they became one of the largest exporters of frozen meat from Australia. The Sydney works can deal with 3,500 carcasses of mutton per day, and have a storage capacity of about 40,000 carcasses. The Musgrave Wharf works can deal with about 120 head of cattle per day, and have storage for about 700 to 800 tons. In 1901 Messrs. Birt and Co. built their own killing and freezing works at Mooraree, eight miles from Brisbane (now called Murarrie), and at these works there is plant to kill and freeze 150 head of cattle and 600 sheep per day, with storage for 1,030 tons. Mr. E. Owen Cox is managing director for Australia.

Redbank and the Burdekin River Meat Works.—Reference has now been made to the Queensland beef freezing works possessing historic interest. Turning to modern times, Messrs. John Cooke and Co. erected the Redbank works on the upper Brisbane river, one of the largest in the State, in 1902. Two large Hercules machines are installed at the Redbank works, which until lately were under the charge of Mr. J. H. McConnell. The daily freezing capacity of the works is 1,500 sheep and 300 cattle; storage capacity, 100,000 carcasses. During the South African war Mr. Cooke supplied about 90 per cent. of the Australian meat issued to the British troops. As beef was largely called for in that connection, Mr. Cooke erected the Redbank freezing works, and with the aid of the Burdekin Meat Works Co., situated on the Burdekin river, North Queensland, in which he had a large shareholding interest, supplemented by the annual output of the Gladstone freezing works, which he regularly acquired, a large export trade to Africa, the Philippines, Siberia, Japan, Mediterranean ports, and the United Kingdom was built up, the only drawback being the long-continued scarcity of

THE MURARRIE FREEZING WORKS, BRISBANE, QUEENSLAND (MESSRS. BIRT AND CO., LTD.).

To face p. 50.

supplies resulting from the severe drought which broke up in 1902.

The importance of Queensland as a contributor of frozen beef to Great Britain is noted in the number of freezing works dotting the course of the Brisbane river. The works of Thomas Borthwick and Sons, Ltd., expected to start freezing early in 1912, are situated on the south side of the river, six or seven miles from Brisbane, and between the city and Moreton Bay. The buildings are within 200 feet of the river, and ocean steamers will load meat at the company's wharf. The site which, it is stated, has been acquired by the American Meat Trust for freezing works is about half a mile from Messrs. Borthwick's works.

A recent estimate of the capacity of the freezing works in Queensland gives 180,000 head of cattle treated during a six months' running season. This is on a basis of a daily slaughtering capacity of 2,230 cattle for the ten works in the State. In addition, over 1,000,000 sheep can be handled for freezing whilst the works are going at their full capacity for cattle.

New South Wales.

New South Wales Fresh Food and Ice Co.—Turning now to New South Wales, and considering the export of frozen mutton, it may be mentioned that these works at Darling Harbour, Sydney, opened in 1861 by Mr. Mort, are the oldest in Australasia. Mr. Mort invested £250,000 in the works; it was in 1874 that the company was formally incorporated under its present title. In 1898 were erected for the New South Wales Fresh Food and Ice Co., Ltd., in Liverpool Street, Sydney, works designed to treat about 5,000 sheep daily or an equivalent in beef, etc., and to store about 85,000 carcasses of mutton as well as large quantities of butter, ice, fish, etc., in addition to which about 32 tons of ice were manufactured daily in the same building. In arrangement these freezing rooms and stores were very much after the style of those at Deniliquin, mentioned hereafter, but considerably larger and much more heavily piped with ammonia coils, this being done to enable the company to treat meat, when desired for shipment, in the shortest possible time, which is frequently very essential when completing the loading of a steamer with a special brand of meat. The company do not slaughter at their works, but only freeze. The direct expansion system of refrigeration is employed. The company have a very extensive local connection in the supply of refrigerated provisions. At these works are frozen the sheep slaughtered at the Riverstone establishment, thirty miles from Sydney, owned by Messrs. B. Richards and Sons, who have shown regularity and courage in their export trade, having taken their chances year in year out in selling their mutton on the London market. Mr. A. E. Pitt is the London agent. Some years ago the Riverstone Meat Co. had several stalls in the Smithfield market annexe.

Australian Chilling and Freezing Co.—The Australian Chilling and Freezing Co., Ltd., a London concern of which Sir Montague Nelson is chairman, first opened works at Aberdeen, on the Hunter river, in 1891, and in February, 1892, the s.s. *Port Douglas* took the first shipment of 13,000 carcasses of mutton. Mr. W. A. Benn, of Sydney (prominently connected of late years with the frozen rabbit business), was for many years manager of the company. The Aberdeen works at the start offered three forms of contract: first, to purchase delivered, fat wethers weighing 47 lbs. and upwards, dressed, with shanks off, and kidneys and kidney fat removed, at 1*d*. per lb. cold weight, all offal, including fat, to belong to the company, skin and wool to seller; second, partial sale, the company to make an advance of $\frac{3}{4}$*d*. per lb. on sheep, and in the event of the mutton selling on average above $3\frac{1}{8}$*d*. per lb. the company returning shippers 75 per cent. of surplus, the offal to belong to the company, the skins and wool to owner; the third form, to consign on owner's account, the company treating the meat, and bagging and shipping it, as well as paying all charges for a consolidated rate of 2.20*d*., giving shippers an advance, and keeping offal and by-products as before. These works have had to contend with variable seasons and low London values for the excellent class of mutton and lamb exported, and their operations have benefited New South Wales and sheep growers more than the shareholders. Mr. R. C. McAdam is the manager of the company.

Pastoral Finance Association.—The era of modern well-equipped freezing works in the parent State of Australia began in 1891. In that year a freezing plant was installed at the Pastoral Finance Association's premises, Kirribilli Point, Sydney Harbour. The first directors were Messrs. J. H. Geddes (manager and managing director), J. F. Burns, Hon. R. H. Roberts, Hon. G. H. Cox (chairman), Russell Barton, and J. B. Christian. The daily freezing capacity (Linde machinery) at the start was 1,500 carcasses, now it is 2,500; storage capacity was at first 50,000 carcasses, now it is 75,000. The total quantity of meat shipped from 1891 to (March) 1909 was 2,750,000 carcasses. The present directors are Messrs. Russell Barton (managing director), J. B. Christian, Hon. G. H. Greene, and W. F. Jacques; Mr. H. Chilton-Young is the manager. Owing to the severe drought the works closed down from 1899 to 1904. No slaughtering is done at this establishment.

Chilling Up-Country.—An important feature of the freezing industry which was about the beginning of the nineties advocated by some of the most able and practical men connected with sheep raising in New South Wales, was the slaughter of the sheep as near to their pastures as possible, chilling the carcasses there, and railing the chilled mutton to port for freezing. Mr. Robert Hudson, of Melbourne and Sydney, was the first man to introduce this chilling at inland centres. Works were started at Narrandera (Riverina) and at Tenterfield in the north. Refrigerator cars were attached to the trains for the conveyance of the meat to Sydney for freezing. The works were well equipped, the engineering work being undertaken by Messrs. J. Wildridge and Sinclair, Ltd. But neither of the works was operated very successfully, as they had to compete with the Homebush market, and carriage and freight costs were against the enterprise. The Tenterfield works were dismantled shortly afterwards, but the Narrandera works are still in operation, and are even to-day being refitted to deal with sheep for export. A carload of mutton from Narrandera was offered by auction by Goldsbrough, Mort and Co. in 1890, this handselling of the up-country scheme representing $1\frac{1}{2}d.$ to $1\frac{3}{4}d.$ per lb.—a poor start! The late Mr. J. H. Geddes was enthusiastic in recommending this system. The late Alexander Bruce, Chief Inspector of Stock, New South Wales, also highly approved of up-country chilling, which was thoroughly discussed by Mr. Cuthbert Fetherstonhaugh in his pamphlet "The Meat Export Trade," published in 1894. The authors acknowledge indebtedness for information to this pamphlet and to Mr. Bruce's "The Meat Trade of Australia," 1895. On this basis the Young Meat Chilling and Export Co. was started in 1892 with a guarantee of 200,000 sheep and 3,000 cattle yearly. The mutton was killed and chilled at Young, and was then sent down in a refrigerated car, run on board a specially constructed craft at Darling Harbour, and towed across to Kirribilli Point, where the truck was unloaded at the works right into the freezing rooms. A similar plant on a rather larger scale was fitted up at Dubbo, and one also at Gunnedah. But the risks and costliness of this method put it out of court; a sheep cost $2\frac{1}{2}d.$ per lb. to kill and send from Young to London. The sponsors for this up-country chilling took their cue from North America, where it was, and is, practised extensively (but not for *freezing* afterwards).

Riverina Frozen Meat Co.—It may be mentioned that in 1892–1893 there were about 56,000,000 sheep in New South Wales, and a surplus of about 13,000,000 wethers of fattening ages, and it was desirable that steps should be taken to increase freezing facilities. In 1892 Mr. John Cooke, who had gone to Melbourne from New Zealand in 1889, promoted the Riverina Frozen Meat Co., which built its works at Deniliquin, New South Wales, 200 miles from Melbourne. This was the first concern to demonstrate the practicability of safely carrying frozen carcasses long distances and in ordinary insulated wagons in summer. The main idea was to supply country-killed and frozen mutton and lamb, driven short distances from their own pastures to the works and thus spared the deterioration resulting from long railage of the live animals to Melbourne. The first directors of the Riverina Co. were

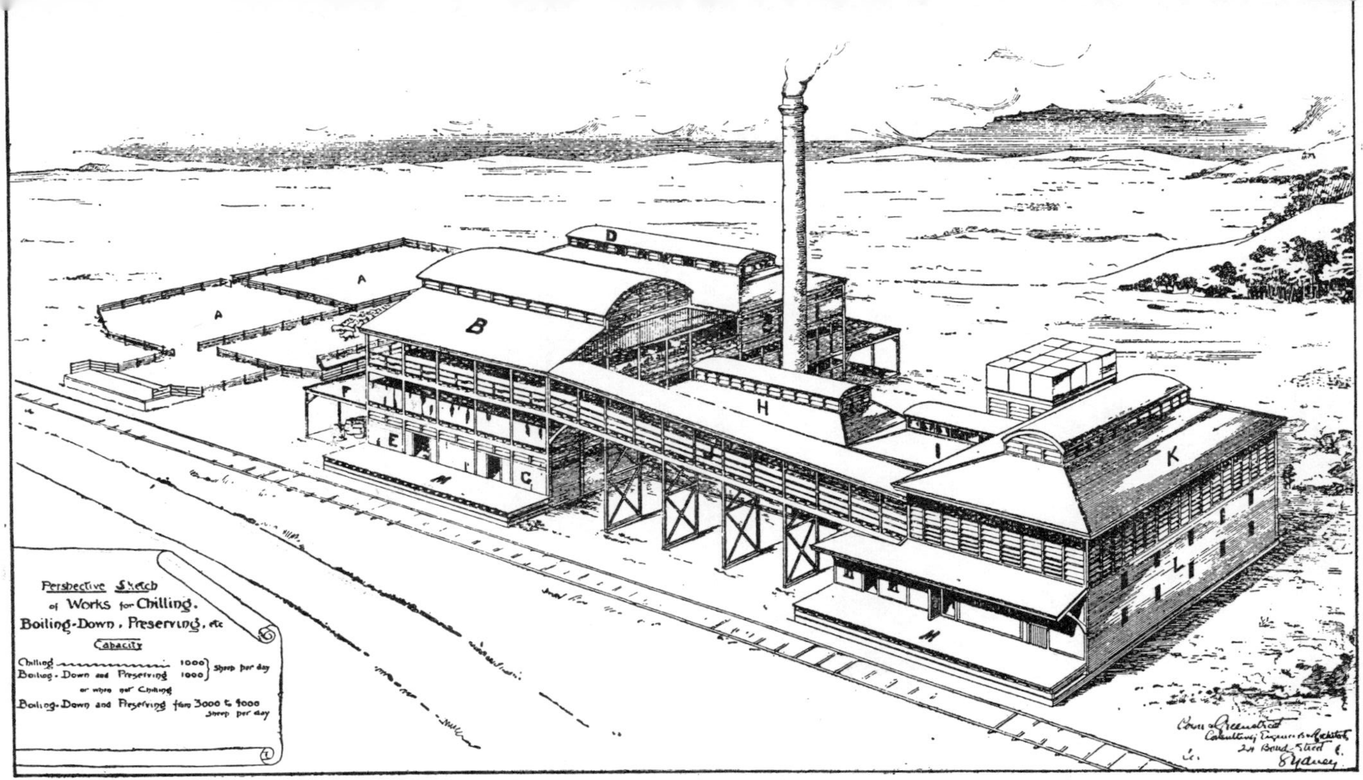

BIRD'S EYE VIEW OF THE RIVERINA FREEZING WORKS, DENILIQUIN, NEW SOUTH WALES.

A. Receiving and Drafting Yards.
B. Slaughter-house.
C. Digester-house.
D. Preserving Rooms.
E. and G. Stores for By-products.
F. Cooperage.
H. Boiler-house.
I. Engine-room.
J. Inclined Passage.
K. Cooling Rooms.
L. Chilling Rooms.
M. Loading Platforms.

To face p. 54.

Messrs. John Cooke, Albert Austin, R. O. Blackwood, S. Frazer, L. Kiddle, J. Raleigh, and R. B. Ronald. Dalgety and Co., Goldsbrough, Mort and Co., the Australian Mortgage Land and Finance Co., and the Australian Mortgage and Agency Co. co-operated with Mr. Cooke in securing the necessary capital. The first shipment of meat was despatched from Melbourne in 1895 per s.s. *Maori*. The long series of droughty seasons in the Riverina reduced surplus fat stock to such an extent that the works have never been kept going steadily, and, indeed, have perforce remained idle for several consecutive years. Mr. Frank Coxon, the well-known Australian consulting and refrigerating engineer, says of the Deniliquin works that they were at the time of their erection far and away the most complete in Australasia or elsewhere for the economical handling and freezing of sheep, and although certain after-provisions were made for the treatment of a limited number of cattle, that was of small moment, the Riverina not being a large beef-producing district. Everything, he adds, went exceedingly well with the Deniliquin works until the great drought set in, and it is a noteworthy fact that, notwithstanding that those works are situated about 245 miles from Melbourne, the port of shipment, shipmasters and others who have handled the meat declare that it is as good as the best frozen and hardest meat shipped from that port.

Graziers' Meat Export Co.—The other company formed in 1892 to build works inland was the Graziers' Meat Export Co. of New South Wales, capital £250,000. It was Mr. Fetherstonhaugh who, in spite of the apathy of the squatters, worked up this concern, and the names associated with its initiation, in addition to his, were W. A. Cottee, T. F. Knox, George Maiden, W. F. Lawry, F. W. Bacon, W. H. Armstrong, and George Mair. Works were put up at Sandown (Parramatta River), Werris Creek, Carrathool, Nyngan, and Forbes. Sandown was a freezing establishment, the other four being up-country chilling depôts. The career of the Graziers' Co. was disastrous, and about 1901 the whole of the works were acquired by a syndicate of London capitalists called the Austral Freezing Works, Ltd., the country works being dismantled or sold, experience having proved that a central works at Sydney, where the supplies of fat stock from nearly all parts of the State are marketed, and where large and regular freezing operations are possible, is essential from every point of view. Sandown only is now worked—by Messrs. John Cooke and Co., who took it over from the Austral Co. in 1902.

Sandown Freezing Works.—The Sandown undertaking has been extended from time to time, and can now treat 6,000 sheep and 150 bullocks daily,

From the above short history regarding country-killed frozen meat the conclusion may be drawn that, although the principle was apparently a sound one, it has since been demonstrated that such meat, whilst admittedly a better article of food, costs more and does not command an enhanced price over that railed alive to the seaboard and treated there.

In August, 1911, the farmers of the Byron Bay (north coast) district formed "the Byron Bay Co-operative Canning and Freezing Co.," capital £50,000, with excellent prospects of founding an important concern.

Victoria.

Newport Freezing Works.—Victoria was early in the field, for this company, formed in Melbourne after the *Strathleven* shipment, erected works at Newport, near Melbourne, the second freezing works erected in Australasia. The Victorian Government afterwards acquired Newport, which first assumed commercial export shape when Mr. John Hotson secured a lease of the works. In 1893 Nelson Brothers, Ltd., and Mr. Hotson came to a working arrangement to freeze for export regularly at Newport. In 1896 Mr. Hotson sold out his interest to the Austral Co., and the firm of John Cooke and Co. has ever since used the works as its chief base for preparing and shipping its well-known "Champion brand" of mutton and lamb. Fuller details of the formation and operations of these works appear in "The Work of the Pioneers," Chapter II.

Portland Freezing Works.—These works were built in

THE FREEZING WORKS OF MESSRS. THOMAS BORTHWICK AND SONS (AUSTRALASIA), LTD., AT PORTLAND, VICTORIA, AUSTRALIA.

To face p. 56.

1895, and were taken over in 1901 by Messrs. Thomas Borthwick and Sons, Ltd., from the Portland and Western District Freezing Co. The daily capacity is 1,200 lambs; storage capacity 25,000 carcasses of mutton. Messrs. Borthwick also have works near Melbourne.

Western Freezing Co.—These works at Geelong also date from 1895, and were likewise (in 1901) acquired by London capitalists; the purchasers, Messrs. W. and R. Fletcher, Ltd., have operated the works regularly up to the present time. Both Messrs. Fletcher's works and those of the Geelong Harbour Trust **(Corio Freezing Works and Abattoirs)** are well designed and compact establishments. At the latter works, which were opened in 1909, an excellent innovation has been introduced. A stream of water flows at the back of the slaughterers, who, in killing, throw the offal behind them, and the water conveys it away to a grating, where the fat is collected.

Government Cool Stores.—This large undertaking, belonging to the City Council, established in the heart of Melbourne, is under the charge of Mr. R. Crowe, and turns over a considerable quantity of frozen lambs, poultry, and sundries. The Victorian Government have decided to erect works of their own near the docks at a cost of £75,000.

W. Angliss and Co.—One of the large freezing works of Australia is that belonging to Messrs. W. Angliss and Co., at Footscray, near Melbourne. The Imperial Freezing Works were opened in 1905. Messrs. Angliss also own works in Bourke Street, Melbourne. Mr. A. E. Pitt, 64, West Smithfield, E.C., is the London agent of Messrs. Angliss.

Wimmera Co-operative Freezing Works.—The opening of the Wimmera Co-operative Freezing Works at Murtoa, Victoria, on September 28, 1911, marked the commencement of yet another effort to make up-country freezing establishments a payable proposition in Australia. These works, which constitute one of the biggest co-operative schemes in the State, cost £45,000, and have storage accommodation for 40,000 carcasses, and are capable of treating 2,500 a day. The first shipment—10,000 lambs—was made on October 21.

Four other freezing works in Melbourne occasionally available for meat export are: The Melbourne Ice Skating and Refrigerating Co.; Messrs. J. Bartram and Sons' works, Flinders Lane; Sennitt and Sons Proprietary, Ltd., Miller Street; and the Victorian Butter Factories Co-operative Co., Flinders Street.

South Australia and Western Australia.

Freezing works in Adelaide, South Australia, were suggested as early as 1891. The Government of the State started freezing by building State refrigeration works at Port Adelaide, and has been much criticized for so doing. It is an axiom that the Government of a country should not interfere with private enterprise. But where private enterprise is lacking in a young country, the State is justified in stepping in to promote and conduct an industry required for the country's development. About 1894, Mr. D. J. Gordon, M.H.R., then commercial editor of the Adelaide *Register*, strongly advocated the establishment of the frozen lamb export trade, and he called a meeting for the discussion of the subject. Many South Australian leading pastoralists who were present threw cold water on the proposal, and said that South Australia would never be a lamb-exporting country. Capital to float a private company formed to build freezing works at Port Adelaide could not be obtained. Now was the occasion for the Government of the Colony to step in.

Port Adelaide Works.—The influence of the late Hon. Thomas Price, leader of the Labour Party in the South Australian Parliament from 1901, and afterwards Premier, and of Sir John Cockburn, Premier in 1889—1890, was successful in getting the Port Adelaide works erected; this establishment, equal to handling 300,000 sheep and lambs yearly, is one of the largest in Australasia. Mr. Gordon's spirited advocacy of the lamb export trade has been justified by the considerable number of lambs which have been and are now put through these works. Exporting was begun in 1895, and to June 30, 1911, 2,468,076 animals had been frozen for export, and for the last season 241,533 carcasses were shipped, of which number 195,000 were lambs. In 1906 it was apparent that the works needed enlargement and reconstruction in order to deal effectively with the rush of lambs that had to be handled smartly during the short lamb-freezing season. An investigation of the latest methods adopted by the New Zealand freezing companies was made by Mr. C. F. G. McCann on behalf of the Government of South Australia, and as the result of the special enquiries made by this gentleman (who is now Trades Commissioner for South Australia in London) the Port Adelaide works were considerably enlarged and modernized, especially in the direction of economical treatment of the by-products. The only other exporting freezing works in the State is the works formerly owned by the Adelaide Ice and Cold Storage Co. This was purchased by the South Australian Government in 1910 for £35,000. So the State has the monopoly of freezing for export in South Australia.

Western Fresh Food and Ice Co.—This company completed a small works at Fremantle, Western Australia (which also has inland freezing establishments at Perth and Kalgoorlie), in 1906; but the only frozen meat export taking place thence has been a tentative shipment or two of lambs sent to London. The Western Australian Government are going to erect works at Wyndham and other ports on the north-west coast, and if the intentions, as stated, of the Bovril Australian Estates, Ltd., an important London company, are carried out, the cattle owners of North-Western Australia will have an outlet for their stock in the shape of a meat freezing works on the northern coast line of Western Australia.

CHAPTER IV

NEW ZEALAND FREEZING WORKS

THE frozen meat industry in New Zealand was started by the large sheep owners with the assistance of the various land, finance, and mercantile companies, and it may be said that but for the assistance of these companies there would not have been any chance of the industry being floated. The majority of the sheep farmers in the colony were so involved, and the price of stock was so low, at the beginning of the eighties, that the pastoral industry was largely in the hands of these companies. The early shipments of frozen meat were on account of the growers, and were consigned to the various financial companies in England doing business with the Colony. The freezing companies came to be termed "farmers' companies," many of them simply freezing on behalf of the owners.

The progress of the frozen meat trade in New Zealand can probably be best reviewed by referring to the establishment, one after another, of the freezing works in that country.

New Zealand Refrigerating Co.—The first enterprise of the kind was the New Zealand Refrigerating Co., which was formed in 1881 with freezing works at Burnside, near Dunedin, and a few years afterwards at Oamaru. This concern, though the pioneer company, never reached any great proportions. For some time the directors endeavoured to carry on purely a freezing business; in later years they developed into a buying company. The Dunedin works were the first in New Zealand. The first directors were Messrs. John Roberts, C.M.G., (Murray Roberts and Co.), W. I. M. Larnach, E. B. Cargill, E. I. Spence (Dalgety and Co.), Robert Wilson, A. C. Begg (Robert Campbell and Co., Ltd.), and James Shand. The company froze the meat cargo for the s.s. *Marsala*—the third shipment—8,506 carcasses; this shipment, made

PICTON FREEZING WORKS OF THE CHRISTCHURCH MEAT CO., LTD., NEW ZEALAND (SHOWING THE OLD HULK, *Edwin Fox*).

To face p. 60.

in 1883, being the first from New Zealand by steamship. The company also loaded, in 1883, the steamers *Sorrento*, *Fenstanton*, and *Ionic*, with respectively, 8,295, 7,840, and 9,591 carcasses, and also killed for the *Dunedin's* second voyage 8,295 sheep, these being railed to the *Dunedin* and frozen on board. The *British King* also sailed in the same year, 1883, with some 10,000 carcasses. In 1884 the New Zealand Refrigerating Co. froze and shipped 72,420 carcasses up to July, and up to July, 1885, it froze and shipped 77,370 carcasses and 54 cattle. During 1885, 39,370 carcasses of sheep were frozen on board the s.s. *Elderslie* at Oamaru. The freight paid on the mutton carried by this vessel was 2¼d. per lb.

Mr. John Roberts remained a director of the New Zealand Refrigerating Co. till 1905, when it was wound up and absorbed by the Christchurch Meat Co. Mr. Thomas Brydone had joined the board in 1884, and these two gentlemen always took a leading part in the development of the trade. The Otago District thus started the frozen meat trade in New Zealand, but now lags far behind, being outstripped by Canterbury, and Southland, and the North Island.

Canterbury Frozen Meat Co.—The second company to be formed was the Canterbury Frozen Meat and Dairy Produce Co. with a capital of £20,000. The names on the circular convening the meeting on November 11, 1881, which led to the formation of the company, were :—John Grigg, of Ashburton, John Tinline, of Amuri, and John Macfarlane, of Coldstream, Rangiora. Mr. John Cooke, then Canterbury manager of the New Zealand Loan and Mercantile Agency Co., was largely instrumental in the formation of the company ; he wrote to the gentlemen named, and obtained their permission to summon the meeting in their names. Messrs. Frederick Banks, William Chrystall, John Cooke, John T. Ford, and John Grigg, were the first directors, Mr. Frederick Waymouth being appointed secretary, *pro tem.* The promoters, naturally, knew very little about mechanical refrigeration then, and had to grope their way from the very start. There was the inevitable "battle of the sites," and Lyttelton itself being impossible, they selected Belfast, an excellent location, as it proved. A Haslam refrigerating machine of 40,000 cubic feet cooling capacity was ordered. The company was fortunate in getting as its architect and engineer Mr. Frank Coxon, now of Sydney.

The daily killing and freezing capacity was only 250 to 300 carcasses of mutton, and it was predicted by many business men that the export of 2,000 carcasses weekly, if maintained, would soon deplete the flocks of Canterbury and bring about something like a meat famine. As the freezing works of Canterbury have now a weekly killing and freezing capacity of fully 150,000 sheep and/or lambs, while the province must have exported nearer thirty than twenty million carcasses since 1883, such prophecies now seem very absurd. Slaughtering began on February 16, 1883, and in April the first shipment was made from Belfast on the s.s. *British King*. Curiously enough, this vessel was built at Belfast, Ireland, and she was commanded by Captain Kelly, of Belfast, while Mr. John Cooke was born in Belfast, a somewhat remarkable chain of circumstances.

For about five years the Canterbury Frozen Meat Co., which was merely a freezing company not operating in stock or the export of meat on its own account at all, had the field to itself. Another company was promoted, and some capital raised, but it did not even buy a site and eventually refunded to the applicants the money subscribed. One of the leading features of the Canterbury Company was strict adherence to a high standard of quality, including an absolute embargo on the freezing of old ewes. Nothing but prime wethers and maiden ewes and prime lambs were accepted by the company for treatment, and the directors resisted all pressure to relax this rule; but in 1890 ewe mutton was taken for freezing, and some time after that second quality lambs were frozen and shipped. The company's leading brands are "C F M Co" and "Diamond." It took Belfast six years to record 1,000,000 sheep frozen. In 1887 the works were enlarged and completely remodelled; in 1889 by-products were taken in hand; in 1893 a Hall's CO_2 machine was installed; in 1896 a Hercules machine was added; and by

THE GEAR MEAT PRESERVING AND FREEZING CO.'S WORKS, WELLINGTON, NEW ZEALAND.

To face p. 62.

1908 the works had a capacity of 6,000 sheep and lambs per day, and storage equal to 120,000 carcasses. In 1899 the company established works at Fairfield, Ashburton. Works at Pareora, South Canterbury, were opened on April 7, 1904; these have a capacity of 4,500 sheep and lambs per day, and storage space equal to over 100,000 carcasses.

The question of freight greatly embarrassed this company's early operations. In May, 1887, the company, together with Nelson Brothers and the Southland company, signed a contract with the Tyser Line, and with the advent of the *Balmoral Castle*, the first Tyser steamer sent out, the difficulty about tonnage gradually disappeared. The capital of the company is now £225,000, and during the twenty-five years of freezing the company has paid in all 191 per cent. dividend on the ordinary shares. The present directorate is: Messrs. John C. N. Grigg, James Gough, Sir George Clifford (chairman), George Humphreys, and R. H. Rhodes.

Gear Meat Co.—The next freezing concern to be registered in New Zealand was the Gear Meat Preserving and Freezing Co. of New Zealand. This was formed in November, 1882, for the purpose of acquiring the butchering and meat preserving business of the late Mr. James Gear, at Wellington, in the North Island. The slaughtering capacity at the start was 500 sheep and 40 cattle per day; at the present time it is over 6,000 and 100 respectively. The refrigerating system originally employed was cold air compression, afterwards replaced by CO_2 compression machinery. The first directors were Messrs. P. A. Buckley, J. Duthie, R. Greenfield, W. H. Levin, J. R. Lysaght, J. McKelvie, N. Reid, J. S. M. Thompson, and James Gear, managing director. The present board are Messrs. D. Anderson, H. Beauchamp, H. D. Bell, J. R. Blair, A. K. Newman, N. Reid, and W. H. Millward, chairman and managing director, and the secretary is Mr. W. H. Tripe. Mr. Gear died in 1911, at the age of seventy-five, and was chairman of directors up to the time of his death. With a consistent policy and under able management the Gear Co. has had a very successful career. It possesses one of the most complete freezing establishments in Australasia.

Nelson Brothers.—The operations of Nelson Brothers have been alluded to; their works were first erected at Tomoana, and afterwards at Waipukurau and Woodville. The company now has two establishments in the Hawke's Bay district—Tomoana and Gisborne. Woodville has been sold to a bacon-curing company, and Waipukurau is dismantled. The daily killing capacity and the sheep storage of these works are respectively 6,100 and 160,000. Messrs. Nelson Brothers also have works at Hornby, Canterbury district. The company built the Ocean Beach works at the Bluff, now owned by Messrs. Birt and Co., Ltd., of London. It is generally considered that the most carefully conceived plan for carrying on the frozen meat trade was that proposed by Mr. William Nelson. Many of the New Zealand sheep growers had no faith in the permanency of the new industry, and some were actively in opposition, so Mr. Nelson, who formed clear ideas as to buying arrangements in New Zealand and a selling organization in London, entered into contracts with farmers in various districts. In 1887-88 Nelson Brothers made forward contracts with sheep farmers, giving 2*d.* per lb. for the carcass unfrozen, sellers also getting the full value of the skin and fat.

The Christchurch Meat Co.—In 1888 it was made quite clear that many producers were dissatisfied with the system of consigning their meat to London for sale, and that they wished to determine their risks by selling their stock alive to a freezing company or a speculative exporter. Up to that time freezing had been carried on by the larger growers and stock owners. The farmer, in contradistinction to the sheep grower or squatter, realized that he was not obtaining for his stock a price equivalent to what was being realized by those who were freezing and shipping sheep. He himself had not the large supplies to draw from to enable him to contract forward for freezing space, or to make regular shipments to London. The late Mr. James Watt, and the late Mr. Peter Cunningham, in conjunction with Mr. John Cooke, were mainly instrumental in forming the Christchurch Meat Co., Ltd., a concern established to conduct the business on new methods. The old Templeton preserving

THE FREEZING WORKS OF MESSRS. NELSON BROS., LTD., AT GISBORNE, NEW ZEALAND.

To face p. 64.

works and property were acquired, the name being changed to Islington. Thus the Islington freezing factory came into existence. The first directors of the Christchurch Meat Co., Ltd., were Messrs. J. M. Watt, John Cooke, David Morrow, Joseph Murgatroyd, Mcgregor Watt, Peter Cunningham, and S. K. Bassett. The present directors are Messrs. H. A. Knight (chairman), Thomas Teschemaker, John Roberts, C.M.G., and the Hon. H. F. Wigram. Mr. Wm. Murray is manager, and Mr. W. O. Campbell is secretary.

A word or two may here be said about the London representation of the Christchurch Meat Co. In 1896 the company decided to send Mr. John A. Randall to supervise its interests in London. Mr. Randall had a thorough knowledge of the trade from a farmer's point of view; he remained in London for about a year, until Mr. Robert Galloway, the company's manager at Timaru, was sent to London as the representative. On Mr. Galloway leaving to join the firm of Gordon, Woodroffe and Co. in 1900, Mr. Randall again took charge. On Mr. Randall's early and regretted death in 1901, Mr. A. W. Pottinger took the position, which he held till 1905. From 1905 to 1910 Mr. William Henderson represented the Christchurch Meat Co. at West Smithfield, and on his term of agreement expiring he returned to New Zealand, when he was succeeded by Mr. F. T. Boys, the company's secretary at Christchurch; Mr. Boys is the London manager at the present time; the office is at 64, West Smithfield, E.C.

At the company's start the slaughtering capacity was 500 sheep per day, and now it is about 15,000 and 100 cattle at the Islington, Smithfield, Picton, Oamaru, and Burnside works. Cold air refrigerating machinery was originally used, but the ammonia compression system is now employed. The successful starting of the company took place at the beginning of the era of the modern, well equipped, freezing works. The Christchurch Meat Co. was really founded in the interests of the small farmer who could not afford to consign his meat to London, or whose stock available for freezing was not considerable enough for him to undertake the export trade. The company pooled these farmers' lots.

Mr. John Cooke sends the following remarks concerning the new conditions then introduced: "The rivalry created between the two freezing companies was very beneficial to producing interests, as not only did it ensure full values, but it proved a great stimulus to breeding and fattening; indeed, the example shown to the rest of New Zealand is admitted to have been one of the most important factors in the great progress of the Colony during the last twenty years. Prior to 1883 there were periods of great depression in values of stock and landed property, but the frozen meat industry quickly provided a huge outlet, and gave a permanence to pastoral and agricultural enterprise which cannot be disputed. I remember seeing shorn sheep sold by the score because the price was so low that they had hardly any value per head. The pelts after the wool was taken off were so valueless at one time that the cheapest method of disposing of them was by burying them in pits immediately they left the puller's beam.

"One of the initial obstacles to the speedy expansion of the trade was the scarcity of refrigerated space, and the greatest trouble was experienced by freezing companies (1) in getting sufficient insulated tonnage, and (2) in obtaining a reasonable rate of freight. The rate of freight was originally 2½*d.* per lb., including freezing on board ship, but when steamers were introduced a reduction to 1¾*d.* per lb. was secured, at which it stood for some time. The total charges for treatment, freight, insurance, and selling, were originally in the neighbourhood of 3*d.* per lb., a figure which was prohibitive if the trade was to assume any magnitude, and having regard to the undoubted prejudice which frozen meat had originally to contend with at the hands of the British consumer.

"Generally speaking, the shipowners from the outset recognized the wisdom of friendly and active co-operation with the freezing companies, but they made one serious blunder which producers and freezing companies deeply resented, but which, fortunately, was quickly remedied. They allied themselves with certain manufacturers of refrigerating machinery under an agreement whereby the manufacturer was to supply their steamers alone with his plant, while the shipowner was to use

THE ISLINGTON WORKS OF THE CHRISTCHURCH MEAT CO., LTD., CANTERBURY, NEW ZEALAND.

To face p. 66.

that special plant exclusively. By that attempted restraint it was expected that outside refrigerated tonnage would be excluded from New Zealand, and that further reduction on freight would be prevented.

"One of the great ambitions of the directors of the Canterbury Frozen Meat Co., actively led by the late Mr. John Grigg, was to get the rate of freight reduced to the "round 1*d*. per lb.," and when this was accomplished, with the assistance of the late Mr. W. H. Tyser, there was great rejoicing in pastoral circles. This brought the consolidated charges down to under 2*d*. per lb., and was the means of stimulating production all over the South Island in a remarkable way."

The first venture undertaken by the Christchurch Meat Co. was killing the stock at Islington, and freezing on board the sailing ship *Wellington*. This shipment, which realized remarkably good prices, consisted mainly of lambs, and was handled by Messrs. W. Weddel and Co., who had just started in business when the cargo was entrusted to them. The cargo was sold splendidly. The prices realized would make shippers' mouths water to-day; from 6*d*. to 9*d*. per lb. was fetched for lambs! The Christchurch Meat Co. saw that if it was to carry out the wishes of the farmer, it must be in a position to do the work thoroughly, and, therefore, special attention was paid to working up the by-products. Up to this time the pelts and the largest portion of the viscera were simply buried, in many works even the blood was allowed to run to waste. The Christchurch Meat Co.'s first concern was to utilize these by-products, and to work up a scheme which would result in the company making the greater proportion of its profits out of these and the skin and fat, looking on the meat mainly as a by-product of the works.

Within two years of the start of the company's operations, that is, in 1891, Mr. Gilbert Anderson was asked to take up the position of managing director, and it was under his management that the scheme just mentioned was carried out, and that the general organization of the company's business was brought up to an excellent standard. Special attention was given to grading the sheep, not only for quality, but for weight. This enabled the c.i.f. business to be put on a proper footing, and the standard and grades which were established at these works have been extensively adopted throughout the New Zealand freezing industry.

The Christchurch Meat Co., in 1893, took over the works and business of the South Canterbury Freezing Co. at Timaru. In two years' time these works were enlarged, and named Smithfield. Another absorption took place in 1899, when the Wairau Freezing Co.'s business was taken over, this necessitating the erection of new and modern works at Picton. In 1905 the company absorbed the New Zealand Refrigerating Co., with its works at Burnside, Dunedin; and Oamaru. The Christchurch Meat Co. has always been progressive, enterprising, and modern. In 1906 Mr. Anderson retired from the company.

Wellington Meat Export Co.—The Wellington Meat Export Co. was incorporated in September, 1881, as a farmers' freezing company. The late William Dilnot Sladdon, who joined the company as manager, later added the c.i.f. principle to the business; Mr. F. D. Sladdon, his son, is the present secretary. This company has shown considerable enterprise in the utilization of new machinery, the directors being the first to introduce ammonia refrigerating machines into use at freezing works. The company's works have been recently rebuilt in brick, and raised to a killing capacity of 6,000 sheep per day with a storage capacity of 150,000, the freezing capacity at the start being only 300 sheep per day. The producer gas system has been introduced for power, driving not only the refrigerating machines but also the by-products machinery. Hercules ammonia compression machinery cooling on the direct expansion system is now installed, the plant used at the beginning of operations being the Haslam air compression machine. The first directors were Messrs. W. C. Buchanan, W. Booth, George Beetham, J. T. Dalrymple, H. H. Lang, J. R. Lysaght, J. E. Nathan, Chas. Pharazin, and D. Peat. The present directors are Messrs. W. G. Foster

THE NGHAURANGA WORKS OF THE WELLINGTON MEAT EXPORT CO., LTD., NEW ZEALAND.

To face p. 68.

(managing director), W. C. Buchanan, M.P., W. H. Beetham, Charles Elgar, E. Newman, M.P., J. W. Marshall, and J. Campion.

Longburn Freezing Co.—The Longburn Freezing Co. was established in 1895; Longburn is 80 miles from Wellington, on the main line of railway. The original directors were Messrs. C. Bull (chairman), J. McLennan, R. S. Abraham, J. O. Batchelor, D. Buick, and — Howard; Mr. J. Beale was manager. The original slaughtering capacity was 800 sheep and 25 cattle per day. Now the works can slaughter 1,000 sheep and 80 cattle daily, and the storage space is equal to 25,000 carcasses. Haslam machines were first installed, Linde plant now being used. In 1896 the company was taken over by the National Mortgage and Agency Co. of New Zealand. Mr. J. Anderson is the manager of the Longburn Freezing Company.

Auckland Farmers' Freezing Co.—The Auckland Farmers' Freezing Co. was established in 1903, the works being completed and killing started in March, 1905. The capital of the company is £75,000, of which £43,830 has been issued, and the directors are Messrs. J. Barugh (chairman), J. E. Makgill, H. E. Worsp, G. Goodwin, S. Wing, S. J. Ambury, and L. J. Bagnall; with Mr. H. G. Stringer as manager and secretary. The daily killing capacity is 1,600 sheep and 100 cattle, and the storage space is equal to 45,000 carcasses. Haslam ammonia compression machines are employed. The Auckland Farmers' Freezing Co. in 1906 bought out the Auckland Freezing Co., which was established in 1884. The farmers of the province thereupon decided to build works of their own, and, negotiations with the proprietors of the old company having failed, they erected works at Southdown, near Auckland. In 1906, as stated, the Farmers' Co. took over the Auckland works of the old company. The stock and offal are frozen at Southdown and railed to the ship's side, and the local freezing is done at the Auckland works.

Wanganui Meat Freezing Co.—The Wanganui Meat Freezing Co., Ltd., was established in 1891 at the mouth of the Wanganui river. The first directors were Messrs. E. A. Campbell (chairman), A. Burnett, H. Comfoot, J. B. Murray, A. J. Parsons, G. S. Robertson, William Ritchie, and John Stevenson. Mr. C. M. Cresswell, the secretary, states that when operations were started, and for a few years afterwards, the output of the works was nearly all sheep, with a few cattle. Now more lambs than sheep are put through, and the quality of the lamb is improving year by year, owing mainly to the use in the district of Southdown rams. Twenty years ago the sheep round Wanganui were nearly all Lincoln, since then Romney and Leicester rams have been largely used for general purposes and the Down ram for lamb raising. The Wanganui works have a killing capacity of 3,500 sheep a day, with a storage capacity of 60,000 carcasses.

Gisborne Sheepfarmers' Frozen Meat Co.—The Gisborne Sheepfarmers' Frozen Meat Co., Ltd., started work in 1902. The first directors were Messrs. C. A. de Lautour, (chairman), F. B. Barker, W. R. Barker, John Clark, William Cooper, W. K. Chambers, F. Hall, E. M. Hutchinson, P. T. Kenway, and W. D. Lysnar. The board at present is composed of Messrs. C. A. de Lautour (chairman), F. B. Barker, W. R. Barker, W. K. Chambers, Charles Gray, E. M. Hutchinson, F. Hall, John Clark, F. Holden, and C. J. Parker. Mr. W. F. Cederwell has been manager from the beginning. The slaughtering capacity at date of establishment was 800 sheep per day; now it is 4,500, and 150 cattle. A Hercules refrigerating machine cooling on the direct expansion system is installed. The company was formed on co-operative lines; in 1908 £30,000 was paid in wages.

Waitara Freezing Works.—The Waitara Freezing Works were purchased in 1902 by Messrs. Thomas Borthwick and Sons, Ltd. The directors of the old company when taken over were Messrs. G. Riddell, E. H. Godsal, J. Hine, A. A. Fantham, H. Goode, and G. Bailey. The works were destroyed by fire in 1904, and were rebuilt on an enlarged plan in the following year. The daily capacity is now 150 cattle and 750 sheep, and the storage capacity is equal to 40,000 carcasses.

Hastings Freezing Works.—The Hastings Freezing Works,

THE FREEZING WORKS OF THE GISBORNE SHEEP-FARMERS' FROZEN MEAT CO., LTD., NEW ZEALAND.

To face p. 70.

Paki-Paki, Hawke's Bay, were erected by Messrs. Borthwick in 1905. The daily capacity is 30 cattle and 2,000 sheep, with a storage capacity of 30,000 carcasses of mutton.

The Wellington Farmers' Meat Co.—This company was formed in 1909. The Board consists of the following gentlemen :—Messrs. J. C. Cooper (chairman), R. Clephane, F. B. Lowes, J. R. Franklin, R. D. McKenzie, T. Hodgins, and George Pain. The works are situated near Masterton, Wairarapa, and the daily slaughtering capacity is equal to the handling of over 2,000 sheep.

The Nelson Freezing Co.—This company began shipping in 1908. The directors are :—Messrs. George MacMahon (chairman), A. Drummond, F. W. Fairey, Frank Hamilton, D. T. J. Rouse, and J. S. Wratt. The works are at Stoke, and the capacity is given as 1,000 sheep per day, with storage equal to 30,000 carcasses.

The Ocean Beach Works, Bluff (owned by Birt and Co., Ltd.), were erected in 1891 ; the North British and Hawke's Bay Freezing Co. (Napier Works) were built in 1888 ; the Patea Farmers' Co-operative Freezing Co., began exporting meat in 1904 ; the Southland Frozen Meat and Produce Export Co., formed in 1884, has works at Mataura and the Bluff ; and the Tokomaru Sheepfarmers' Freezing Co. began shipping in 1911.

Such are the particulars of the establishment and progress of New Zealand's meat freezing works. A full list of these works in the Dominion will be found in Appendix VII. Reference may now be made in a general way to the process of development under which the meat works of to-day have attained their fine equipment and completeness.

Refrigerating Installations.—Of course, the meat works in the early eighties were primitive affairs. Freezing for export was quite a speculation, and there were many "doubting Thomases" in New Zealand, either actively or passively antagonistic to the new industry. The scale of shipments was small, both on account of the limited capacity of vessels and limited trade in England. So, everything was elementary, and the design and equipment of the freezing establishments were crude. As time went on and the market value in England of frozen meat fell, it was naturally found necessary to effect economy in the general organization of the works. About 1890 it was plain that the actual freezing was costing too much. The cold air machine had been used successfully, but it was expensive in the matter of fuel consumed. British engineers were not alert in meeting the felt want, and New Zealanders had to go to America to gain information. The British refrigerating machinery manufacturers, as soon as they saw their business being threatened, set themselves to supply an economical freezing machine, with the result that the Haslam and the Linde ammonia machines, and Hall's CO_2 machine, are the types of refrigerating plant used in New Zealand to-day. The American Hercules ammonia machine is found at some works. A Linde "Disc" machine installed in the new works of the Wellington Meat Export Co. was the first departure from the cold air principle. This was followed by ammonia machines cooling on the direct expansion system being fitted at Islington, and by Hall's carbonic acid machines cooling on the brine system at Belfast.

In 1891 there were seventeen freezing works in New Zealand, with a total freezing capacity of 3,665,000 sheep a year, and in 1911 the number of works had increased to 31, with a capacity for dealing with 82,000 sheep per day.

The Operations at a New Zealand Meat Works.—In New Zealand the invariable practice is to have the whole works complete in a series of buildings. The stock are brought by rail or road to the drafting yards. All stock suitable for freezing are carefully drafted, and animals not suitable for freezing are either sent back to the farms or killed for tinning. The stock after leaving the yards are driven into carefully constructed abattoirs, where they are readily handled by the slaughtermen under the very best approved sanitary conditions. The stock are then killed and dressed, and the offal is at once removed to the buildings for treating the by-products. The carcasses are carefully cleaned, dressed, graded, weighed, and passed into the cooling room. In most of the New Zealand slaughterhouses the cooling room is constructed in such a manner as to allow for a rapid current of air, which quickly cools down and " sets " the carcass. In the northern portion of New Zealand, and in most of the Australian works, the cooling has to be assisted by artificial means—brine pipes or cool air. As soon as the carcasses are " set," they are conveyed—invariably by overhead rail—to the freezing rooms, which are long narrow apartments where the carcasses remain suspended from the rails till they are hard frozen. The actual time taken in freezing varies according to the lowness of the temperature applied, and ranges from 24 to 48 hours. As soon as the carcasses are frozen thoroughly hard, bags are put on, with numbers corresponding to the tickets—indicating the grade and quality—which have been attached to the carcasses when graded. Immediately after bagging, the carcasses are fit to go into the store room, where they are stacked up one on top of the other, the various marks and numbers being kept separate until ready for shipment. The frozen sheep and lambs are loaded direct into insulated railway vans and conveyed to the steamer. In almost all the New Zealand ports the steamers are provided with loading port-holes. Canvas awnings are spread between the dock and the steamer, and the carcasses are passed rapidly from the vans along wooden shoots through the portholes into the refrigerated holds of the vessels.

This rough sketch of the operations at a New Zealand freezing works conveys but a poor idea of the thorough methods and scientific management now practised in the meat freezing industry of the Dominion. The figures and facts given by Sir Joseph Ward in Chapter XXII. show at a glance the splendid results flowing from the meat freezing industry, and these results—the direct outcome of the operations at the freezing works—indicate a very considerable effectiveness in the general system and management of the frozen meat industry in New Zealand.

CHAPTER V

THE SOUTH AMERICAN FRIGORIFICOS

The story of the rise and development of the Argentine frigorifico, or freezing works, could be made romantic, so abounding with stirring events has it been. The tale of the beginnings of the freezing industry in South America has been told in the chapter of this book in which the authors have endeavoured to immortalize some of refrigeration's pioneers. From these beginnings, founded partly with British capital, great successes have arisen, though the movements of the Argentine meat exporting companies have not been uniformly forward nor financially successful year by year. But the results achieved, both in the dividends to shareholders and the establishment of a splendid industry, helpful to both the estanciero and the meat exporter, form a record of which all persons engaged in the business may well be proud. The descent upon Argentina of the North American "Trust" houses has been the most startling event in the later stages of the Argentine meat export industry, and it is a subject of frequent discussion how far-reaching that important happening is destined to be in future developments.

Before describing the various meat freezing enterprises in Argentina and their development, it may be well briefly to review the growth of the industry in that country. The first period of the freezing industry in Argentina may be said to have closed in 1899, up to the end of which year 442,000,000 kilos. of mutton and 29,000,000 kilos. of beef were exported by the three great concerns (Sansinena's, River Plate Fresh Meat Co., and Las Palmas [J. Nelson and Sons]) which held the field without competition. The shipments of mutton year by year showed wonderful expansion. Frozen beef was shipped irregularly up to 1895, in which year this section of the Argentine meat trade was begun in earnest. It is worth giving the figures to show the rate of increase: 1895, 1,587,000 kilos.; 1896, 2,996,000; 1897, 4,241,000; 1898, 5,867,000; 1899, 9,079,000. Increases were shown in the shipment of this article right away from 1895 to 1906. During this period lambs were of no account; in 1897 12,000 were shipped to England, but after that the shipments were reduced, and thus did not compete with the increasing Australasian trade in this article.

The Live Cattle Trade and its Stoppage.

The closing of the English ports to Argentine cattle in 1900, owing to the outbreak in the Republic of foot-and-mouth disease, was a great stimulus to the frigorificos. The serious disturbance caused by the crisis is seen in the fact that the respective values of live stock and frozen meat exported in 1899 were $8,482,511 and $2,665,073, gold. The import of Argentine live stock into Great Britain was prohibited by an Order under the Contagious Diseases (Animals) Act after April 30 of that year. For the four years 1896 to 1899 Argentine fat cattle imported into English markets had averaged 80,000 head; sheep, 380,000; and the sudden cessation of these considerable imports had, of course, a dislocating effect for the time upon the various industries at the export end, also causing much embarrassment to the cattle salesmen and "carcass butchers" in Great Britain. The shippers in the United States in the following year increased their despatches of live cattle and sheep heavily. In 1903, from March 1 to June 13, British ports were again opened to Argentine live stock, and 110,000 animals were shipped in that period. Great pressure had been brought to bear upon the Government to take this step, which, however, in the opinion of practical persons, was unwise. Upon the discovery of foot-and-mouth disease in cattle sent to South Africa and Great Britain, the ports were again closed, and, taking all things into consideration, it is doubtful if they will ever be again opened to Argentine live stock.

The high-water mark of the importation of live stock from all countries for slaughter in British markets was, for cattle,

643,000 head, in 1890, and for sheep, 1,056,000, in 1895. The dead meat trade, preferable in every way, has forged ahead steadily, and imports of North American live stock will probably soon be a thing of the past.

Although the above remarks are somewhat of a digression, it is well to place on record the circumstances which led to the stoppage of the Argentine live stock exports to Great Britain. This undoubtedly placed the frigorifico men in a stronger position—it gave them cheaper stock to buy at home and an expanded market in Great Britain. Frozen meat production increased by leaps and bounds ; the export in 1900 to Great Britain was 76,338 tons, and in 1903 this had grown to 131,000 tons. The triennial period, 1901 to 1903, Dr. Bergés (of the national bacteriological institute of Buenos Aires, the chronicler of the records of Argentina's meat export industry) terms Argentina's " golden age of the freezing industry." It was so, indeed, to the shareholders of the three companies previously mentioned, which still held undisputed possession in 1902. Not alone was the cessation of Argentine live stock shipments to Great Britain in their favour ; drought in Australia crippled meat exporters there, the war in South Africa attracted meat imports, and labour troubles in New York and Chicago reduced North American live and dead meat exports. Never was so favourable a group of factors present in one year, and it was these circumstances which led to the establishment of other freezing concerns in the Argentine.

River Plate Fresh Meat Co., Ltd.

To revert to the earliest days, the commercial beginning of the great Argentine meat industry was the shipment by the s.s. *Meath* of 7,500 frozen sheep from the Campana works of the River Plate Fresh Meat Co., Ltd., in 1883, an enterprise which owed its inception to the late Mr. G. W. Drabble. [The date of the earliest attempt (experimental) to export Argentine frozen meat was about the year 1877, when the salting factory " San Luis," in San Nicolas, shipped trial lots in the holds of the s.s. *Le Frigorifique* and *Paraguay*, repeating the experiment by the s.s. *Teviotdale* in 1882.]

THE LATE MR. GEORGE W. DRABBLE.

THE LATE SEÑOR EDUARDO OLIVERA

DR. EMILIO FRERS.

To face p. 76.

Originally the River Plate Fresh Meat Co. was promoted for the freezing, shipping to England, and sale there of Argentine sheep, but gradually, as the business advanced, the freezing of beef was commenced, and from the year 1896 onwards the export of both rapidly developed. The weight of the beef shipped by the River Plate Fresh Meat Co. soon exceeded that of mutton. The company's shipments of mutton and beef from the commencement of the business (1883) to 1910, twenty-eight years, totalled 14,141,588 carcasses of mutton and lamb, 1,440,595 quarters of chilled and 2,178,987 quarters of frozen beef. The first shipment of mutton was despatched to London on November 23, 1883, the carcasses averaging 38 lbs.

The Campana works are fitted with very complete plant and machinery for dealing with the various by-products arising out of the business, and improvements and additions are being made constantly. The chilling and freezing plant is capable of dealing with 800 cattle and 3,000 sheep per day. This is a great contrast to the early equipment of the works, as originally the plant consisted of a small engine room, cold storage chambers, two digestors, and a slaughter pen. All the work was then manual. The offal was given away or destroyed. It was not till 1884 that beef freezing was started in Argentina: frozen beef and pork were shipped that year, and in 1886 the first lambs were despatched. The River Plate Fresh Meat Co. also established a plant at Colonia, Uruguay, with the idea of shipping from the two works, but the enterprise did not pay, and the machinery was removed in 1888.

The company's first engineer and manager in the Plate was Mr. John Angus, and a large portion of the present complete and splendid works at Campana grew up under his management. In 1893 he went to Buenos Aires to act as manager of the company, and held the position till 1905, when he retired. In 1899 Mr. G. W. Drabble died, and Mr. (now Sir) Henry Bell became chairman, a position he held for three years, his successor being Mr. Charles Drabble, who was in turn followed by Mr. John A. Wood as chairman, Mr. Drabble remaining a director. Mr. Wood had been manager and secretary in London ever since the company began operations, Mr. Sidney

Young, who has been with the company since 1884, succeeding him in that capacity. The London offices of the company are at Cecil House, Holborn Viaduct, E.C.

The following notes concerning the establishment and development of the works at Campana of the River Plate Fresh Meat Co. are furnished by Mr. Wood :—

"The River Plate Fresh Meat Co.'s freezing works in the Argentine were at Campana, on the river Parana, about sixty miles above Buenos Aires. The initiation and development of an entirely new trade was naturally surrounded with difficulties, and if it had not been for the ability and resource of the then chairman of the company, Mr. George W. Drabble, who initiated and brought out the company, and also for the support he received from the leading shareholders, the company would undoubtedly have succumbed in the early years of its existence. In spite of difficulties, the company gradually developed, and is now one of the largest exporters of meat and relative by-products from the Argentine. According to the figures given at the annual general meeting in March, 1910, the imports over the twelve months ending December 31, 1909, were just 50,000 tons, and a further increase was mentioned as probable for 1910. The handling, shipping, and distributing, either retail or wholesale, of these quantities of meat and by-products involve necessarily a large organization, which organization has been specially built up to meet the requirements of the company's trade.

"Chilled beef was, as the result of long experiments carried out by the River Plate Fresh Meat Co., actually shipped by the company on a large scale in the year 1901. The development of this chilled beef business, which was first successfully carried out by the company, has been a great factor in the development of the Argentine trade, and was rendered possible by the improvement in cattle stocks in the Argentine, which enterprising estancieros had been carrying out for some years. This improvement in cattle and sheep stocks has been continued, and the supplies of good-class cattle and sheep available for export are greater now than at any time.

"The system on which the River Plate Fresh Meat Co. has

been worked has been for the company to control and work with their own men the whole business as far as possible, from the buying of the live stock in the Plate, and its shipment, to the delivery to the retailers or consumers, the aim all through having been to meet as far as was practicable the consumers' requirements, and the control of the company being in the hands of those who were acquainted with what was required on this side enabled that object to be fairly well attained.

"So far as my personal part is concerned, I have from my position had to initiate, supervise, and control the various developments of the company's business, and the successful carrying out of the work has only been possible owing to the assistance of a most capable staff both in the Plate and on this side. I have, of course, paid many visits to the Argentine in connection with the business of the company.

"I may, perhaps, mention that in the earlier days of the River Plate Fresh Meat Co. the development of the industry—especially in the mutton trade—was handicapped by the quality in those days of the stock purchased in the Argentine being inferior to what was obtainable at the same time in New Zealand and Australia. In recent years, however, the quality of the Argentine stock leaves little to be desired."

The San Nicolas Works.

Contemporaneously with the formation of the English River Plate Fresh Meat Co., Mr. Eugenio Terrasson established meat works at San Nicolas on the Parana river. His first shipment was despatched in 1883 in the *Loch Ard*, and was composed of hindquarters of mutton, a Bell-Coleman machine being installed on the vessel. In 1884 Mr. Terrasson brought out the prospectus of La Compañia de Carnes Congeladas, capital $250,000, gold. There is no precise record of the actual formation of this company. The San Nicolas frigorifico worked without interruption till 1898.

Dr. Pierre Bergés recorded in a pamphlet published in 1908 that the San Nicolas works had a frontage of 115 metres and occupied an area of upwards of 4 hectares. The depth of the river was sufficient to allow of the loading of the frozen meat into a liner straight from the refrigerating chambers. In 1890, he adds, "this frozen meat warehouse had three large cold chambers, able to contain each 4,000 frozen sheep. In 1898 it had preserved 163,103 sheep, and to-day it is to be sold for £50,000."

Thus far Dr. Bergés; but harking back to 1895, when the works had been shut up for some time and were under mortgage, three English houses became interested in the property, Messrs. W. and R. Fletcher, Ltd., the Liverpool Cold Storage and Ice Co., Ltd., and Mr. Hudson, the Newcastle shipowner, taking over the works. About 1898 the works were let to the three Argentine frozen meat companies, the River Plate Co., Sansinena's, and James Nelson and Sons, for a minimum period of five years, at a rental equivalent to 15 per cent. on the paid-up capital of £40,000, viz., £6,000 per annum. These companies promptly shut up the frigorifico, which was not operated after that date. Since 1903, when the agreement expired, the land, plant, and machinery have belonged to Messrs. W. and R. Fletcher, Ltd.

In 1884 La Congeladora Argentina was founded by the Argentine Rural Society to export frozen meat. The capital was $1,000,000 paper. In 1885 the first shipment of 1,000 cattle and 10,000 sheep was made from Zarate. Dr. Pierre Bergés says that the society did not prosper, and it lost all its paid-up capital.

James Nelson and Sons, Ltd.

In 1886 Mr. Hugh Nelson, a partner in James Nelson and Sons (a firm of cattle salesmen in Liverpool, Dublin, Manchester, and London, founded in the early Victorian era, the partners being the late Mr. James Nelson and Messrs. William [the present baronet], Hugh, and Edward Nelson), went out to Argentina and built Las Palmas freezing works at Zarate. A company was formed called Nelson's River Plate Meat Co., and in 1889 this was changed to Nelson's (New) River Plate Meat Co., with an extension of capital. In 1892 the company of James Nelson and Sons, Ltd., was registered in England to amalgamate the businesses of Nelson's (New) River Plate Meat Co. and James Nelson and Sons. In 1893 the Las Palmas Produce Co., Ltd. was registered in Argentina. This is the South American section of James Nelson and Sons, Ltd., who hold all the shares. These various concerns were based on the Zarate frigorifico and the business proceeding therefrom. Sir William Nelson in 1904 retired from the joint managing directorship, held with Mr. Edward Nelson, of James Nelson and Sons, Ltd., this position now being held jointly by Messrs. Edward Nelson and T. C. Nelson. The Zarate works cover 168 acres, and the pasturages and lairages nearly 3,000 acres. The slaughter yards are capable of dealing with 1,000 cattle, 5,000 sheep, and 250 pigs a day, and the refrigerating chambers, which altogether number 70, have a total capacity equal to 7,000 cattle, 90,000 sheep, and 2,000 pigs. Principally, Linde refrigerating machinery is installed, and the plant which was provided in 1907 to deal with pork products alone cost £40,000. Mr. Philip Holmes has been secretary to James Nelson and Sons, Ltd., and the earlier company since 1889. The London office of the company is at 57, Charterhouse Street, E.C.

Compañia Sansinena de Carnes Congeladas.

The well-known and popular brand of frozen meat in English markets marked **S** represents the outturn of the Compañia Sansinena de Carnes Congeladas, popularly styled in England the Sansinena Co. Messrs. S. G. Sansinena and Co. had a boiling-down works at Barracas al Sud in the early eighties, and in 1885 they erected a freezing plant on the site of the present La Negra works, and began to export mutton to Great Britain. The business continued, at first on a small scale, until 1890, by which time Messrs. Sansinena were shipping 25,000 sheep and lambs a month. In 1891 the business was turned into a company under Argentine law, headquarters in Buenos Aires, with a capital of 20,000 shares of $100 (gold). The original board was: Ernesto Tornquist (president), Robert M. Ramsay,

Samuel H. Pearson, Santiago Luro, and Francisco Sansinena (managing director). The first shipments to England were made to Messrs. James Nelson and Sons, and in January, 1887, Messrs. Sansinena established themselves in Liverpool, their London office being opened in 1888. The success of the Sansinena Co. is a high tribute to the sagacity of the directors and the excellent management of the Buenos Aires and European managers. Mr. Miles Pasman, who has lately retired from the Board of Directors, contributed very largely to the success of the company by his capable and vigorous administration in the position of managing director. The late Mr. William Cook held the position of general manager for Europe from 1887 to 1904, and the late general manager for Europe, Mr. John J. Ward—who retired in 1910—was in the New Zealand frozen meat business in 1884 and joined the Sansinena Co. in 1887. Mr. W. Dyson Barnitt is now the European general manager, and Mr. A. G. Rose is secretary of the London office, Nos. 13–16, Long Lane, West Smithfield, E.C.

The Sansinena Co. has always had far-reaching aims ; during the period from 1891 to 1899 it did a considerable export business in frozen meat with France, where it had depôts at Havre, Dunkirk, and Paris. The prohibitive duty and regulations imposed in 1899, however, brought this to a conclusion. Brazil was also selected for trial, and shipments were made there, but without startling success. In 1902, following the splendid financial trading results, the company built new works at Cuatreros (Bahia Blanca). In 1905 Belgium was attacked, but the campaign was no more successful than in France. In this connection the following extract from the *Review of the River Plate* of January 1, 1909, is of interest :—

" In the matter of destination of frozen meat exports, Great Britain continues to be practically our only client. South African trade has dwindled down to 35,662 quarters of beef and 10,804 carcasses of mutton (1908)."

In 1906 the " Sansinena Distributing Syndicate, Ltd.," was formed with a capital of £200,000, of which £125,000 was paid up, the Sansinena Co. contributing £50,000. This concern ran shops in the chief centres of South Africa, but the depression

THE LA NEGRA FRIGORIFICO OF THE COMPAÑIA SANSINENA DE CARNES CONGELADAS.

SLAUGHTER-HOUSE FLOOR AT THE LA NEGRA FRIGORIFICO.

To face p. 82.

which hung over those Colonies must have set the distributing company a heavy task. The Sansinena Co. has a very extensive establishment in the United Kingdom: it has warehouses, stores, or offices in London, Dublin, Glasgow, Cardiff, Liverpool, Birmingham, Manchester, Newcastle, Bristol, Leeds, Hull, Sheffield, Leicester, Burton, Wolverhampton, and Derby.

A recent enterprise on the part of the Sansinena Co. is the outbidding of Messrs. Swift early in 1911 for the purchase of the Frigorifica Uruguaya. The Sansinena Co. paid £300,000, or £10,000 more than the price offered by Swift's, for this undertaking, which is described later on in this chapter. The capital of the Sansinena Co. was in 1911 increased from $3,000,000 gold to $4,500,000 gold, this being for the purpose of the Uruguaya purchase.

Twentieth Century Companies.

Reference has been made to Argentina's "golden year," 1902. It was then that the estancieros, aghast at the spectacle of the frigorificos making their 50 to 100 per cent. profit, naturally came to the conclusion that they would take a hand at the game themselves. La Société Anonyme de Viandes Congelées La Blanca was founded by Argentine capitalists in 1902 at Buenos Aires, almost all the proprietors being leading estancieros. The capital was fixed at £300,000; the works are situated on the river Riachuelo, and operations were begun in 1903. In 1908 the works were taken over by the American Trust companies for £340,000. In the same year the Cuatreros factory, erected by the Sansinena Co. as a second string at Bahia Blanca, was set going. The next freezing works to be started was that constructed at the port of La Plata by the La Plata Cold Storage Co. The establishment of the La Plata works at Puerto La Plata, excellently situated on deep water, was brought about in this way. There was some difficulty in getting full supplies of frozen meat about 1902—1903 for South African requirements. Australia was out of the trade at that time owing to the drought, and the Sansinena and other Argentine companies would not undertake to sell at all freely, fearing to interfere with their connection in England. The Imperial Cold Storage Co. had a concession placed before it of the site at La Plata by Messrs. Zavala and Co., of Buenos Aires, Mr. Daniel Kingsland, Mr. Joseph Moore, etc., just at this time, and the company closed with the offer. Works and plant were erected, and the system of business established largely on lines suggested by Mr. John Cooke, of Melbourne, and Messrs. W. Weddel and Co., of London. But when the La Plata works got into working order the imported meat trade in South Africa was dying, and so the exports were directed to England. The system of selling c.i.f. to London was practised and developed, but the Imperial Cold Storage Co., having no longer any interest in this trade, wished to sell, and so it came about that these works passed in 1907 into the hands of Swift's, of Chicago, for £350,000, thus marking the beginning of a great revolution in the frozen meat trade. As the head office of the company was always at Cape Town, it is permissible to call La Plata a British company.

The Smithfield and Argentine Meat Co. was formed in 1903 with a capital of £200,000, the shares being distributed amongst English and Argentine capitalists, some leading Smithfield Market men subscribing to the enterprise. The freezing establishment is near Zarate, and work was started on February 24, 1905. This company, which mainly exports chilled beef, includes amongst its directors, Messrs. Assheton Leaver (W. and J. Biggerstaff), chairman, and P. J. Poels. The London office is at 58, West Smithfield, E.C.

Next we come to a purely Argentine company, the Frigorifico Argentino, the capital of which, £250,000, was put up entirely by local people. Operations were begun in June, 1905, the works being on the river Riachuelo. The success of this company in the chilled beef trade has been most marked, the uniformly good condition in which its consignments are landed giving it a strong hold on the retail trade. Mr. S. McC. Rough is manager for the United Kingdom of this company, which has its London offices at 40—44, Holborn Viaduct, E.C.

ARGENTINE FROZEN MUTTON FOR LONDON: INTERIOR VIEW IN THE LA PLATA COLD STORAGE COMPANY'S WORKS.

To face p. 84.

Evolution of Argentine Mutton.

A word is not out of place here as to the evolution of the class of mutton exported from Argentina's frigorificos.

At the time when meat freezing in Argentina was started, the sheep offering were very unpromising for the industry. The merino was the national sheep, though a few South Down animals had been introduced in 1825, and the Lincoln Long-wool was imported with great success several decades later. But merino mutton was shipped by the freezing companies for years after the beginning of the exporting business, and the pioneers were much handicapped in selling such an inferior article against the well-grown, meaty, New Zealand carcass. On referring to some of the old London price lists, one notes that in 1884 New Zealand mutton was quoted at 3*s.* to 3*s.* 4*d.* per stone, whilst River Plate was priced at 2*s.* to 2*s.* 8*d.* From the same source of information it appears that in 1886 (June) a considerable improvement both in breed and condition was observed by London salesmen in the latter class of mutton. The necessity to improve upon the lean, light-carcassed, merino sheep in order to put up a serious competition with New Zealand cross-bred mutton in Smithfield was no doubt a considerable factor in the bettering of the estancieros' flocks in Argentina by a beginning being made in the great import trade in high-class rams from British pedigree flocks.

Other South American Contributors.

Argentina, though the principal, is by no means the only frozen meat exporting State of South America. British territory in the South Atlantic—the Falkland Islands—did a considerable business for some years; Uruguay contributes frozen mutton, lamb, and beef, to the needs of Great Britain; and down at the southern tip of South America there are a couple of freezing works on the Straits of Magellan, both in Chilian territory. Venezuela is the latest South American principality to enter the trade.

To detail first the connection of the Falkland Islands with the industry, it may be said that the Falkland Islands Co.

entered the field in 1886 with a shipment of 30,000 sheep in the steamer *Selembria*. The vessel arrived at London on July 15, 1886, and she delivered two other lots, 45,000 carcasses, in 1887. The sheep were about 64 lbs., "rather lean." When first offered, the mutton made $4\frac{1}{2}d$. to $5\frac{1}{4}d$. per lb., or about $\frac{1}{4}d$. under the price of New Zealand mutton. Later, the Falkland Islands mutton made a lower average market price, $3\frac{1}{4}d$. to $5\frac{1}{2}d$. per lb. It was understood that these shipments were disastrous in every way. It is said that 7*s*. 6*d*. in the £ of the capital embarked was lost.

After these three shipments the Falkland Islands Co. dropped out, and Messrs. Spearing and Waldron took up the running. The Waldrons had been connected with sheep breeding in the Falklands from the earliest times. The firm chartered the *Hengist*, a sailing ship of about 1,500 tons, which had been engaged in the New Zealand frozen meat trade. This vessel made her first trip in 1890, and continued to bring a shipment yearly till 1895; in her six voyages the vessel conveyed about 100,000 carcasses. The *Hengist* loaded and froze sheep at two ports, San Carlos on the east and Port Howard on the west. Two lifeboats were lashed together and a platform was put on them both. The mutton was placed on this and transferred thence to the 'tween decks of the ship and frozen on board; then it was stowed below and shipped at intervals. Some of the shipments sold fairly well, but the sheep were too big and coarse for Smithfield buyers, and the lack of grading told against them—the mutton could not compete with that from New Zealand. The net return to the farmers was about 1*d*. to $1\frac{1}{2}$d. per lb., 5*s*. to 7*s*. per head. If 20,000 sheep could have been got at one port, the enterprise, it is said, would have paid. Messrs. Spearing and Waldron shipped on their own account, and also as agents for some of the farmers who preferred to consign and take their own risk. On the last two trips of the *Hengist* some of the meat was transhipped to the s.s. *Hornby Grange*. The end of the Falkland Islands frozen meat trade was the wreck of the *Hengist* in the Straits of Magellan. The trade, as above described, lasted from 1886 to 1895, and in all 169,973 sheep were frozen and shipped to London. The

Falkland Islands frozen sheep first shipped were too old. Two gentlemen, Messrs. Windsor and Wolff, were mainly responsible for the Falkland Islands mutton export business in the early days.

Another start was made by the Straits of Magellan Frozen Meat Co. in 1896, Messrs. Spearing and Waldron being largely interested in that company. The sailing ship *Oneida* was dismantled and turned into a freezing hulk in the Straits of Magellan, and is still there. Shipments of mutton from Patagonia, frozen on this hulk at Punta Delgada, were made to London in the Grange Line boats in 1896, 1897, and 1899—70,000 carcasses in all. As a matter of fact, the Falkland Islands and Patagonian mutton was never much fancied at Smithfield. One shipment in 1895, according to the market circulars of that year, went as low as 2⅝*d.* to 3*d.* per lb.

The third stage of meat freezing for export in this part of the world was reached with the establishment of freezing works at Rio Seco in 1905. The works are situated on the northern shore of the Straits of Magellan, ten miles east of Punta Arenas, which is 80 miles from the mouth of the Straits. The proprietors are the South American Export Syndicate, in which Messrs. Houlder Brothers and Birt and Co. are largely interested. The first shipment of frozen meat was made in April, 1905. The mutton and lamb come under the "Shell" brand. The London office is at 102, Fenchurch Street, E.C.

On December 27, 1906, a number of ranch owners and merchants met and resolved to erect a freezing works at Puerto Sara, San Gregorio, on the Chilian side of the Straits of Magellan, 60 miles east of the Rio Seco works. They put up £39,000, and a month later the Compañia Frigorifica de Patagonia was formed, the head office being in Punta Arenas. The ranchers of Gallegos were asked to join, but they declined. Haslam's refrigerating machinery was installed in the building, the foundation stone of which was laid at San Gregorio by Don Pedro Montt, President of Chili, on February 25, 1907, and the works were completed on February 20, 1908. The president of the company was Mr. Rodolfo Stubenrauch, and the other directors were Messrs. Alejandro

Menendez, Pablo Van Peborgh, Luis Bonvalot, and Mr. Frank H. Townsend, the works manager being Mr. David Anderson. The paid-up capital of the company is £65,000. The average output for the season is about 150,000 carcasses. The shipping season of these Straits of Magellan works is extremely short, February to May, as stock cannot be moved in winter.

It does not require much imagination to picture the whole of the Atlantic seaboard from Monte Video—possibly from the Venezuelan coast—to Magellan Straits dotted with freezing works for the provision of meat for the Old World. At present there is a wide gap between Bahia Blanca, where the southernmost freezing works (Sansinena's) in Argentina is placed, and the two works on the northern shore of Magellan Straits (Chili). But this gap of a thousand miles will doubtless be bridged by-and-by. Port Madryn is a rising settlement on the Patagonian coast, and at Puerto Gallegos (lat. 51° S.), just opposite the Falkland Islands, there are canning works, founded in 1898, owned by the Patagonian Meat Preserving Co., of London. A freezing plant was added, and the Puerto Gallegos frigorifico, under the name of the New Patagonia Meat and Cold Storage Co., Ltd., is expected to begin operations shortly. A fair number of sheep are available between this point and the mouth of the Santa Cruz river, about 100 miles farther north. With the development of the railway system over the Patagonian portion of the Argentine Republic, and the movement of the sheep farmers north from the districts of Punta Arenas (the region talked of as being full of millionaires), and south from Argentina proper over the river Negro, conversation with men who know the country leads one to believe that sooner or later there will be works right up the coast. The difficulty lies in the dryness of the climate of Patagonia—the rainfall is small. Including three canning works—one in Tierra del Fuego—there are in all five companies in Patagonia preparing mutton for export.

La Frigorifica Uruguaya, which was formed in 1902, had an original capital of $500,000 gold, later increased to $700,000 (£140,000). The promoter and first chairman was

THE SAN GREGORIO FREEZING WORKS, PATAGONIA.

To face p. 88.

Señor Manuel Lessa, a prominent financier. Slaughtering was started in December, 1904, and the first shipment of frozen meat was despatched to London in the s.s. *Sussex* in March, 1905. The works are fitted throughout with modern appliances, and Linde ammonia compression refrigerating machinery is driven by triple expansion Sulzer steam engines. The original capacity of the works was for the production of 50 tons per day, and storage of 1,000 tons, but during the 1909 season the works were extended, and can now produce 120 tons daily, with a storage capacity of over 2,000 tons. In 1910 the company paid 12 per cent. dividend. As mentioned above, this frigorifico was purchased in 1911 for £300,000 by the Compañia Sansinena, and is now being doubled in capacity.

Uruguay may be expected to be exploited vigorously in the interests of cattle freezing. The *Review of the River Plate* states that "The statutes of the new frigorifico 'Frigorifico Montevideo,' have been approved, the capital being two million pesos, but the company may commence operations when 400,000 pesos have been subscribed."

Beef from Venezuela.

The Venezuelan frozen meat venture, begun in 1910, geographically belongs to this South American chapter. The works at Puerto Cabello, from which the s.s. *Star of Victoria* took the first shipment of frozen beef on August 7, 1910, 7,121 quarters (400 tons), are in the latitude 10° N. One would think that peculiar difficulties surrounded the preparation of frozen meat in such a torrid clime; probably the only other freezing works in the world matching the Venezuelan one for tropical situation is the establishment of the Queensland Meat Export Company at Townsville, 19° S. The proprietors of the Venezuelan plant are the Venezuelan Meat and Products Syndicate, Ltd., domiciled at 16, Finsbury Circus, London, E.C. It appears that the cattle available for slaughter at Puerto Cabello can be frozen with profit to meet the special demands for beef from small-boned, light-weight beasts for the northern markets of England. The beef will be shipped mainly to Liverpool, for distribution in the Manchester, Liverpool, and Glasgow districts. Probably these cattle—which have so far been killed for hides and fat, in the absence of demand—are bought cheaply, and the beef will, no doubt, prove useful as a secondary quality article. It is expected that a great improvement in the available fat stock will result from using for grazing purposes the vast cattle plains of the Orinoco. The hides and offal will be brought over to Liverpool to be worked up; 1,000 tons of beef can be shipped about every ten weeks. The works are now in full working order, and chilled beef from them occasionally comes to Smithfield market *viâ* Southampton.

Three Distinguished Argentine Statesmen.

In the development of Argentina's rural economy, agriculture, and pastoral progress, there are three names that stand out pre-eminently, and seeing that without the great pastoral resources of Argentina the meat freezing industry of the country could never have reached its present stage, some account of the efforts of these three men is not out of place here. These are the late Don Eduardo Olivera, and—still happily with us—Don Exequiel Ramos Mexia, and Dr. Emilio Frers.

Don Eduardo Olivera, who passed away in September, 1910, was born of a family of estancieros eighty-four years ago. Having completed his university career, he travelled, as a young man, through Europe, studying in various countries the science and practice of rural economy. Returning to his country more than half a century ago, he speedily made his mark both in the Press and in political circles by his intelligent and stimulating campaign for the improvement of agricultural and pastoral methods. This was at a time when the majority of landowners heard with indifference any propositions for the improvement of their live stock and better cultivation of their land. Production was limited to wool, jerked beef, hides, and tallow. Wheat was then, and for many years subsequently, imported into the country.

In 1866 Olivera achieved the work with which for all time

his name will be most associated, by founding the Argentine Rural Society, which may now claim to be the most important institution of its class in the world. At first secretary, and afterwards president, of this society, Olivera was for many years its most active member, and had the honour of being elected its honorary president for life. Amongst many other public offices held by him, he was Postmaster-General, Deputy, Senator, and interim Governor of the State of Buenos Aires, and during all his life held a prominent position in the affairs of his country. As a pastoralist, Olivera was a well-known breeder of merino sheep, and formed by selection a type called the "Argentine merino." Outstanding from all his work for the nation's welfare, the offices he so honourably filled, and his contributions to rural legislation and progress, there rises the man himself, whose virtuous life and sixty years of disinterested service to his country will remain a tradition and an example for future generations. Genial in manner and simple in his life, he was beloved of all, and no surname was necessary to individualize the man who was known throughout his country as "Don Eduardo."

Señor Exequiel Ramos Mexia, the member of a patrician Buenos Aires family, is at the present time the national Minister of Public Works, and in that office, which he has held through two Presidencies, he is displaying the same intelligence and statecraft that have marked him in the various offices he has held related to the country's rural economy. For many years president of the Argentine Rural Society, twice Minister of Agriculture, president and presiding genius of the drainage works of the Province of Buenos Aires (the drainage, at a cost of three-and-a-half millions sterling, of sixteen million acres of rich land subject more or less to inundation, and probably the greatest undertaking of its nature in the world): it is not possible in a brief space to enumerate the many services to Argentina's rural industry for which the country is indebted to Señor Ramos Mexia.

The Sanitary Law regarding contagious diseases in animals was initiated and carried through by him, and many improvements in the handling of live stock in transit and inspection of meat for export are associated with his name. On his own fine estate of Miraflores he is a breeder of Shire horses, pedigree Shorthorn cattle, and Lincoln sheep; the organization there and intelligent cultivation of the land are a reflection of the ability and thoroughness he displays in public affairs. Señor Ramos Mexia is an admirer of the British and their commercial methods, and includes among his personal friends more than one distinguished statesman of that nationality. Like many of his countrymen, he has put aside the personal convenience afforded by his private fortune and condition, to give his services to his country.

Dr. Emilio Frers is the son of one of the first presidents of the Argentine Rural Society, and has himself held that office twice. Perhaps the best illustration of the esteem in which he is held is the incident that occurred when the portfolio of Agriculture was first added to the Cabinet. General Roca was at that time President of the Republic, and Dr. Frers belonged to the opposite political party. General Roca drove to his house to tell him that the country's interests were above party politics, and that he had come to the man recognized by all as the right citizen to organize the new Ministry and be the first Secretary of State for Agriculture. Dr. Frers accepted, and fully justified the choice that had been made.

Dr. Frers is an able economist, a gifted writer, and a citizen whose integrity has won for him universal esteem. His last office was that of president of the Centennial International Exhibition of Agriculture held in Buenos Aires. He has given much of his time to the welfare of the small agriculturist, the labourer, and the immigrant. He would not inaptly be described as the Cobden of Argentina. Essentially a citizen with an austere though kindly view of the duties of life, he does not court publicity, and, as he once remarked, his connection with each office he held began by his first refusing it. On his estancia "La Estrella," where he breeds Hereford cattle and merino sheep, he is beloved by his men; and it is there, surrounded by his family and his books, taking his share in the modest county council of the district, that he is happiest.

CHAPTER VI

IMPROVING FLOCKS AND HERDS IN THE SOUTHERN HEMISPHERE

Evolution of Mutton Sheep in Australasia.—Pedigree sheep were imported into New Zealand quite early to improve the wool. The Border Leicester, English Leicester, Cheviot, and various types of Down rams imported, crossed with the merinos from the stations, established the type from which "prime Canterbury" meat was obtained. In 1892, ten years after the meat export trade was started, the sheep men in the Colony were working on well-defined lines as regards crossbreeding. The sheep in the Colony numbered 18,000,000, 8,000,000 in the North Island, and 10,000,000 in the South. Of this number there were 6,000,000 merinos, and the crossbred sheep were estimated as follow: Lincoln and crosses, 5,750,000, Border Leicester, and English Leicester and crosses, 3,000,000 Romney Marsh and crosses, 1,150,000, and Shropshire, Southdown and crosses, 1,500,000. The various English breeds were used in New Zealand according to the nature of the country, class of wool, fat lamb export, and the ideas of the breeders themselves. The question of the most saleable weight of the mutton carcass had to be considered. When the trade began in 1882, carcasses averaging 80 lbs. were worth over 6d. per lb., but by 1892 the most acceptable weight at Smithfield was 55 lbs. to 60 lbs.; it is now 48 to 56 lbs. The first cross out of a merino ewe by an English ram was the most suitable meat in London. As a general rule, the New Zealand sheep breeders inclined to the use of the Lincoln ram when they desired to raise the standard of their flocks as regards weight of fleece, and to the English Leicester and Shropshire if the carcass wanted improving.

In Otago, it may be said, there has been less change in sheep breeding of late years than in any other part of the Dominion. The Border Leicester, Romney, and Lincoln, and their crosses, are still to be found in all parts. On back-country stations there is a more extended use of the Corriedale. The Corriedale was established in the first instance by crossing the merino with the Lincoln or Leicester ram. The produce were carefully culled, and these "in-bred half-bred" sheep were bred from until the breed became well fixed. The Corriedale is now a well established breed in New Zealand. The merino was only suitable for hilly or very dry country, and the wool from the English breed was found coarse.

The Downs have not made much headway, though the Shropshire maintains its position in front of the Southdown. In Canterbury the popular breed of sheep is the English Leicester. The Lincoln has almost disappeared, and the Romney has still only a few supporters. The Corriedale has considerably increased in numbers on the front hills in the northern district. Persons interested in this subject are referred to a pamphlet issued in 1899 by the *Christchurch Press* on "Sheep Breeding in New Zealand."

In New Zealand crossbreeding is largely a question of lamb production. In Australia no great revolution from the original merino type of sheep has taken place, such as has occurred in New Zealand and Argentina. The frozen mutton from Australia is still largely from merino stock; Australian mutton is sold as "merino and/or crossbred," to quote from a form of contract. All New Zealand mutton and lamb shipped to Great Britain are from crossbred sheep, and so are practically all Argentine.

Systematic tests have been conducted in Australia to discover the best cross for fat lamb raising. The general results of these experiments seem to point to the Shropshire-Leicester merino cross as producing the best results, though crosses with the Dorset-Leicester merino worked out almost as well. At twelve weeks old lambs from the first-named cross weighed 62 lbs., and from the other cross 60 lbs.

Mr. Grigg's Communication.—To elucidate this subject the authors asked Mr. J. C. N. Grigg, of Longbeach, New Zealand, to place on record his views concerning the evolution of the mutton sheep in New Zealand.

JOHN GRIGG, OF LONGBEACH, AND THE STATUE ERECTED TO HIS MEMORY AT ASHBURTON.

To face p. 94.

"At the beginning of the freezing industry in New Zealand," writes Mr. Grigg, "the country had large numbers of four, six, and eight tooth wethers, kept simply for their wool; the bulk of them were merino and first cross in the South Island and Lincoln and Romney Marsh in the North Island. The 'first cross' was really a cross between the English Leicester ram and merino ewe, or the Lincoln ram and the merino ram. Before freezing started wool was the main profit, as the profit from boiling down the surplus sheep was small and prices for store sheep and fats ruled very low; therefore the whole returns of a man's property depended on the wool and the increase in his stock, and even wool on the average was lower about 1875 to 1881 than of late years.

"When freezing started, the English Leicester increased as a stud sheep more than any other breed, because it was found that the English Leicester-merino cross was the neatest and best sheep for freezing. (The English Leicester is the same in New Zealand to-day as it was when freezing started, though the old type of English Leicester [a low-set, medium-sized sheep of good quality] of forty or fifty years ago has disappeared in England). A fairly large number of merino wethers were frozen and shipped in the early days of freezing, but did not as a rule leave much margin of profit, as merinos looked very dark in colour, and sold at low prices compared with first and second cross mutton; the meat of the latter looking bright red and the fat whiter than merino.

"The Border Leicester is a very fine sheep, but it is not a breed that corrects want of shape and quality when crossed with coarse breeds. It is rather liable to run away on the leggy side. But the breed is a very useful one, and is used very extensively in Southland and Otago, where a hardy and easily fattened sheep is the first essential. A few South and Shropshire Down flocks were in existence in Canterbury when freezing started; notably John Deans of Riccarton and Samuel Garforth had South Down flocks, and my father had a flock of Shropshires. The first cargo of frozen sheep and lambs which left New Zealand in the sailing ship *Dunedin* in 1881 contained some Down cross sheep and lambs.

"From 1890 the Shropshire was used very extensively, not so much in the North Island as in Canterbury. My father was one of the first Shropshire breeders in Canterbury, and I use the breed now as well as Southdown on three-quarter-bred Leicester and Romney ewes.

"The English Leicester ram on Romney Marsh ewes produces a very useful sheep. If any lambs from this cross are not frozen and are carried over for a year, the females make good mothers, and the wethers are shapely and fatten readily as 'two-tooths' after having cut a useful fleece; the meat from the carcasses is fairly bright. Where wool is thought more of than quality of meat, the Lincoln ram is used on Romney ewes. The Romney Marsh breed is the foundation stone of successful sheep farming in most parts of the North Island of New Zealand.

"The Southdown is the most symmetrical sheep in the world, and full of short, good quality meat. Now that the South Island draws on the North Island for a large number of its breeding ewes, mostly Romney Marsh cross, the Southdown is rightly becoming more popular in Canterbury. The two earliest breeders of Southdowns were Mr. John Deans, of Riccarton, and Mr. Samuel Garforth, of Speydon. The breeding ewes of Canterbury are neater and smaller-boned sheep, as a rule, and carry finer fleeces, than North Island ewes. The merino foundation is still in evidence in a large proportion of the Canterbury flocks; lambs from these ewes, fattened on the short sweet pastures in a clear and rather dry climate, make a model lamb in meat and weight for the London market.

"Practically the whole sheep breed of the North Island flocks are Romney Marsh cross; the bulk of the ewes there are of this cross. The Romney of to-day in New Zealand is not the fiddle-headed sheep of thirty years ago, and it is undoubtedly a profitable and popular breed. Full credit must be given to the Romney Marsh breed for the large percentage of lambs reared in the North Island under a heavy rainfall.

"A very fine good quality ram, either Southdown, English Leicester, or Shropshire, is more necessary to-day in Canterbury than ever it was, to give shape and finish to the freezing lamb,

"BAPTON VICEROY," SHORTHORN BULL (REGISTERED NO. 90770, VOLUME 52).

Sold by Mr. J. Deane Willis, of Bapton Manor, in June, 1906, to Mr. F. Miller, for Señor Manuel Jose Cobo, of Estancia La Belen, Argentina. The price given, 3,000 guineas, was the highest ever paid for a Bull shipped to Argentina.

To face p. 96.

if we are to keep up the old quality. For the reason that as the merino ewe becomes less and less in number, we lose on the mother's side one of the main elements of our supremacy in quality of meat. Thousands of acres of hill land that used to carry merinos are now carrying 'in-bred half-breds,' generally called 'Corriedales,' most of them Lincoln-merino cross. The first and by far the oldest, and I might almost say best, in-bred half-bred flock is the one still in existence started by the New Zealand and Australian Land Co. on the Levels, near Timaru. Since the Levels was sold to the Government, the flock has been carried on at Moraki. These sheep have proved on many runs in Canterbury more profitable than merino flocks. Heavy culling every year is necessary in this made breed between two sheep so opposite in character as a Lincoln or Leicester ram and a merino ewe. Ewes from the hill in-bred half-bred flocks are now coming down on to the plains to be used as farmers' breeding ewes for the lamb export trade, and taking the place of the old half-bred English Leicester or Lincoln merino, the latter, without doubt, one of the most beautiful sheep ever produced for mutton and wool combined. As in-bred half-breds in many cases are the mothers of the fat lambs to-day, it is more necessary to use a very high-class Southdown, English Leicester, or Shropshire ram to keep up the quality and shape of the sheep.

"The New Zealand and Australian Land Co., and John Little, of Waikari, North Canterbury, were amongst the first breeders of in-bred half-breds, now named 'Corriedales.' Many others are breeding in-bred half-breds to-day, notably Sir George Clifford, of Stonehurst, whose flock is of Lincoln-merino foundation.

"To-day the farmers of Canterbury find the lamb trade much more profitable than mutton, as they get for a lamb five to six months old the same price as they would get for a wether if they kept him another year. Therefore, the bulk of the stock in farmers' hands in Canterbury are breeding ewes. The English Leicester is still the favoured sheep of Canterbury for breeding fat lambs; from this cross the lamb is of nice quality, and the pick of the ewe lambs are kept for breeding.

Where farmers on heavy land mix grain growing with lamb fattening, the Southdown or Shropshire is used, and all the lambs are sent away fat. There is no doubt that from a Romney, Lincoln, or Leicester cross ewe the Southdown half-bred lamb is of the best freezing quality. Finally, no country in the world has a finer record of natural increase in sheep per annum ; with a total number of 23,000,000 sheep, New Zealand in 1910 exported 1,968,254 sheep and 3,522,333 lambs (without reducing her total very much), and fed her 1,000,000 inhabitants at the same time. The total number of ewes bred from were about 10,000,000. I believe the average percentage of lambs reared is about 90 per cent., this shows clearly New Zealand's splendid advantages as a pastoral country."

Argentine Imports of Pedigree Stock.—The improvement of the flocks and herds in Argentina, rendered necessary for the successful establishment and prosecution of an export trade in meat, began at an early stage by the importation of pedigree stock from Great Britain. Argentine buyers for many years have been the great supporters of Great Britain's most flourishing landed industry, pedigree stock breeding ; their determination in securing the best animals practically regardless of cost and the extensive scale of their operations have resulted in a vast improvement in the marketing stock slaughtered for the frozen and chilled meat trades. In the thirty-one years, 1880—1910, Argentina imported from all countries for breeding purposes 18,709 cattle and 77,505 sheep. From 1885 to 1908 an analysis of the imports from Great Britain gives 12,094 Durham cattle and 55,338 Lincoln sheep. It is not easy to estimate the monetary value of these imports of pedigree stock, but the figures, £70 per head for cattle, and £15 per head for sheep, may be taken to represent the value realized very approximately by British breeders for their exports. Applying this valuation to the cattle and sheep exported from all countries to the Argentine Republic for breeding purposes for the period named above, the following totals—representing, roughly, the f.o.b. values—are arrived at :—

Cattle	18,709	£1,309,630
Sheep	77,505	£1,162,575

"RIBY DERBY CHAMPION," LINCOLN RAM (REGISTERED NO. 9722, VOLUME 16).

· Mr. Henry Dudding, Riby Grove, Grimsby, on July 12, 1906, to Mr. F. Miller, whose portrait appears here, for Ianuel Jose Cobo, of Estancia La Belen, Argentina. The price paid, 1,450 guineas, was the highest ever paid for a Ram shipped to South America.

To face p. 98.

The average f.o.b. valuation of the 5,518 pedigree bulls shipped from the United Kingdom to Argentina from 1903 to 1909, inclusive, was £100 each. The figures of exports of British pedigree stock for the complete period 1880—1910 are not available, but the fact that British breeders exported 15,476 cattle and 71,359 sheep to the Republic from 1885 to 1910 (twenty-six years) shows what a preponderating share of the business has fallen to them.

That the expenditure of these large sums by the enterprising Argentine estancieros has been remunerative no one can doubt who is acquainted with the good quality of the beef and mutton now imported into Great Britain from the Republic. In the export beef trade the ideal to work up to is the marketing of chilled beef, an article highly superior to frozen beef from a selling point of view. That the managers of the Argentine frigorificos are able to purchase cattle good enough for chilling may be placed to the credit of the policy under which the estancieros and other importers have since 1880 spent a sum approaching two and a half millions sterling in improving the cattle stock of the country. Reference is made in another chapter to the benefit to the English and Scotch pedigree stock breeders which has accrued by this demand from South America. It is a curious reflection, but obvious, that, whilst the British breeders in building up the pure-bred stock export trade have brought English agriculture in this particular section to a high condition of prosperity, they have prepared a rod for the backs of the British farmer! For, although to the public of Great Britain, the £10,500,000 worth of frozen and chilled meat imported during 1910 was a blessing, it was obviously regarded in quite another light by the farmers.

CHAPTER VII

THE STOCKRAISERS' MARKET

This is a very important and practical department of the subject. It would not be difficult to outline the system which regulates the handling of the frozen carcass through the various stages, from discharge at the London or Liverpool wharf to the final destination at the retailer's shop, but something more than generalization is wanted here.

The Argentine Way.

Before following the meat from ship to shop, it is well to take a step backwards and refer to the methods by which the shipment of frozen meat is worked. To take the simplest way first—the South American. Meat shipped from Argentina is the property of the freezing works, which, in all but a few instances, have their own offices in London, and depôts, and a complete system for the sale of the meat at various ports and important marketing centres throughout England. Messrs. James Nelson and Sons have about 1,500 retail shops, and two of the other Argentine companies own shops. In the case of all the South American frozen meat shipped to Great Britain, the officials in England, or the regular agents of those companies which have not English offices, take charge of and realize the goods in their own shops, on the market, or ex store. There is much less forward selling in the South American than in the Australasian trade. Much of the Uruguayan, Patagonian, and Venezuelan meat is sold forward, and Argentine has been, and is occasionally thus sold now. One important difference has marked the Argentine selling system as compared with the Australasian. In the former trade the meat is, as a rule, turned over quickly ; the holders have averaged the market values and sold steadily right along, and have used the cold stores merely as receiving depôts for their meat. Continuous supplies have enabled the Argentine companies to develop distribution pretty well on retail lines, and owing to regular and continuous imports into Great Britain the Argentine houses have been able to avoid, to a great extent, the embarrassing accumulations and temporary scarcities which have so frequently caused disaster to those engaged in the necessarily more speculative Australasian trade, in which, unfortunately, there has always been a lack of continuity in supplies. The advantage of an extended season enables Argentine shippers to export practically uniform monthly quantities. (Australasian works all have a more or less lengthy " closed down " period.) Mention may be made of the freight contracts for definite quantities, to cover long periods, made in the Argentine trade, say, one to three years, and shippers are under penalty to ship these specified quantities at regular intervals, be the British market good or bad, or pay dead freight.

Australasian Methods.

The Australian and New Zealand meat export business is worked in two ways. First, there is the old-fashioned consignment or commission system, by which the grower or the merchant ships on owner's account. All the trade in the early days was conducted on this basis. It was then very commonly the practice for exporters of small lots to send the meat through the banks or wool houses ; such consignees, knowing nothing of the frozen meat trade, sent the documents to Smithfield salesmen. Nowadays in New Zealand the owner of the stock takes the risk of the London market to a limited extent ; in Australia the grower does not do this, preferring to sell his stock to the freezing companies, which, in order to keep their factories going, have to buy from the graziers and pastoralists extensively. The Australian producer nowadays is not anxious to become a direct shipper to the British market on consignment, but in the early days of the Queensland export a very large proportion of the beef was sent forward at the growers' risk.

We may take it that the Australian meat producer, as a rule, sells to the shipper, who either consigns or sells forward, according to circumstances. The London offices of the Australasian banks receive a small amount of business, but the great bulk of consignments is sent direct to the houses which lay themselves out specially for this trade. The advances to the shipper are calculated in much the same way as in any other trade, and generally bear a fixed relation to the London value of the meat at the time of sailing. The charges on account sales are generally on one of two bases, either a consolidated rate (a method very little used now), which covers all expenses from the ship's rail in London up to rendering account sales, or charging all actual out-of-pocket expenses, such as lighterage, storage, fire insurance, cartage, pitching and market tolls, railage, interest, port dues, etc., plus brokerage. The Smithfield salesman's commission is 2 per cent., and the bank or agent usually charges 1 per cent. for his work.

Buying and Selling Forward on C.I.F. Terms.

With the introduction of the grading process, about 1890, purchasing frozen meat forward became possible. Large retailers in London and the Provinces who have regular outlets for meat of a certain quality and weight at once saw that they could partly cover their requirements for many months in advance by means of contracting to buy on a cost, freight, and insurance basis. This system has had to be adopted with the leading lines of perishable food produce handled by the large stores and " multiple shop " companies. It has been applied quite scientifically to the frozen meat trade, and represents one of the most important and interesting developments. The volume of business passing during the last ten or fifteen years on this basis has been very considerable, but is apt to fluctuate widely, according to the requirements of the " multiple shop " companies, and shippers' costs. In the disastrous 1909 season, when frozen lamb fell 50 per cent. in price on the rates of the previous year, forward buyers dropped money heavily, and it must be noted that a proportion (though quite a small one) of the c.i.f. purchasers are purely speculators. The c.i.f. system was started in New Zealand, and receives its most scientific development there. In Australia, speaking of meat shipments as a whole, grading has not been so thoroughly mastered as in New Zealand, and without reliable grading for quality and weight the forward trade is not seen at its best. Australian trade is severely handicapped in this respect, and lower prices are paid to cover faulty grading. A larger percentage of Australian than of New Zealand meat is now sold on c.i.f. terms. Whilst it is true that Australian grading of frozen meat, as a whole, is not as reliable as is desirable, nearly every one in the trade will accept the brands of some of the Commonwealth shippers as meritorious examples of the grading system.

How the C.I.F. Trade is Worked.

The *modus operandi* in the c.i.f. business is as follows. The representatives in London (termed in the trade "agents") of the freezing companies, having certain lines of mutton and lamb of specified weights on their books, go round to their c.i.f. buyers on Smithfield and telephone to their country clients in the endeavour to fix up contracts; the cable is freely used in the business. When a sale is made, a formal contract passes between seller and buyer; there are several different forms current. The contract fixes the time for shipment, either by the month or the steamer being specified. On the arrival of the carrying vessel, the buyer pays net cash against documents (bill of lading and insurance policy), and takes possession of the meat, provided that the documents are in order, as to brand, port of shipment, date of bill of lading, terms of insurance, range of weights, etc. If the bill of lading bears date the first day of the month following, or the last day of the month preceding, the specified month, there is a breach of contract. Forward sales have been made for as long as six months ahead, but a more common plan is to sell month by month. Under this system, the shipper has to stand the cost of freight, insurance, and exchange, and the buyer in England bears all costs of landing, storing, and market tolls, also loss of weight up to an agreed percentage.

Ex ship sales are practically a department of the "forward" system of doing business; the term may be applied to contracts made after the carrying vessel has sailed, but is principally used for sales made after the ship has arrived at London. The agent, having documents in his hands representing consignments sent to his care, may make a contract when he takes delivery from the vessel.

Allowances Off Actual Weights.

An interesting question which may be referred to at this point is that of allowances off actual weights made in the frozen meat trade. The c.i.f. or ex ship buyer pays on colonial bill of lading weights, which represent a deduction of from 5 to 6 per cent. off the "hot" or "green" weight. Occasionally meat is weighed "cold," while some of the Argentine companies weigh the carcasses in a frozen state, in lots of 20, in order to arrive at the bill of lading weight. Sellers guarantee that the loss on weight when the carcasses are weighed in store at the time of delivery shall not exceed an agreed percentage, usually 2 per cent., from the bill of lading weight.

At Smithfield when selling off the hooks the weights are taken as a rule without the wrappers, and buyers claim allowances from these weights to cover the loss in cutting up and the turn of the scale. These allowances vary slightly in accordance with the terms of sale. Until recently abatements had been accorded only in the London trade, but they have now been introduced at Liverpool and elsewhere, as far as Australasian meat is concerned. The allowances in question from the gross weight are 2 lbs. on each New Zealand sheep, 8 lbs. on every five Australian and South American sheep, 1 lb. on each lamb, and 2 to 3 lbs. on each quarter of beef, according to the nature of the wrapper. The allowance on Australian beef was reduced from 2 lbs. to 1 lb. per quarter by concerted action of the agents at a time when supplies were in one or two hands, and that reduction was accepted for many years. Latterly the old scale has been reverted to owing to the keen competition amongst sellers. This London "bate" formed the ground for strife between the North American "Beef Trust" houses and Smithfield a long time ago. When these firms first opened their business in Great Britain, they allowed the usual "bate" of 1 lb. off their chilled beef quarters, but when they got firmly settled down, and were sufficiently strong to dictate to their customers, they withdrew this allowance. The battle was sharp, but as the American refrigerated beef had become by that time absolutely necessary to the Smithfield salesman, the position of the Americans was impregnable. Their victory, however, left a bitter feeling, traces of which are observable to this day. The origin of the market allowances on mutton is obscure. Some people trace it to the time when Scotch mutton was sent to Smithfield with the kidneys left in the carcasses; as kidneys had no value, apparently, the salesmen removed them or allowed the buyer 1 lb. off the consignors' weight. But frozen sheep, as a rule, contain no kidneys! The allowance is made partly to cover the butcher's loss by wastage in cutting the meat into small joints. As far back as one can discover from research, the Smithfield custom was to allow 1 lb. draft on every quarter of beef and 3 lbs. per side, besides tare.

The Great Grading Question.

As grading of the carcass is the foundation of the c.i.f. trade, some remarks on the system may be made. The earliest references appear in the year 1890. At that time grading and classification on more exact lines were suggested to facilitate mercantile handling of New Zealand mutton and lamb. "Forward" sales on c.f. and i. basis had increased, and this system of carrying on the trade necessitated the employment of more precise standards of quality and weight. In the nineties the "multiple shop" principle of trading began to be developed in the large cities of England, and the proprietors of these businesses found it as convenient to contract for forward delivery of their frozen meat—in the case of meat retailing concerns—as the New Zealand farmers found it convenient to secure a market on the spot by selling outright to the freezing companies. As the multiple shop companies grew, the necessity to cover requirements some months ahead and to guard against sudden variations in values became more pressing, with the result that the c.i.f. system became incorporated as a fundamental part of the New Zealand frozen meat trade. It has had many critics, and had it remained as it began—only a speculative affair (on the part of buyers)—the "forward" trade, and possibly the trade as a whole, would not have assumed the volume and regularity it now possesses. As the introduction of grading is one of the great historic events of the frozen meat trade, the following extract from a market review of 1890 is of interest :—

"In view of the vast extent of the trade and its established character, some serious efforts should be made to grade and classify the exports from New Zealand in a more thorough manner than has hitherto been done."

It was found by importers that the brand was not such a guarantee of uniformity and quality as forward buyers required. As a record, there may be inserted here the grades in force at the New Zealand Refrigerating Co.'s works at Dunedin in August, 1890 :

A.	Sheep 55 to 70 lbs.	
B.	,, 50 to 54 ,,	Crossbred wethers
C.	each weighing from 71 lbs. upwards	and
D.	ranging from 40 to 50 lbs.	maiden ewes.
MER.	Merinos 45 lbs. and upwards	

This company was the first concern to sell c.i.f. and the first to grade for weight. It always failed to grade satisfactorily for quality.

At the beginning of the trade the only thing necessary in this connection was that the meat should be graded to quality, the demand in the first instance being largely for heavy weights in mutton. As the trade developed it was found that the requirements of Smithfield were turning to lighter weight

"PRIME CANTERBURY LAMB."

To face p. 106.

sheep; buyers paid a higher price for light mutton, say, under 64 lbs., than for the heavy carcasses which had been ordinarily shipped. The tendency for the favourite weight of sheep to grow steadily less and less is largely due to the insistent demand of the lower classes for variety on their table. The small joint sells first because the wife of the English artisan and labourer is not skilled in making tasty dishes out of cold meat. In 1887 the most favoured weight was 64 lbs., to-day it is from 48 to 52 lbs. The first grading suggestions from London favoured the somewhat arbitrary system of classifying the mutton carcasses into 5-lb. grades. After consideration it was found more commercially convenient to adopt the stone grade of 8 lbs.; this brought the New Zealand mutton grading into correspondence with the weight measure regulating sales of live and dead meat in the London trade. The grades on this basis of the Christchurch Meat Co.—practically, also, of the Canterbury Frozen Meat Co.—are as follow. The New Zealand (North Island), Australian, and South American systems are different. But as these Canterbury grades have much currency at Smithfield, it is convenient to give them.

Sheep	Under 48 lbs.	Weight brand	1
,,	48 to 56 lbs.	,,	7
,,	56 to 64 lbs.	,,	3
,,	64 to 72 lbs.	,,	9
,,	Over 72 lbs.	,,	5
Lambs	Under 36 lbs.	,,	2
,,	36 to 42 lbs.	,,	8
,,	42 to 50 lbs.	,,	4
Tegs	Over 50 lbs.	,,	T

These weight grades are mainly for the South Island of New Zealand; the North Island favours to some degree this gradation: under 50 lbs., 50—55, 55—60, 60—65, 65—70, and over 70. The Wellington Meat Export Co.'s grades for mutton run: under 55 lbs., 55—65, and 65—70. Other exporters adopt slight variations on these classifications. It would certainly be convenient if one standard could be adopted for the whole of the freezing works of the Dominion.

It may be stated that the 8-lb. butcher's stone of the London trade is not accepted at Liverpool or any other large centres of the United Kingdom, where meat transactions go by the pound.

The sheep and lamb carcasses are first graded for quality, and then for weight. Beef is also weight-graded, but on broader lines, the favourite range of weights being from 160 to 220 lbs. per quarter: under 160 lbs., 160—180, 180—200, 200—220, and over 220.

Probably the first c.i.f. transaction on record, the authors learn on inquiry, was a sale in 1888 of 2,000 Dunedin sheep to Messrs. W. and R. Fletcher, Ltd., by Messrs. A. S. Paterson and Co., of Dunedin, through their London agents, Messrs. W. Weddel and Co.

Argentine Grading.—With regard to the grading of Argentine frozen meat, each company grades its own meat in its own way—there is no general classification, as there is with Australasian mutton and lamb. The shipments of mutton and lamb from Argentina mainly represent first quality; secondary, or relatively inferior, meat is shipped separately under certain marks, but the classifications used in the Australian trade—"g.a.q." "f.a.q.," etc.—are not recognized in the Argentine trade. The weight grades of the Compañia Sansinena de Carnes Congeladas for mutton are: 40 to 46 lbs., 47 to 56, 57 to 64, 65 to 72, and over 72. The River Plate Fresh Meat Company's classifications are as follows. Mutton; 40 to 48 lbs., 49 to 54, 55 to 60, 61 to 68, 69 to 75. Lamb; under 30 lbs., 31 to 36, 37 to 40, 41 to 44, 45 to 50. Beef is graded as systematically as mutton, but both chilled and frozen are graded more for quality than for weight. The two Patagonian companies grade for weight and quality, and grading is also practised at the Uruguayan and Venezuelan works.

Rates and Freights.

In one way or another the grower cashes his meat partially or, in cases of outright sales, entirely at the time of shipment. His banker or agent at the port of shipment makes no difficulty under ordinary circumstances about advancing 75 per cent. of the then market value of the meat. Companies handling a large quantity, of course, make special arrangements with their bankers. Farmers shipping through freezing companies are sometimes charged a "consolidated rate," covering everything from works to sale at the Central Markets, London. This inclusive rate may be roughly stated as $1\frac{1}{8}$*d.* to $1\frac{1}{4}$*d.* per lb. on a parcel of mutton or lamb shipped from New Zealand to London. As the companies do so much buying now, the "consolidated rate" may be considered old-fashioned. The economy now existing in the trade is noted in comparing these figures with those of, say, 1893–1894, when the movement for reducing charges in Australia was initiated. The New Zealand consolidated rate then in force was 1·80*d.* per lb., and that on beef shipped from Queensland was 2·49*d.* The charge to-day on Queensland mutton is $1\frac{1}{2}$*d.*, when squatters ship on their own account. Going back to 1885, the shipper in New Zealand of a 65-lb. sheep was mulcted, on the above basis, to the extent of $3\frac{1}{2}$*d.* per lb. (in 1883, 4*d.*): colonial charges $\frac{1}{2}$*d.*, transport $2\frac{1}{4}$*d.*, and London expenses $\frac{3}{4}$*d.* Selling his meat at 5*d.* per lb., he netted about 8*s.* on his sheep. By 1888 the total charges had dropped to 2·61*d.* per lb. Now they are $1\frac{1}{2}$*d.* per lb.

The freight on frozen meat from New Zealand has been fixed from 1905 on the following scale:—

—	Mutton.	Lamb.	Lamb over 42 lbs.	Beef.	Legs of Mutton.
	Per lb.	Per lb.	Per lb.	Per lb.	Per lb.
From December to May, inclusive	$\frac{11}{16}$*d.*	$\frac{3}{4}$*d.*	$\frac{11}{16}$*d.*	$\frac{5}{8}$*d.*	$\frac{9}{16}$*d.*
From June to November, inclusive	$\frac{9}{16}$*d.*	$\frac{5}{8}$*d.*	$\frac{9}{16}$*d.*	$\frac{1}{2}$*d.*	$\frac{7}{16}$*d.*

New Zealand Brands.

On p. 110 appear the leading brands used by the New Zealand meat works. Many of these are at the same time quality grade marks, such as "Eclipse," "Sun," etc.

The farmers in New Zealand have been in clover for many years past as regards the realization of their meat. Most of

Town.	Name of Freezing Works.	Brands.
Auckland	Auckland Sheep Farmers' Co.	Name in full and "Glas."
Gisborne	Gisborne Sheep Farmers' Co.	Name in full.
	Nelson Brothers	Circle.
Napier	Nelson Brothers	Circle.
	North British and Hawke's Bay Freezing Co.	"N. B. & H. B. F. Co."
	T. Borthwick and Sons	"Paki Paki," "Hastings."
Wellington	Wellington Meat Export Co.	"W.M.E.Co.," "N.G.A.," and name in full.
	Longburn Meat Freezing Co.	"Longburn, N.Z.," and sub-marks.
	Gear Meat Preserving and Freezing Co. of New Zealand.	"G. C."
Masterton	Wellington Farmers' Meat Co.	"W. F. M.," "Taratahi," and "Masterton."
Patea	Patea Freezing Works	"Patea Freezing Works."
Wanganui	Wanganui Meat Freezing Co.	"Wanganui" (red and black), Thistle.
Waitara	T. Borthwick and Sons	"Waitara, N.Z.," "Mount Egmont."
Stoke	Nelson Freezing Co.	"N. F. C.," "Stoke," One Anchor, and Two Anchors.
Picton	Christchurch Meat Co.	"C. M. C.," "Wairau," Crown, and Three Crowns.
Belfast	Canterbury Frozen Meat and Dairy Produce Export Co.	"C. F. M. Co.," Diamond, and Star.
Fairfield	" " "	
Pareoroa	" " "	
Islington	Christchurch Meat Company	"Eclipse," "C. M. C.," Sun, One Crown, Three Crowns.
Smithfield (or Timaru)	" " "	
Oamaru	" " "	
Burnside	" " "	
Hornby	Nelson Brothers	"C. M. C." and "C. F. M. Co." 777 (the brands denote the various Canterbury works where the stock were killed).
Mataura	Southland Frozen Meat Co.	"S. F. M. C.," "M.," "Z.," and Crosskeys.
Wallace Town	" " "	
Ocean Beach	Birt and Co.	"Princeps," "O. B."
Tokomaru	Tokomaru Farmers' Freezing Co.	"Tokomaru," "Waima," and "Tawhiti."

them sell their live stock on the farm, or at so much per lb. at the works. Their market, at high prices, frequently too high (compared with London), has been assured. The disastrous 1909 season would probably have had the effect of lowering the value of sheep and lambs in New Zealand for freezing had not the following year been one of high prices on the English market. Some of the New Zealand freezing companies operate on their own account, while others are 'farmers' companies," freezing only. The Canterbury Frozen Meat Co., formed in 1881, is an example of the latter system. Some companies combine the two methods.

Australian Grades.

Much of the Australian mutton and lamb—and it must be stated that grading in Australia is slowly improving—is sold on two quality standards: "f.a.q." (fair average quality), and "g.a.q." (good average quality). Sydney meat mostly comes under the former, and Melbourne under the latter. "F.a.q." allows but little recourse on account of quality, and buyers want a considerable concession in price for that reason. In the cases of well-known accepted brands, with a reputation, the meat is sold on f.a.q. or g.a.q. of the brand, and the trouble arising from the interpretation of "g.a.q" is lessened. The difficulty in the Australian trade has been the uncertainty as to a standard of quality. In the New Zealand trade there is a recognized basis of quality. Every weight grade carries with it a certain standard, and for a delivery against sale a carcass must carry proportionately as much weight as a sheep in good condition would carry. Without such a standard, how can allowances be assessed? In the New Zealand business only two grades denoting quality are used in the general trade: "Prime Quality" and "Second Quality." With the latter, except in extreme cases, there is no recourse for the dissatisfied purchaser, who can call for surveys in the case of "Prime Quality." With the North Island companies' meat there is a grade "good average quality," carrying with it recourse. The London c.i.f. buyer has to cover himself to the extent of $\frac{1}{16}$*d.* per lb. in accepting "colonial weights"; that allows for shrinkage in freezing and the London "bate."

Allusion has been made above to the c.i.f. form of contract

commonly employed in the "forward" trade (New Zealand). Although buyers and sellers, assisted by the efforts of the Frozen Meat Trade Association, for years hammered away at a "uniform contract," the difficulty of harmonizing the many divergent views expressed was found insurmountable. It is to be hoped that a form that commends itself to the trade will be generally adopted throughout trading centres in the United Kingdom.

To conclude these remarks regarding the "forward" system of handling New Zealand and Australian meat, it may be stated that, notwithstanding the results of the terrible 1909 season, there can be no doubt that this method of selling the meat, based as it is upon the modern commercial ways of working large meat and provision shop businesses, has become firmly established and will extend. There are, it is true, certain drawbacks to it, but these, no doubt, in time will be partly or wholly removed. For instance, there is a considerable difference of opinion, with consequent friction, as to what constitutes the Australian grades referred to above, "g.a.q." and "f.a.q.," and some weakness in the working of the general system is shown in the arbitrations as to quality which are frequently called. It is alleged, perhaps without very good grounds, that the calls for these arbitrations vary in direct ratio to the tone of the market. The c.i.f. trade fluctuates as regards the business done, depending upon the views that are held regarding the immediate future prospects of the market.

Discharging the Meat at the London Docks.

The stockowner, although often he may be less concerned than others financially in frozen meat cargoes on their arrival at the port of destination, evinces a keen interest in the systems under which the shipments are handled on discharge, and the following information will, therefore, be of interest to him as well as to others concerned in the trade.

The vessels conveying frozen meat from Australasia to London berth at Tilbury, Victoria, and Royal Albert Docks, and those from South America in the Victoria, Royal

CAPTAIN NOAKES'S MECHANICAL CONVEYOR DISCHARGING FROZEN MEAT AT THE PORT OF LONDON.

To face p. 112.

Albert, and West India Docks. The two Australian mail lines, which do not carry a great quantity of meat in each vessel, and the White Star steamers, use the Tilbury Docks. The steamers usually break bulk within twenty-four hours of docking, and as a rule the discharging goes forward during working hours without any stoppage till completion. The discharging is done by dock labourers, who are sometimes the employees of the shipowners and sometimes of firms of stevedores discharging under contract. Carcasses of mutton and lamb and quarters of beef are commonly discharged in slings, but the New Zealand Shipping Co., the Shaw, Savill and Albion Co., the Shire Line, and other companies, use patent elevators and shoots for the rapid handling of mutton and lamb.

Noakes' "telescopic elevator" was the first appliance of the kind. This was invented by Captain G. H. Noakes, superintendent of discharge to the New Zealand Shipping Co. in London, and was patented about the year 1900. It consists of a telescopic frame carrying two endless chains, and on the chains are fitted at equal intervals shelves or projections to convey the goods. It is driven off the ordinary ships' winches by means of a rope slung round from the winch end. This machine is portable and very handy, it being possible to pick it up from the quay and place it in the hold ready for work in about twenty minutes, and it is capable of discharging sheep, cases of butter, crates of bananas, or any packages of uniform size, at the rate of over 1,000 per hour.

Captain Noakes has recently designed an improved form of conveyor for the discharge of frozen meat from vessels, which consists of a system of mechanical chutes and endless belt-carriers driven by small electric motors fitted inside the mechanical chute, leading from the ship's deck to the quayside. This apparatus, which has lately been installed and set to work at the London docks, is illustrated herewith. Although the driving is in the upper part of the chute the control is at the bottom, and a man by pressing a button can stop it immediately if necessary. By this system, in conjunction with the elevators for raising the meat from the holds, the carcasses are carried from the hold to the end of the quay for delivery, without the aid of meat slings or hand trucks. The mechanical chute does away entirely with the system of sliding sheep down chutes, as they rest against a projection, and are carried down by the electrically-driven appliances. This prevents exposure (the carcasses are protected by canvas coverings), bruises and abrasions.

The method of discharge at the docks in London depends upon the destination of the meat. There are three general courses open: warehousing at the dock stores, despatch by rail to the country, and barging along the river to the "up-town" stores. Taking the Tilbury Docks first, farthest down the river, if meat is to go into the dock stores at Victoria Docks it is barged up. If intended for the country it is forwarded by railway, or if for the up-town, riverside stores, barging is resorted to. If the meat is for the dock stores at West Smithfield, the railway and also insulated vans are employed. At the Victoria and Royal Albert Docks the same procedure is followed, except that meat intended for the dock stores at these two docks is transferred there direct, either by railway or hand truck, according to the distance from the ship's berth, and that intended for the dock stores at West Smithfield is forwarded in insulated vans. The handling of meat at the West India Dock is practically the same, meat for the dock stores at Victoria Dock being conveyed thither by insulated vans. The railing, vanning, and trucking involve the landing of the meat on the quay alongside which the vessel lies, and meat for barges is usually delivered from the other, or water, side of the vessel, alongside which the barges lie. The railway wagons run alongside the quay, and the dock stores have connection with the Great Eastern, Midland, London and North Western, and Great Northern lines. Meat by steamers in the Tilbury Dock intended for forwarding by railway to London is dispatched at frequent intervals, and trucks for the country (insulated and iced in the summer), having received the full consignments, or such portions as are available, are sent away to their destinations. Transit is fairly rapid; a line of meat loaded direct from the vessel on to the railway can be delivered at Manchester or Cardiff within twelve hours.

Conveying to Store.

A very large quantity of the Australian, New Zealand, and South American meat is taken in barges to the Lambeth stores of the Colonial Consignment and Distributing Co., Ltd., the various stores of the Union Cold Storage Co., the Blackfriars Cold Storage Co., Ltd., the Thames Cold Storage Co., etc. The barges are brought up the river on the next tide after the day's work is finished. The great aim of the lighterman is to get a full load, so as to ensure the meat travelling in the best condition. Considerable delay takes place at times in the conveyance of meat by this method, particularly in foggy weather, and advocates of storing at the docks suggest that the barging system is far from an ideal method of conveying frozen meat. But some of the meat that goes into the dock stores is moved in this way, as mentioned above. The secretary of the Colonial Consignment and Distributing Co., Ltd., states that his company's barges are "carefully insulated with hair felt, and with full cargoes it is an unknown thing for damage to occur in transit." All the barges conveying meat from ship's side to up-town stores are insulated and are passed by Lloyd's inspectors. Formerly, however, this was not the case, and much damage occurred to meat in transit in hot weather through the imperfect conditions which ruled. All the London public cold stores are on the Thames, excepting those of the London Central Markets Cold Storage Co., which are under and adjoining Smithfield, the West Smithfield cold store of the Port of London Authority, and one or two others indicated in the London cold storage map, Appendix VI. Meat for the first-named store is barged to that company's depôt at Poplar and carried thence by insulated motor vans.

Having warehoused his mutton, lamb, or beef at one of the London cold stores, the owner arranges further steps in his campaign according to his business and the state of the market. His one great aim is to clear the meat within four weeks, so as to save the second month's "management rate" charge. The object of the merchant is to deliver to the market only such quantity as he may expect to dispose of every day, but if meat is left over unsold at the close of business at Smithfield it is seldom taken back to store, as it will remain in sufficiently good condition on the hooks for the next day's business. Occasionally, however, meat unsold is taken back to store, especially in the case of the cold stores handy to the market. In hot weather, naturally, this course is adopted more freely. Those tenants who rent storage space under their stalls from the London Central Markets Stores, which space communicates with the premises above, generally pop their left-over stuff down into their cold rooms.

In putting forward these details concerning an exceedingly technical section of this subject, the authors desire to mention that the business systems of the various firms of merchants, importers, and salesmen, are not all framed on the same lines. All that must be expected of this chapter is a more or less rough and, it is feared, incomplete outline of the methods under which the Australasian and South American producers get their stock to the market.

CHAPTER VIII

THE FUNCTIONS OF THE MEAT INSPECTOR

In this chapter is given information concerning the general principles and methods which govern the official examination of frozen and chilled meat in the exporting countries, and details are added referring to the practice of British Medical Officers of Health in dealing with this meat on its arrival at the ports of Great Britain. Reference is also made to the totally inadequate measures taken to protect the British public from the consumption of diseased meat, bred and slaughtered in the United Kingdom. The more one regards the meat inspection systems in vogue in different countries, so widely differing in principle and detail, the more necessary does it appear to work for an international standard of meat inspection, respecting which proposal it is still hoped there will be a conference in Paris in 1912, or at some early date.

New Zealand.

New Zealand has an excellent official system of inspection, which is rigidly applied to exported frozen meat. The Dominion spends a very considerable sum annually in paying the salaries of a staff of specially qualified inspectors, twenty-four of whom write M.R.C.V.S. after their names. Official inspection in New Zealand is carried on under the General Meat Inspection Act of 1900. The inspectors have by this Act full control over the sanitary and general conditions of the freezing works of the Dominion of New Zealand. The Government of New Zealand has, from early days in the industry, applied itself consistently to the question of having veterinary examination made of meat intended for the export trade. Concerning this,

Sir Joseph Ward, as Prime Minister, made himself responsible for the following statement :—

All meat as exported from New Zealand is absolutely guaranteed by the Government to be healthy, wholesome, and thoroughly fit for human food.

Australia.

As to Australia, the inspection and supervision of meat for export is undertaken by the Federal Government. Prior to February 1, 1911, the various States were working on individual lines, and though the regulations in force in the States were most thorough, the want of uniformity led to some confusion. But at the date mentioned, the Federal Government assumed control of Australian meat inspection, and now all meat exported from the Commonwealth is inspected on a uniform basis under the Commerce (Trade Descriptions) Act, 1905, and the Customs Act, 1910, prior to export, and meat found to be in every way satisfactory is marked " approved for export." Second grade meat is marked as " passed for export," while emaciated and diseased meat is, of course, refused any permit for export. Here attention may be drawn to the fine distinction between "approved" and "passed" in the Commonwealth inspection regulations. Whilst New Zealand and Argentina, in common with usual practice, each have one standard of fitness, Australia has two. A large staff of veterinary surgeons is stationed at the abattoirs and meat works throughout the Commonwealth, and the regulations under which they work are drastic. The examination is ante- and post-mortem. The central point of the Commonwealth regulations is that the exportation of any meat is prohibited unless it has been certified by an inspector under the Commerce Act to be fit for export. Under the Commerce Act there is prohibition of the export of carcass meat in a diseased state, and " disease " includes any defect, inferiority, or abnormal condition in the meat which renders it unsightly or unfit for human food.

Australia can afford to go still farther than European countries in the raising of her standard. Her proportion of

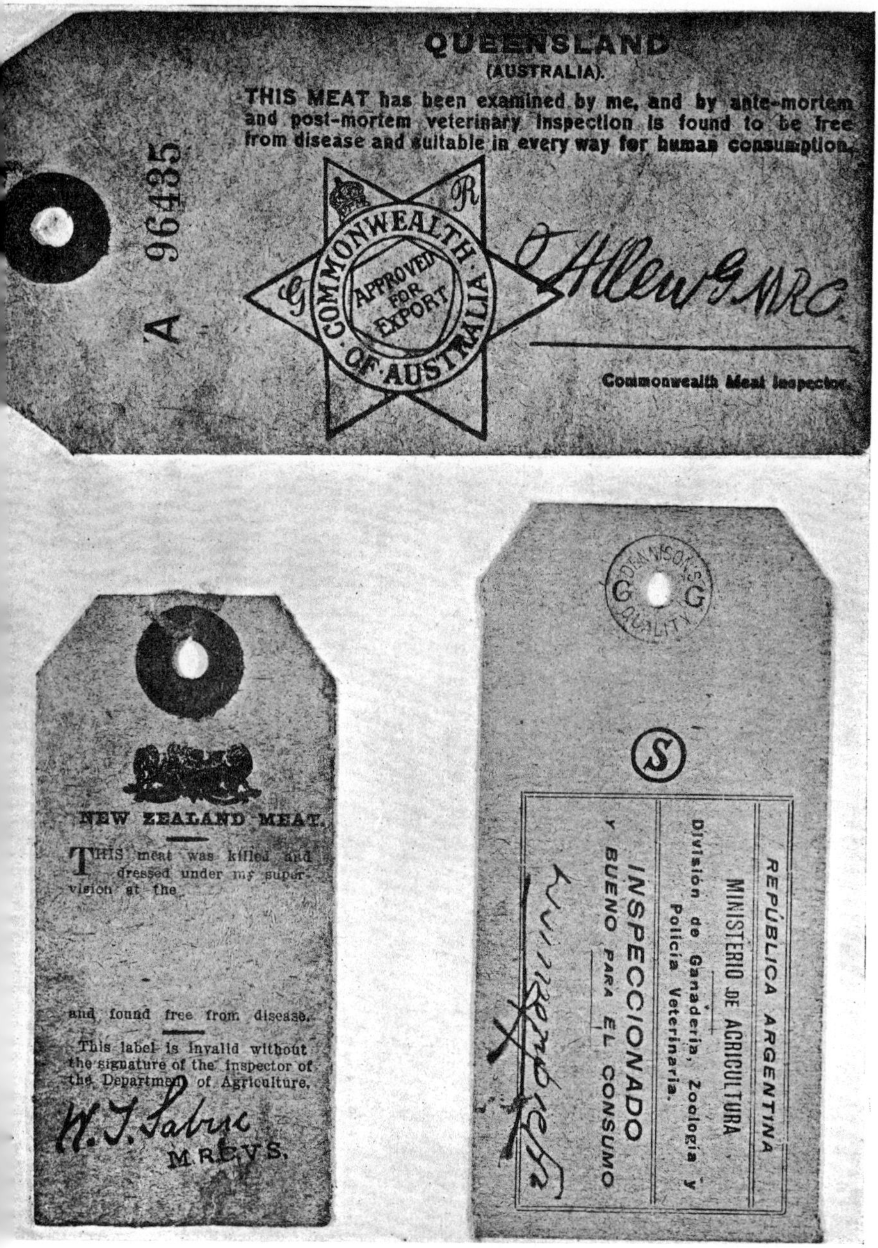

ACSIMILE REPRODUCTION OF AUSTRALIAN, NEW ZEALAND, AND ARGENTINE GOVERNMENT INSPECTION LABELS.

To face p. 118.

stock diseases is wonderfully small, and as boiling down is available for rejects, the Australian freezing works have in the case of animals not up to freezing mark a second string to their bow which the British farmer and butcher lack. No meat can be exported from Australia unless it has passed the veterinary's health tests.

Argentina.

The measures taken by the Government of Argentina to ensure that frozen and chilled meat exported shall be in all respects sound, free from disease, and of first class quality, are most thorough. Before stock can be moved in the Republic, official permission has to be obtained. Attached to every frigorifico is a Government inspector who has an office in the works. He has to look into all the processes, inspects the stock before killing, and the meat before export, and has the right to reject anything of inferior quality. The following statement by the manager of one of the Argentine works shows how thorough is the inspection :—" Two or more Government inspectors are billeted at each works ; they are there all their time, scrutinize everything, examine live stock, slaughterings, the meat, walking about the whole time, condemning anything and everything they are not absolutely satisfied with, and nothing leaves the works without the Government inspectors' certification." The vessels carrying the frozen meat from Argentine ports are under Government supervision as to cleanliness, disinfection, and hygiene. Of late, as a complement to this careful system of inspection, the Government of the Republic have enacted that a brand in aniline dye, which has the force of a Government certificate of soundness, shall be placed upon all frozen meat exported to Great Britain. The Act under which the official supervision is undertaken is the Animals Sanitary Law, No. 4155.

The certificated labels that are attached to meat exported from the three countries whose official inspection systems are above explained are reproduced herewith. They are very useful and necessary documents; but, after all, speaking of Great Britain's importation of frozen and chilled meat, the Local Government Board's inspector is the real tribunal.

Inspection of Meat Killed in the United Kingdom.

Several references have been made in this book to the inefficient methods of inspecting home-killed meat in the United Kingdom. Under regulations officially observed, there is a competent veterinary staff employed at British ports to see to it that no diseased or unsound meat from overseas is landed for consumption. One would think that an equally satisfactory examination of stock slaughtered throughout the United Kingdom would be made. But what do we find? Dr. Collingridge, Medical Officer of Health for the following observation when referring to the Australian the City of London, in his report dated April 25, 1911, made official inspection clauses:—"It is unfortunate that while such regulations are made and enforced abroad, *there are no provisions for the compulsory inspection of meat at the time of slaughter in this country.*"

A flood of light is thrown upon this subject by Mr. W. G. Barnes, Chief Veterinary Inspector and Superintendent of Abattoirs, Islington Cattle Market, in his paper read before the Royal Sanitary Institute Congress at Belfast on July 28, 1911, "Meat Branding and Uniformity of Inspection." In many of the rural districts, meat inspection, Mr. Barnes stated, "is a theory instead of a practice." The following quotations from the paper are startling indeed:—

"The Britisher is daily eating the flesh of diseased animals, and is unaware of it. He takes it for granted that the law of the country provides for the inspection of all butcher meat intended for his food, and consequently he does not deem it necessary to make further investigation into the matter, but is content to remain in ignorance of the fact that more than half of the meat consumed in this country is never seen by an inspector, and that in many parts of the country the system of meat inspection is scandalous. As illustrative of this, it is known that carcass butchers from towns some distance from

London can afford to pay from 30 shillings to £2 more per head for a doubtful class of cattle, pay carriage to one of these places where there is little or no inspection, outside London, and then send the carcass to London for sale for human consumption ; or, if it is too risky for sale in London, sell it locally. . . . Only a short time ago, in conversation with a butcher who has a large trade in a town not far from London, I was informed that no inspector ever visited his premises for the purpose of examining the meat. It was only visited three or at most four times a year by the inspector of nuisances, and that was with regard to its sanitary condition. . . . A pure meat supply can only be secured by a universal compulsory abattoir system and general inspection of all animals intended for human food, also branding and marking of all meat. This system would afford the highest degree of protection to the consumer and would probably be the indirect means of causing a great lessening in disease."

Do not such facts as these concerning the supplies of home-killed meat to British consumers throw into strong relief the scientific veterinary inspection of chilled and frozen meat in Australasia and South America, as detailed in the early part of this chapter, and can there be a more forcible point to urge in support of the case for frozen meat ?

An International Standard of Inspection.

Meat is not an interchangeable article of commerce, but is subject to strict rules of inspection in most countries before it can obtain entry. These conditions formed no hardship so long as trade was carried on only in fresh meat from the adjacent countries. The immense development of the refrigerated meat trade and its conditions of handling render these conditions antiquated and unworkable. There are no insurmountable difficulties to be overcome in placing refrigerated meat in the same category as the articles of commerce which may be consigned to any country. In exporting countries where refrigeration of meat is properly carried on, it is killed under Government veterinary inspection. Inspection to be thorough and complete has to be ante- and post-mortem, and can only be done at the time of slaughter. Refrigerated meat has to be stripped of all internal organs before freezing and shipping. It therefore follows that the inspection, however rigid, can only be guaranteed by an assurance and certificate that the animals were healthy and the meat wholesome. A certificate and guarantee to be valid must be given by a Government authority, and must represent conditions well and clearly understood. A complete harmony exists among all the training institutions for veterinary science of all countries as to principle and methods. Veterinary inspection is on a strictly scientific basis, and there are an international standard and regulations common to all lands where the inspection of meat is considered of vital importance.

To free refrigerated meat from present trade disabilities in European States, all that is required is that the standard which already exists should be recognized as international. Further, that an agreement be arrived at by the various countries that, provided an exporting country will guarantee that all the conditions of inspection are being complied with, the veterinary certificate of such country shall be recognized as complete. Refrigerated meat would then be available for all countries, and would only be subject to local inspection as to refrigerated condition and soundness. It would then take its place along with other articles of commerce—fresh markets would be opened up, and its price would be regulated by the world-wide demand.

Public Health Regulations.

A development which had bearing on some minor section of the frozen meat trade, and made necessary certificates of examination from the Governments of exporting countries, was the passing in England of the Public Health (Regulations as to Food) Act, on August 28, 1907. On September 12, 1908, the Local Government Board issued two sets of Regulations :—

(1) " The Public Health (First Series Unsound Food) Regulations, 1908," and

(2) "The Public Health (Foreign Meat) Regulations, 1908."

The first became operative on October 1, 1908, and the second on January 1, 1909. The Regulations are now in full working order, and have strengthened the hands of the sanitary authorities at the various ports. These Regulations have had a very considerable influence upon the department of the frozen meat trade discussed in Appendix II., for they were put in force mainly to check irregularities in American meat oddments and to enable the sanitary authorities in Great Britain to have an effective weapon wherewith to fight tuberculosis in pig carcasses and boxed pork.

Before the issue of the Regulations inspection of meat imported into Great Britain had been casual and unmethodical. The port sanitary authorities and the medical officers of health, as well as the Board of Trade expert officers, found difficulty in dealing with certain classes of imported frozen meat, and it was considered desirable to give the inspectors further powers. For instance, boneless beef, arriving in boxes with the pieces of meat frozen into a solid mass, was impossible of identification and examination by the inspectors as to the separate pieces. This boned beef trade from New Zealand had grown to considerable proportions ; it was a profitable means of disposing of old cows, and the form in which the meat arrived and its cheapness allowed it to be used for mincing and other purposes. As the conditions under which this boneless beef was prepared and shipped were inconsistent with the requirements of British inspectors, acting under the Public Health Regulations of 1907, viz., that frozen meat must arrive in separate pieces, the trade was suspended for several years. But of late the inspectors have modified the severity of their views, and boned beef, frozen in separate pieces, is imported to a small extent, mainly at the port of Liverpool. It is held in many quarters that had the export of boneless beef been stopped altogether it would not have been a matter for much regret, for, even allowing that this low-grade meat was a valuable adjunct to New

Zealand's carcass and quarter trade, it adds no lustre to the Dominion's export of produce.

The Regulations referred to, in full working order, no doubt, render more precise and (in some instances) equitable the methods of the sanitary authorities with regard to any meat seized for unsoundness. Merchants and importers of frozen meats, at the time when the Regulations were being framed, took the opportunity to press upon the local Government Board their view that identical instructions should be sent to inspectors throughout the country, regarding the general question of inspection of imported meats, and that the owners of meat seized by the inspectors should be supplied with data, such as the name of the vessel, the brand, etc. But this reasonable suggestion has not been adopted. The officials at different ports in England have varying methods of seizing meat and of treating meat so seized; it is a confusing and unfair state of things to traders that different "standards" and different practices should exist in the ports and wholesale markets of Great Britain as regards meat inspection.

The Regulations make three classifications, the first of which applies to "scrap" meat, and the other two to pigs and parts of pigs. An important reference is to the "official certificate" which the Regulations required. This is defined as a certificate, label, mark, stamp, or other voucher, declaring that the cattle or pig from which the meat is derived has been certified by a competent authority in the place of origin to be free from disease at the time of slaughter, and that the meat has been certified by the like authority to have been dressed or prepared, and packed, with the needful observance of all requirements for the prevention of danger arising to public health from the meat as an article of food. On the authors' application in September, 1911, for information about these certificates, which are obviously required in the interests of shippers from various countries, Dr. G. S. Buchanan, Medical Inspector to the Local Government Board, supplied copies of accepted official certificates, as follow:—Denmark, the Netherlands, Canada, New Zealand, Belgium, and the Commonwealth of Australia; Dr. Buchanan wrote that the list of accepted certificates did not include any from the United States or South America. These Regulations have been dealt with above very fully, because they appear to be the only example of official enactment in Great Britain concerning meat inspection drafted on thoroughly scientific and uniform lines of procedure.

CHAPTER IX

THE SHIPOWNERS' BURDEN

THE importance of the part that the shipowner plays in the frozen meat trade of to-day is universally recognized, and most of the pages of this book bear evidence of the great stake he has in the industry. It was the shipowner who was always ready in the pioneering and experimental days of the eighties to risk his capital in meeting shippers' and merchants' wishes, and to-day he is constantly called upon to extend the facilities he offers for refrigerated overseas traffic. The freezing works and the "refrigerated fleet" are the two great agencies engaged in the preparation and transportation of the farmers' meat to the markets of the Old World, and skill, capital, and organizing power, both in freezing and transport, have all along been required to ensure that the Australasian and South American sheep, driven to the freezing works and landed on the London Docks as frozen meat, have suffered the minimum of depreciation in quality and condition.

Given the necessary information from behind the scenes, what a book could be written on the beginnings and development of the refrigerated produce trade, from the shipowners' point of view! If the walls of board rooms had ears, and shipping clubs' discussions were recorded for the benefit of the trade historian, it would be possible to write a useful and entertaining narrative. The personalities of the owners and managers of the great shipping lines that

> Give the Pole the Produce of the Sun,
> And knit th' unsocial climates into one,

are in many cases distinctly striking; many of us remember the graphic, good-natured articles that Mr. Hope Robinson used to write on them in the pages of *Fairplay*. What a tale could be told, for instance, of the men who built up the "Austral-

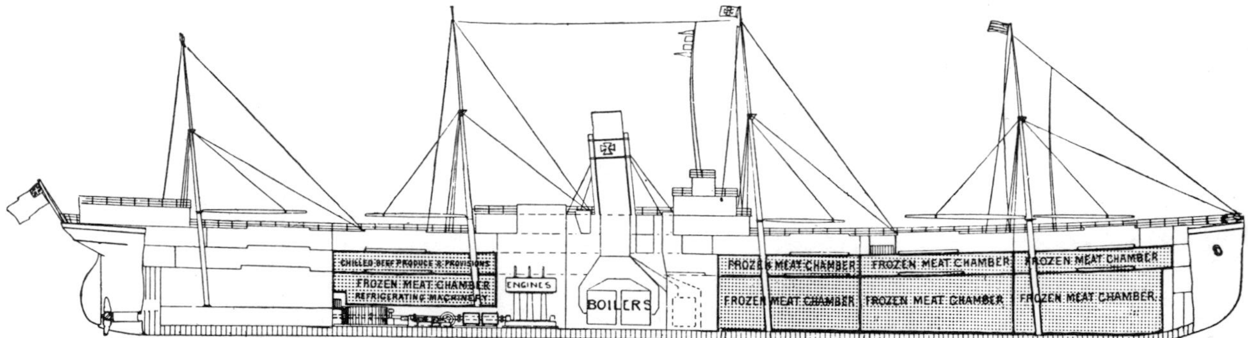

SECTION OF THE HOULDER LINE S.S. *Oswestry Grange*, SHOWING THE TYPICAL FITTING OF A FROZEN MEAT CARRIER.

To face p. 126.

asian Shipping Conference" and their work; will that chapter ever be given to the world from the precincts of Leadenhall Street? The man at the head of a shipping line must, to be successful, have great business gifts as well as force of character. He must be resourceful to the last degree, incredibly audacious, when necessary, in making contracts and adjusting business with clients; all the commercial facts from pole to pole he masters concerning production and transport of merchandize, and at a pinch, as the frozen meat trade shows, turns himself into a merchant filling the holds of his own steamers. The part the British shipowners—of refrigerated and other vessels—played in the South African War stands out brilliantly amongst many failures in other directions. Incidentally, they made heaps of money!

"What Frozen Meat has done for the Shipowner" might well form a sequel to "The Shipowners' Burden." The 1910 figures showed 214 vessels engaged in the conveyance of the South American and Australasian frozen meat; the carrying capacity of these vessels was 14,225,500 carcasses of sheep. Of these steamers 43 had a carrying capacity exceeding 100,000 carcasses each. As the total import of frozen meat for 1910 was the equivalent of 18,056,844 carcasses, it seems that, assuming that each ship of the "refrigerated fleet" came home fully loaded, the portion of frozen meat allotted to each vessel to carry during the year was the equivalent of about 84,380 carcasses. During 1911 the "refrigerated fleet" was much augmented, and the insulated space of the vessels on the Australasian and South American lines was at the end of that year equal to 44,000,000 cubic feet (see Appendix VIII.), say, 16,000,000 carcasses.

Then there is the great trade in butter, cheese, fruit, and rabbits from Australia and New Zealand to consider. What a welcome addition to the freight refrigerated produce must have been, and what potent influence it must have had on the expansion of the great fleets, as well as upon the construction of the vessels! In 1888 there were 57 vessels in the trade with a total carrying capacity of 955,000 carcasses; by 1891 there were 87; in 1894, 100; in 1895, 108; and in 1898, 131.

The carrying capacities in those years were, respectively, 2,267,000, 3,367,000, 3,816,000, and 5,460,800 carcasses.

It would have been interesting to discover that there were one or two master minds applying themselves with a prophet's foresight to the task of adapting their ships to the new trade; men who, with the genius of perspicacity, fully grasped the situation with all its possibilities. One cannot discover any such figures, and the progress of the refrigerated carrying trade seems to have been a natural evolution. Improvements in refrigerating machinery, insulating materials, and appliances or stowage and handling, have come along as the need suggested.

The shipowners' attitude in the early eighties was this:—They said "we are the servants of the public, and if there is meat to carry we must carry it." In this way they have adapted themselves to the different departments of the refrigerated business, science revealing the particular conditions requisite for the safe carriage of each separate kind of refrigerated produce. A great deal of capital is invested in the "refrigerated fleet"; roughly, it takes £20,000 merely to fit a New Zealand liner with refrigerating machinery and insulate her holds, and to this has to be added the cost of upkeep, and the loss from the permanent reduction in dead weight and cubic carrying capacities. Regarding the weight of cargo carried, a vessel which, without refrigerating machinery, could carry, say, 4,000 tons, could with it only take 3,000 tons of ordinary cargo.

The Australian Lines.

The Australian shipping lines were not backward in taking up refrigerated produce, and the companies running regular services perceived quite early the possibilities of the new trade and what it meant to the shipowner. The prospectus of the Orient Steam Navigation Company, Limited, dated 21 May, 1880, contained this paragraph: "The export of fresh frozen meat is likely to yield an important addition to the company's earnings. A number of applications for space have already been received, and the necessary refrigeration machines are about to be fitted in the steamers." The first steamers fitted were the s.s. *Cuzco, Orient,* and *Garonne* in 1881. On p. 34 mention is made of the pioneering part taken by the Orient line vessels in fitting up their steamers for the carriage of frozen meat, and note should be taken of the fact that this company, in installing Haslam machinery on its ships in 1880, was the first shipping line in the world to go into the refrigerated meat trade on a regular basis. The Peninsular and Oriental Company entered the trade in 1887, and their first vessel to carry meat was the s.s. *Victoria*. The s.s. *Hornby Grange* was Messrs. Houlder's first refrigerated vessel to enter, in 1890, the Australian trade. The Aberdeen Line took up frozen meat freight in 1892, and the s.s. *Australasian* was the first steamer to be fitted. The first Federal Line steamer to carry meat was the s.s. *Maori King* in 1893. The White Star vessels entered the Australian trade in 1899, the s.s. *Medic* being the first steamer to bring meat for the company from Australian ports.

The late Mr. William Haverside Tyser, the founder of the Tyser Line, entered the refrigerated shipping trade in 1886—1887 (see p. 67). The first ship which he fitted up and loaded was the *Balmoral Castle*, next coming the *Ashley Brooke*, and others following. The *Balmoral Castle*, by the way, was the first vessel to sail to Australia in ballast, a practice subsequently extensively followed.

The New Zealand Services.

In 1881 the Government of New Zealand offered a subsidy of £20,000 a year for the establishment of a service of refrigerated fifty-day boats, but the inducement was too slight to attract shipowners. The honour of opening the frozen meat trade, which has been of such enormous value to the Dominion, belongs to the Shaw, Savill and Albion Co. Full particulars appear at pp. 40 to 44 of the enterprise of the New Zealand and Australian Land Co., and the successful co-operation of the shipping company as regards the voyage of the sailer *Dunedin*. The New Zealand Shipping Co. (which had been founded by New Zealand merchants at Christchurch in 1873) despatched the second vessel with frozen meat to London, this shipment being the company's own venture. In 1882 the company fitted the sailing ship *Mataura* with Haslam's cold air refrigerating machinery at a cost of several thousand pounds. On the first voyage outwards quantities of fish, poultry, and game were carried from London. These goods were delivered in New Zealand in excellent order and a return cargo of frozen beef and mutton, at a freight of $2\frac{1}{2}d.$ per lb., was landed in London in perfect condition. This cargo had to be frozen on board. Captain Greenstreet, now the skipper of the s.s. *Remuera*, was in charge. The first steamer fitted up for refrigerated cargoes by the New Zealand Shipping Co. was the *Fenstanton*, belonging to Watts, Milburn and Co.; it was a tramp steamer taken up on time charter. In January, 1883, the company despatched the first steamer of its regular service from London, the *British King*, 3,356 tons, fitted with Haslam's machinery sufficient to carry 8,000 carcasses of sheep. The *British Queen*, a sister ship, followed in March, 1883, and the White Star steamers *Ionic* and *Doric*, vessels of 4,750 tons, were also chartered so as to keep up the monthly service until the company could put on steamers of their own. A fifth steamer being needed, the Cunard liner *Catalonia* was taken up in 1883 for one voyage, and even for that one voyage it was fitted up with complete refrigerating machinery.

In the same year the company inaugurated a direct steamer service, building five 15-knot steamers of about 4,500 tons register each, viz., the *Tongariro* (August, 1883), *Aorangi* (October, 1883), *Ruapehu* (November, 1883), *Kaikoura* (September, 1884), and *Rimutaka* (October, 1884), all fitted with Haslam's refrigerating machines. Three of the company's sailing ships were also fitted with refrigerating machinery for the carriage of frozen meat, etc. After the New Zealand Shipping Co. had received delivery of the above-named five steamers, the Shaw, Savill and Albion Co. continued to employ the *Ionic* and *Doric* in the trade, and also chartered the *Coptic*, besides building the *Arawa* and *Tainui*, so that the two lines, each with five mail and passenger steamers, commenced a regular fortnightly steam service between London and New

THE NEW ZEALAND SHIPPING CO.'S S.S. *Remuera.*

To face p. 130.

Zealand, which has been maintained without a break ever since. At the present time all the foregoing steamers of the New Zealand Shipping Co. have been replaced by new vessels, bearing the same names but of about double the size.

Whilst the Shaw, Savill Co.'s own steam fleet was in preparation in 1883—1884, the chartered steamers *Triumph*, *Bombay*, *Victory* and *Florida* were fitted. But these vessels were soon replaced by the mail steamers *Arawa*, *Tainui*, *Coptic*, *Ionic* and *Doric*, and these were supplemented in 1889—1890 by the cargo vessels the *Mamari*, *Maori*, *Matatua*, *Rangatira* and *Pakeha*. The original *Arawa*, *Tainui*, *Mamari*, *Matatua*, *Rangatira* and *Pakeha* have gone, but bigger successors, of the same names, do similar work to-day. The old *Arawa* was fitted for 25,000 carcasses, the modern *Waimana*, built in 1911, is insulated for 100,000.

The s.s. *Elderslie*, launched in 1884, was the first of a fine fleet of steamers owned by the Shire Line and Messrs. Turnbull, Martin and Co. To Mr. John Reid, of Elderslie, near Oamaru, the credit of this enterprise must be accorded, and he certainly must take rank as a pioneer of the trade. He persuaded Messrs. Turnbull, Martin and Co. to build this and other steamers, and was always in the front supplying sheep freely from his fine estate. The other Shire liners followed quickly, and competed with the steamers of the New Zealand Shipping Co. and Shaw, Savill and Albion Co. Freight on the meat for the *Elderslie* was reduced to 2¼*d*., by the *Fifeshire* to 2*d*., and by the next Shire Line steamer to 1¾*d*. Competition and the increased tonnage available caused further reductions in the rates steadily onwards.

Sailing Ships as Meat Carriers.—It is noteworthy that the early work done by the fleet of sailers in carrying frozen meat was excellent, and these craft proved quite suited to the trade in the early days. Certainly the 90 to 110 days' passage, instead of the mail steamers' 40, from New Zealand, was against them, but, then, the c.i.f. business and "catching the market" were rudimentary points in the eighties. The sailers did their refrigerating remarkably well, and the early fleet was considerable. Messrs. Shaw, Savill and Co. (which combined with the Albion Co., to form the Shaw, Savill and Albion Co., incorporated in 1882) had the *Dunedin*, *Lady Jocelyn*, *Lyttelton*, *Invercargill*, *Oamaru*, *Northumberland*, *Wellington*, *Westland*, *Timaru*, *Marlborough*, and *Hinemoa*. These ships carried anything from 10,000 to 15,000 carcasses, and were fitted with either Bell-Coleman or Haslam machines freezing on the cold air blast system. Meat cargoes were brought home in these boats with commendable success, but their being superseded by steamers was in the natural course of evolution which has settled the fate of the older class of vessel in modern trade. The Shaw, Savill Co. fitted up the sailer *Edwin Fox* as a freezing hulk, and sent her out to New Zealand in 1885. Subsequently the vessel was transferred to the Christchurch Meat Co., and is still used as a hulk at Picton (see illustration, p. 60). The *Timaru*, after her active career as a frozen meat carrier was over, was used as a freezing store on the South African coast, and was wrecked in 1907. The *Marlborough* and *Dunedin* were lost in 1890. Another old sailing ship, formerly employed to convey frozen meat from New Zealand, is now the property of the Royal Mail Steam Packet Co., and is stationed in the River Plate as a hulk, and is called the *Rothay*. Before the company purchased her the vessel was called the *Duleep Singh*, and was sent as a freezing hulk to Gibraltar in 1890 by Messrs. Wills and Co.

The Evolution of the Frozen Meat Carrier.

The evolution of the frozen meat carrier is a most interesting point to refer to. From the 10,000 carcasses sailing ship, "adapted" for the trade, through special sailers, like the *Hinemoa*, there is the development to adapted steamers like the *British Princess*, *Selembria*, the early Shire boats, etc., to the half-and-half (refrigerated, and/or general cargo) steamers largely used by the New Zealand Shipping Co., and the Shaw, Savill and Albion Co., to the very highly specialized steamers that have been built recently for the South American frozen and chilled meat trade.

Early refrigerating installations on board ship were, no doubt, somewhat crude, and, with engineers untrained in the new branch of their profession, the difficulties met with must have been many. Captain Whitson's experiences on the *Dunedin*, recorded in another part of this book, are an instance. But the engineers did their work well, as the meat was brought in sound condition in the great majority of cases. Up to the beginning of 1884 only nine cargoes had been delivered in "unsatisfactory condition." Some of the damage to the meat was done in transit between the works and the ship, and there was the fault of insufficient freezing in the case of the meat brought to London by the *Mataura* in 1883. The trade circulars of the time reported as follows: "Imperfectly prepared for the voyage; about 4,000 carcasses had to be destroyed on arrival."

By those who remember the large number of cargoes which arrived in bad condition in the nineties, the contrast of the cleaner record of the shipments made in the previous decade may seem remarkable. As a matter of fact, if an exact comparison of percentages of unsound arrivals were made over the two periods, it would probably be found that the later term did not show a larger percentage of casualties; the volume of the trade was much bigger than in the eighties, consequently the failures loomed larger. However, three reasons have been assigned for the trouble which became so apparent in the nineties. By those well qualified to judge, the cause was said to be partly the fitting of the ships, partly the refrigerating machinery, and partly—perhaps, principally—the indifferent preparation and handling of the meat just prior to exportation. With the old sailers freezing was done on board at the port of shipment, but in the case of the steamers the meat was frozen ashore, and it is thought that insufficient refrigeration of the carcasses before their stowage in the steamers' holds was often responsible for their bad carriage. The reliable working of the old refrigerating machinery installed in the ships on the cold air blast system in the earlier days of the trade is a bright feature of the industry, and some of the old machines continue to do good service right up to the present time. For instance, only to mention four Shaw, Savill boats; the *Waiwera* (capacity 80,000 carcasses), *Tokomaru* (85,000 carcasses), *Kumara* (75,000 carcasses), and *Karamea* (75,000 carcasses), are still carrying meat cargoes efficiently frozen by the old Haslam cold air machine.

Contrasting the early days' difficulties with the smoothness which characterizes the conditions of to-day, Mr. J. A. Potter, general manager of the Shaw, Savill and Albion Co., sends the following remarks:—

"In the absence of freezing works ashore, and the consequent initial freezing and preparation of the carcasses on board, great delay in loading was caused, and even steamers for a time had to perform this work. But soon freezing works ashore began to be erected, and gradually all cargoes were prepared prior to shipment, and sent in a frozen state to the carrying vessel. In the early days of the industry complete or partial failure at times took place due to some breakdown of machinery, imperfect preparation of the carcasses, faulty insulation, etc., but with the growth of the trade experience and science brought knowledge, and to-day it is marvellous with what certainty such a delicate article of commerce can be conveyed in perfect condition from one end of the world to the other. Mutton, lamb, beef, butter, cheese, and delicate fruits, are shipped at the far Antipodes with a practical certainty that they will be put on the English market in the same state of preservation as they would be if offered for sale at the place of production."

The evolution of the modern frozen or chilled meat carrier has been a gradual process, in which details have one by one been revised in accordance with the march of modern invention and the discovery of the more up-to-date appliances employed in mechanical refrigeration. In the early eighties mail steamers were fitted with limited refrigerated accommodation, *e.g.*, in 1884 the *Arawa* and the *Tainui*, which have been already referred to; and it was then thought that about 15,000 carcasses refrigerated capacity was sufficient. Ideas grew, and following the equipment of the early mail steamers came the pressing into refrigerated service of cargo steamers, later coming the boats built especially for cold storage transport, of which the latest examples carry huge cargoes considerably in advance of 100,000 carcasses.

It will enable the reader to gain some idea of the great

THE SHAW, SAVILL AND ALBION S.S. *Rangatira.*

To face p. 134.

progress made with the frozen meat carrier since its first conception thirty years ago, if he remembers that from 1885 to 1888 a 4,000-ton dead weight steamer was a big vessel. Ships were then constructed to carry the maximum quantity of dead weight on the minimum net register. Consequently they had the minimum amount of cabin capacity within their dimensions. When frozen meat became a factor in the steam carrying trade, it was found that, as it required 160 cubic feet, including refrigerating apparatus and insulation, to carry 20 cwts. of meat, it was advisable to increase the number of decks within the dimensions of the vessel. This was effected by putting on a shade deck, which filled up the space between the forecastle and the short bridge and the deck, and making it a continuous deck, thereby forming, as it were, another storey in the vessel, this giving roughly 25 to 33 per cent. more cubic capacity than the old style. On top of this shade deck was put a small forecastle, and possibly a small bridge, which would carry the officers of the steamer, the question of stability being provided for by increasing the beam to counteract the extra height. To-day the modern steamer is roughly eight to nine beams in her length, whereas in the old days she was nine to ten, and possibly a little more.

The Shipowner as Merchant.

There have been times in which the shipowner in his connection with the frozen meat trade has turned merchant. About twenty years ago, say, 1891, the three regular lines carrying meat from New Zealand began to buy, in order to fill their ships. They were forced to do so, otherwise their ships would have sailed for London with empty holds. Even previous to this they had taken an interest in the meat by receiving freight on a sliding scale according to the price realized. The reason for such action could well be understood, and the later development of the same general policy was shown by the tendency of shipowners to finance the erection of works, recouping themselves by freight contracts securing the output. The case against the shipowner actually competing with his clients is very strong, and no circumstances but the most unusual and compelling could justify such a departure from mercantile ethics. There are other instances of shipowners in the Australasian trade acting in this dual capacity. One of these relates to the shipments of frozen beef from Queensland to Vladivostock, in regard to which, perhaps, one may say that as this was a new departure, as well as speculative, and probably temporary in character, the element of competition with the merchant did not arise at least, in a marked degree.

In 1894, owing to exceptionally heavy imports of meat from Australia, the London cold stores became choked with meat, and the vessels as they arrived in the Thames were unable to discharge their cargo. This state of affairs marked one of the most serious crises the frozen meat trade has passed through. The same position was, as a matter of fact, within an ace of occurring in 1909, but by that time the capacity of the stores had greatly increased. On the earlier occasion referred to above, vessels had to be used as stores, and lay for weeks on demurrage. This might suit the shipowner in some instances, but not those having to work a regular service on time tables. Such owners could not get their ships discharged, and their arrangements were completely upset, while, to make matters still more awkward for them, some clients who had sold their meat insisted on delivery. This could not be effected, because the meat was under that of other consignees who were unable to authorize landing, having no store room obtainable. The shipowner in such circumstances stood to be shot at by the consignees who were losing their market, for he could not throw the meat of the others out on the quay.

Freights.

Freights fell, of course, as frozen meat cargo became regular and bulky. On the first shipment from New Zealand, that of the *Dunedin*, the freight was 2¼*d.*; steamers of the same line are carrying meat to-day at ½*d.* The freight on the first meat vessel from the Plate, the s.s. *Meath*, was 2½*d.* In 1891 New Zealand mutton freight was reduced to 1*d.*, by 1894 to ¾*d.*

and by 1897 to $\frac{1}{2}d$. The present rate of freight on frozen meat from Australia to United Kingdom ports, say $\frac{5}{8}d$. per lb., and $\frac{13}{16}d$. to Mediterranean ports, shippers consider too high. It is only the great quantities of meat carried, and the construction of specially designed steamers, that allow of more moderate rates being fixed. Shipowners, when asked, say that frozen meat is not a "good paying freight." An occurrence which would seem to fortify this assertion was the taking out of the refrigerating machinery from the sailer *Hinemoa* some years ago, the owner, Mr. Leslie, declaring that ordinary cargo would pay better. Still, for the splendid and well-equipped vessels of the lines bringing food products from South America and Australasia there can be no doubt that the frozen meat trade has been a very profitable occupation indeed, and responsible to no small extent for the development of the "refrigerated fleet."

Multiplicity of Marks.

One of the greatest burdens the shipowners have to bear, with what philosophy they may—in the New Zealand and Australian trade—is the excessive sub-division of the mutton and lamb consignments into marks. The system has come about through the personal part which the farmers (mainly in New Zealand) take in the business. Where the growers sell their stock to the freezing works, as is the case invariably in Argentina, and usually in Australia, the grouping of the carcasses into separate parcels and the placing of distinguishing marks on them is only governed in the main by the consignees to whom they go. But the way in which the New Zealand trade is done involves the division, to a considerable degree, of the frozen meat shipments from a works. There may be scores of c.i.f. buyers interested in the meat shipped in a liner by one of the freezing establishments, and such lots have to be "delivered to marks." This is bad enough, but where the shipowner—and the subsequent handlers, too—find this splitting up maddening is when the New Zealand farmer sends in his few sheep—frequently less than a hundred—to the works, and insists upon those carcasses being specially marked, and each grade of weight and quality kept separate. It is natural, perhaps, that he should desire that an account sales should reach him, giving the sale price of each class of carcass in his particular lot; but his action places an almost unbearable burden upon the various agents he employs to put his meat on the English market, namely, the shipowner, the storekeeper, and the consignee. In the course of the discussion of a paper on "The Inspection and Distribution of Meat Foods," read by Mr. F. Knowles before the Cold Storage and Ice Association in April, 1910, a speaker said that on one occasion, owing to the multiplicity of marks occurring in a frozen meat cargo arriving at Liverpool, it had taken him a whole week to get 2,000 lambs from a ship, and all that time he had been constantly travelling between Liverpool and Manchester. Carters sometimes abandoned altogether contracts for carrying such meat to store because of the delay in discharging from the vessel owing to the marks difficulty.

To take an actual shipment of mutton from New Zealand in 1909 from one of the North Island Works, for 6,000 carcasses there were no less than 460 different marks. Again, in the case of one bill of lading for 211 sheep, altering the marks and numbers, this is how the division was made :—

		Sheep.
H F C	425/1	4
,,	425/7	27
,,	425/3	8
,,	425/9	3
,,	425/4	7
Pakeha	425/1	3
2 Ships	425	2
H F C	640/1	22
,,	640/7	29
,,	640/3	20
,,	640/9	6
,,	640/5	2
2 Ships	640	16
H F C	705/1	2
,,	705/7	1

		Sheep.
H F C	705/4	4
Pakeha	705/1	2
H F C	680/1	3
,,	680/7	9
,,	680/3	12
,,	680/9	8
,,	680/5	1
2 Ships	680	8
H F C	120/1	2
,,	120/7	1
,,	120/3	1
,,	120/4	6
H F C	720/4	2
		211 C/s

Now, in all the processes through which these carcasses pass from works to market, "H F C 705/7" (one sheep) has got

MESSRS. TURNBULL, MARTIN AND CO.'S SHIRE LINE S.S. *Argylshire.*

To face p. 138.

to be kept separate. That is the theory, but at times it is absolutely impossible to do so, and it is to be feared that "H F C 705/7" gets mixed up with other goods. It is obvious that such a system of business is faulty; it would be so with any description of merchandise, more or less, but with perishable produce like frozen meat it is often fatal. The great thing with refrigerated meat is to save handling, and multiplicity of marks makes for handling. An enormous amount of sorting over of meat cargoes takes place at the point of discharge before "H F C 705/7" and its fellows can be got at, and in this sorting over the meat gets damaged. This matter is mentioned here not only to record an existing feature of the trade, but to urge strongly upon all parties in the New Zealand trade to see if the farmers cannot preserve their own individual interests by other means. There is no doubt that New Zealand meat suffers greatly from the way in which it has to be handled in separate lots; exposure to the atmosphere causes much loss of bloom. The bright ruddy appearance of a newly arrived crossbred sheep is a wonderful market asset. Canterbury, New Zealand, has been the centre of this sub-division system. In 1907 the shipowners began to take action, and since then the question has been constantly discussed, and conferences with merchants have been held. In the Australian trade, where the shipowners were unhampered by contracts, they took action by stamping bills of lading to the effect that they would not be responsible for sub-marks. A "sorting store" at the Dock stores in London was mooted in 1892 and later, in order to deal with this difficulty, but clashings of interests and the small extra expense per carcass involved prevented reform being carried out. In 1899 the marks trouble became very pronounced and shipowners threatened "pooling" the cargoes.

A scheme to adopt special stripes varying in design and colour for use on the meat wraps, so as to provide one distinguishing and easily visible brand for each works was suggested in 1909—1910, and met with the approval of shipowners in principle, but in working out the details many trivial objections were raised which still remain to be dealt with before the shippers will willingly adopt the recommendation. It has much to commend it as a means of lessening the handling at this end.

Many of the criticisms made of London methods of discharge would have no existence but for the continued use of so many distinctive marks in each cargo—especially from New Zealand. The endless sub-divisions employed are an anachronism—relics of the youthful days of the trade when the farmer *was* the freezing company. Argentina, Uruguay, Patagonia, all ship in big lines of 500 to 1,000 or even 5,000 carcasses of one mark. Until Australia and New Zealand can act similarly, they will be handicapped in their efforts to secure the trade of the big retailers.

The Shipowner and the South American Meat Trade.

In 1883, when the first shipments of frozen meat were ready to be despatched from the Campana works, Buenos Aires, there were no shipowners with sufficient knowledge of the possibilities of the new industry to induce them to instal refrigerating machinery and insulation in their steamers at their own expense, so the shippers—the River Plate Fresh Meat Co.—had to put in the machines, etc., themselves. Their initial shipment, in 1883, was brought by the s.s. *Meath*, as stated earlier in this book. The vessel arrived in London in January, 1884, via Antwerp, where she had discharged some meat. The s.s. *Meath* and the s.s. *Wexford*—the next vessel—both owned by Mr. R. M. Hudson, of Sunderland, and running under Messrs. Houlders' contract, continued in the regular Plate trade until 1886, and for many years afterwards as independent units available for chartering to any part of the world. In addition to these vessels were the s.s. *Zenobia* and *Zephyrus*, which were fitted with Haslam cold air machinery. The " Z " boats, owned by Messrs. Turner, Brightman and Co., were, and are still, taken by the River Plate Fresh Meat Co. on long time charters ; excepting the s.s. *Zephyrus* and *Zenobia*, which ran under the Houlder contract until 1893, the various " Z " boats have practically been built for the company. The s.s. *Zephyrus*, by the way, is still bringing frozen meat from Argentina—though not now on account of the River Plate Co.—after twenty-eight years of good service as a meat carrier, with the original Haslam machinery on board. The River Plate Co. put Haslam's refrigerating machinery and insulation into ten of the " Z " boats. From about 1909 the owners of the line have installed the machinery and insulation.

Messrs. Houlder were associated with the conveying of the first shipments of frozen meat from the River Plate. As mentioned above, Messrs. Houlder contracted to bring to the United Kingdom the first shipments from the Campana Works, and their chartered vessels, the *Meath* and the *Wexford*, therefore, had pride of place in opening the South American trade. Referring to the Houlder Line, the s.s. *Hornby Grange* and *Ovingdean Grange* were the pioneer ships in the refrigerated produce trade of the service, these vessels being fitted in 1890 and 1895 respectively. Ever since that year the Houlder Line steamers have held a very important position in the South American frozen meat trade, conveying meat on contract to the ports of the United Kingdom for the various frigorificos. For example, they have had carrying contracts with one of the largest Argentine frozen meat exporting firms for the past sixteen or seventeen years, and with the freezing works at Monte Video from the time of their erection. At the present time, as far as the service to the " outports " is concerned, the Houlder Line is running a refrigerated boat every three weeks to Liverpool and Cardiff, and one also every three weeks to Southampton, London, and Newcastle. At the time this book goes to press the Houlder Line owns the frozen meat carrier possessing a larger refrigerated capacity than any other ship in the world; this is the *Sutherland Grange*, which has an insulated capacity of 397,000 cubic feet.

Messrs. R. P. Houston and Co. (Houston Line) were one of the first firms to insulate and fit their steamers for the conveyance of frozen meat from the Argentine Republic to ports in the United Kingdom ; this was in the year 1884, and they then fitted up four steamers, the first one being the *Hesperides*. Mr. R. P. Houston, M.P., was in Buenos Aires in the early part of 1884, and he arranged a contract for a period with the late Señor

Don Francisco de Sansinena for the conveyance of frozen meat to this country. The first live cattle ever shipped from the River Plate to Europe were carried by a Houston Line steamer and landed at Dunkirk in the year 1884.

The first steamer of the Lamport and Holt line to carry meat from South America was the *Thales*, in 1887. The European ports to which the refrigerated vessels of this service now run are Southampton and Liverpool. With the delivery of the two new twin screw steamers, *Vauban* and *Vestris*, the service will be a four-weekly one, calling at Lisbon, Vigo, and Cherbourg with passengers, and Southampton and Liverpool with passengers and frozen and chilled produce.

Messrs. H. and W. Nelson, Ltd., as managers of two important refrigerated steamship services from the River Plate to England, have played a prominent part in the Argentine carrying trade. Formed in the early nineties, for the purpose of carrying frozen meat from Argentina for Messrs. James Nelson and Sons, Ltd., Messrs. H. and W. Nelson later contracted with other meat companies, and fortnightly and weekly services are run respectively by the two lines they manage, the Nelson Line (Liverpool), Ltd., and the Nelson Steam Navigation Co., Ltd., of London. The latter company was formed in 1910 to acquire from the Nelson Line (London), Ltd., a company registered in the same year, nine new steamers, each of a refrigerated capacity of 330,000 cubic feet, and fitted with ammonia compression refrigerating machinery by the Liverpool Refrigeration Co., Ltd., and also a tenth steamer to complete a weekly service to London. With these boats, which are a most modern and well-equipped type of the chilled meat carrier, were acquired freight contracts with the Swift Beef Co., the La Blanca Co., the Smithfield and Argentine Meat Co., and the Frigorifico Argentino. The first boat run by Messrs. H. and W. Nelson was the *Highland Scot*, in which cold air refrigerating machinery was installed by Messrs. J. and E. Hall.

The Royal Mail Steam Packet Co. first entered the refrigerated trade in 1883, and the first vessel fitted was the s.s. *Tagus*. The mail steamers of the line, which carry meat to Southampton only, take twenty-one days from the River Plate; the cargo

THE ARGENTINE CARGO LINE S.S. *El Argentino.*

To face p. 142.

vessels, despatched to Southampton, London, and Hull, take twenty-eight days. The ships of this company carry meat for the various Argentine meat companies, under long contracts.

The Chargeurs Réunis, a French line which runs from South America to Havre, via London and Hull, carries chilled and frozen meat for the Compañia Sansinena de Carnes Congeladas.

The Argentine Cargo line, controlled by Messrs. Furness, Withy and Co., and Messrs. Birt, Potter and Hughes, has vessels carrying refrigerated meat from Argentina to London. A new line is to be started about June, 1912, which Messrs. Furness, Withy, and Co. and Messrs. Birt, Potter and Hughes, Ltd., will control; this will run from South America to Liverpool in less than twenty-one days. Five vessels for this service are being fitted out, or are in hand, which will each have a cubic capacity of 400,000. Arrangements have been made so that the vessels of this line will be despatched alternately with those of the Royal Mail Co. at weekly intervals. These steamers are intended chiefly for the conveyance of chilled beef.

Messrs. R. M. Hudson and Sons also run a regular line of cargo vessels in the Argentine meat trade.

Shipping Arrangements of the South American Frigorificos. In a previous paragraph appears a note concerning the way in which the company which first began the regular despatch of frozen meat from Argentina, the River Plate Fresh Meat Co., manages its ocean transport. The Compañia Sansinena de Carnes Congeladas have shipped their frozen and chilled meat largely by Messrs. Houlder's steamers, and also by the Chargeurs Réunis boats, though the company uses other lines of vessels too. In the early days Messrs. Houston's steamers were employed. The company despatches its meat from its Argentine and Uruguayan works to London, Liverpool, Cardiff, Hull, Newcastle, and Southampton. Messrs. James Nelson and Sons have shipped meat regularly from the Las Palmas works by the vessels of the Nelson Line; the bulk of the meat from these works is conveyed to Liverpool. The first shipments despatched were in 1888, and the s.s. *Ranmoor*, of the Lamport and

Holt line, was the vessel that brought the initial shipment, which was on account of Nelson's River Plate Meat Co.

The other Argentine freezing works, formed in comparatively modern times, those of the La Plata, La Blanca, Smithfield and Argentine, and Frigorifico Argentino companies, despatch their frozen and chilled meat to British ports by the various lines above mentioned. Their shipments to Continental ports are carried by specially chartered British steamers and by the Italian or Austrian-owned liners.

The two freezing works in the Straits of Magellan, at Rio Seco and San Gregorio, despatch their frozen meat to Liverpool by the Houlder steamers, and by other steamers specially chartered from time to time.

The Venezuelan Meat and Products Syndicate ships its meat from the works at Puerto Cabello to Southampton by the Royal Mail liners, and to Liverpool and the Continent by the new s.s. *Imataka*, belonging to Messrs. Brooker Brothers, and McConnell, Ltd.

Rates of Freight.—The usual plan in shipping frozen meat on the steamers in the Argentine trade is for the shippers to engage a certain space for a definite period. This space has to be filled on each voyage, or dead freight has to be paid. The present rate of freight on frozen meat may be stated as $\frac{3}{8}d.$ per lb. for mutton and beef. Chilled beef pays $\frac{11}{16}d.$, or $\frac{13}{32}d.$ per lb. on the basis of 105 cubic feet to the ton. Frozen meat stows in about 105 cubic feet to the ton, whereas chilled beef occupies from 170 to 200 cubic feet to the ton.

Long contracts are customary in the Argentine trade, usually for five years. It may be mentioned that there is little "ship damage" to meat from South America; this may be accounted for by the fact that the whole shipload of meat frequently represents the loading of one works, or, at most, of three works. The average length of voyage of a cargo vessel bringing meat from Argentina is twenty-six to twenty-eight days to the first English port.

THE WHITE STAR LINE S.S. *Suevic.*

To face p. 144.

CHAPTER X

THE UNDERWRITERS' RISK

It is possible that this subject is not considered as closely by the grower of the meat as, with advantage, it might be. For instance, the New Zealand farmer takes a deal of interest in the marketing at Smithfield of his frozen meat, but shipping and insuring he leaves to his agents. The producer, by getting to note the points where friction occurs, might help the business along. As it is plain that heavy insurance charges must react unfavourably on the profits of the producer, it is a matter for comment that growers of meat in Australia and New Zealand have never attempted to grapple seriously with the conditions which have not only prevented insurance rates falling of late years, but have actually sent them up. More satisfactory conditions of insurance would certainly be the outcome of businesslike and combined efforts for reform by the parties concerned ; not alone by the growers, but to some extent the merchants and agents who put the meat through. Take 1909, when a great outcry was raised in Sydney concerning damaged meat from Australia arriving in Great Britain. The Chamber of Commerce in Sydney took the matter up, and a committee of investigation was formed. Well, the circumstances responsible for the imperfect handling of the meat had been in existence for years ; there was absolutely nothing new, yet the meat exporters and shipping merchants apparently were "surprised" to hear of such a state of affairs. There were few who knew anything at all about the trade who were unacquainted with the weaknesses of the export trade as it was carried on in New South Wales, and the recommendations of the Sydney Chamber of Commerce committee revealed the evils existing. For instance, here are a few sample points urged :—That proper provision should be made at the new abattoirs for weighing and grading meat; that the Government should not allow emaciated or poor-conditioned carcasses to be ticketed with the Government label for export; that the vans carrying lamb carcasses from abattoirs to freezing works should not be overloaded; that the existing methods of carting and lightering meat from the freezing chambers to export vessels in open wagons and lighters was unsatisfactory; and so on. Such weak points had existed in the light of day, yet it was only the tireless wielding of such trenchant pens as that of Captain A. W. Pearse that secured due notice of the misdeeds wrought in the name of refrigerated export.

The issues with which this chapter is concerned are at times so involved that the Courts of Law have to be visited in order to get judicial ruling. From 1880 onwards the insurance of frozen meat has presented peculiar difficulties, and the underwriter has been slow to adapt himself to the situation satisfactorily to all interests in this new section of his business. The frozen meat trade is even yet a new one, and before it settles down quite permanently into its groove in the commercial world all the mercantile methods by which it is handled must become more defined and scientifically sound.

Early Insurance Covers.

The pioneer shipment by the *Dunedin* from New Zealand was insured at £5 5*s.* per cent., as stated in Chapter II., and the first shipment of the *Mataura* in 1883 from New Zealand was insured for £4,000 by seven offices at £7 7*s.* per cent., w.p.a. Insurance was effected then by local insurance companies. A little later the premium, for "all risks" policies, reverted to £5 5*s.* per cent. By 1886 the export of meat from New Zealand had, according to the Otago Marine Underwriters' Association, become so well understood that many of the risks incidental to the introduction of the business had entirely disappeared. Commenting on this, the *Australasian Insurance and Banking Record* in 1897 observed that "consignees had not at that time educated themselves up to their present high standard in the science of claim making." Generally speaking, for the first ten years there was not much to complain about in the carrying of the New Zealand cargoes; from 1882 to 1887, out of 172 shipments only nine were returned as in "unsatisfactory condition," that is to say, thoroughly unsatisfactory condition. But the Australian records do not read so well; from 1880 to 1887, out of 127 shipments there were nineteen cargoes on the black list.

Steadily in the nineties this "unsatisfactory condition" of the meat on arrival increased, as will be seen from the following quotation from the Australian and New Zealand Underwriters' Association report for 1895:—"Frozen Meat.—It would be impossible to omit from this report reference to the subject of frozen meat, the most absorbing, interesting, and perplexing, of the year. Notwithstanding the constant, anxious, and varied attention given to it, the experience of the past twelve months has shown how baffled have been all the attempts made to bring the treatment of this interest into a satisfactory condition. Vast as is the trade, paramount in its importance to the colonies, and of momentous consequence to the carrying companies, it must yet be reported that, with all these interests in combination to make their efforts a success, disappointment and failure has been the result."

Insurance matters continued unsatisfactory till 1895, when the underwriters drew up and adopted a new clause, which cleverly met the difficulty of how to exclude certain cold stores in which damage to frozen meat had occurred. Appended to the clause was a list of "approved" stores. In the clause the period of insurance on the meat after landing was cut down to sixty days as a maximum; "bone taint" was excluded. But the passing of this clause did not improve matters, and claims became heavier and heavier. The underwriters thought they were being victimized by market practices in assessing damage, which in those days was done on the market on the basis of "sound values." The underwriters considered that high "sound values" were engineered so as to throw into strong relief the goods to be surveyed. So they amended the frozen meat clause once again (1897), and fixed thirty days from ship's arrival as the outside period of cover; at the same time they insisted that the meat should be surveyed, and any damage assessed in store and not on the market.

A market annual for 1897 stated that on the frozen meat imported from Australia from February to March and from June to November a depreciation equal to $\frac{1}{8}$*d.* per lb. on 750,000 carcasses had taken place through faulty condition; out of forty-six cargoes, twenty-one contained meat which was condemned. In the following year 20 per cent. of the meat cargoes arrived in a damaged state.

In referring further to insurance premiums, it may be stated that in December, 1895, the sailing vessel *Hinemoa* carried frozen mutton and lamb from Melbourne to London, and then the rate was 90*s.* per cent., with 10*s.* per cent. extra for the proportion frozen on board. Fifteen months later the same vessel lifted another cargo, when the rates were 65*s.* net for mutton and lamb, and 85*s.* net for beef.

In 1897 the *Hinemoa* took another loading insured at 90*s.* per cent., and in March following, the *Opawa*, also a sailer, carried meat to Durban insured at £5 5*s.* per cent. Steamers in the meantime began to cater more for the trade, and with the growth of competition amongst underwriters rates were gradually reduced. In March, 1898, the first frozen meat tariff came into force. The "all risks" rate on mutton to the United Kingdom was then fixed at 47*s.* 6*d.* for works, voyage, and thirty days' storage by P. and O. steamers, 70*s.* by other steamers and regular liners.

Claims, Surveys, and Allowances.

The adjustment of claims has always been a very sore point. When the surveys and allowances were made on the market, the importers' theory was that only when loss was actually made in retailing the meat would allowances be claimed; if the meat was accepted by retail buyers as sound nothing could be claimed. But when it was decided by underwriters, in face of opposition from traders, that surveys must be held in store, the consignees, to protect themselves against possible loss, necessarily made claims upon "estimated losses" before sale. This system has prevailed ever since. Consignees protested against the thirty days' limit, but the insurance people were obdurate. Meat condemned by the sanitary authorities was limited in claims under the new arrangements to the current "sound value," thus reducing what proved to be in some cases excessive insurance to the actual value on a lower market. The cost of survey in store was fixed at £1 1*s*. per 1,000 carcasses "surveyed"; the consignee and the underwriter were both represented at the survey, and the underwriters paid the survey expenses of both parties. This practice existed up to 1909.

In connection with the general question of damage to frozen meat in Australian shipments, it is well to chronicle here the action taken by the New South Wales Freezing Companies' Association in 1899. With a view to checking the damage to frozen meat, which at that time passed all bearing, the Association sent to London Mr. C. C. Tayloe, of Sydney, who urged the extension of the Association's operations to London. This was effected, and on March 25, 1899, Mr. Septimus Merriman, the London secretary of the Association, issued a circular letter informing the trade what had been done. A scheme was worked out showing a contemplated expenditure of £4,000 a year. Surveyors in Sydney and London were to inspect the loading and discharging of the Sydney-shipped meat. The surveys were to include examination of ships' refrigerating machinery and insulation—"supervision of all means of carriage from steamer to store." It was a point of the programme to get shippers to insure f.p.a., and under this it was hoped to bring down rates to 25*s*. and 20*s*. respectively for local companies and Lloyd's. "Shippers must remember," ran the circular, "that with the proposed clause it will be everyone's interest from first to last not to have claims." The proposal of the Association was never entirely put into operation, chiefly because it was urged that steps be first taken to put things right in Sydney. This most rational step did not commend itself to the promoters, and the whole scheme gradually died a natural death.

Allusion may be made to the proposal put forward by importers about the middle of the nineties to form in London the "Australasian Frozen Meat Insurance Co." with a capital of £100,000. The idea was that co-operation introduced into underwriting would provide the panacea required. Many well-known firms discussed preliminaries, but the affair did not take definite shape.

With the opening of the new century improvement took place in the general conditions of handling frozen meat throughout the various stages of the industry. A better class of vessel was brought into service, fitted in a more modern style ; some of the older ships conveying meat were under-powered for the refrigerating work they had to do. New well-equipped cold stores were built, and transport arrangements at the ports of debarkation were also improved. Barges and vans were more efficiently insulated, and lightermen were instructed to proceed to their respective stores (when the tide permitted) after work at ship's side had finished for the day; that is, they did not remain in dock all night waiting for a complete load on the following day. Also on arrival at the store the barges were unloaded immediately. With these reforms the "unsatisfactory cargoes," heard of so much from 1885 to 1899, were seldom recorded. The two great "Lloyd" institutions—Lloyd's, the underwriters at the Royal Exchange, and Lloyd's Register of British and Foreign Shipping—played a prominent and useful part in improving matters. Lloyd's underwriters at that time really took little or no interest in the improvements referred to—in fact, many of them permitted meat to be warehoused in unapproved stores. Any action that was taken was at the initiative of the Institute of London Underwriters, which organization is the mouthpiece of Lloyd's underwriters. The Institute of London Underwriters required the barges and cold stores to be surveyed and reports submitted to them before admittance on the approved list could be granted. Arrangements were made between the shipowners and Lloyd's Register for Lloyd's surveyors at different world's ports to undertake an inspection of the refrigerating and insulating fittings of steamers. This examination being concluded satisfactorily, the ship was from the point of view of freezing machinery and insulation "seaworthy."

In 1908 there were several serious accidents to Australasian steamers and consequent heavy claims on underwriters. As a consequence, insurance rates were put up from 37*s*. 6*d*. per cent. to 45*s*. per cent. from New Zealand and from 50*s*. per cent. to 60*s*. per cent. from Australia. The insurance position and the feeling in the trade one sees in the following extract from the annual review issued by Messrs. W. Weddel and Co. for 1908 :—"Until underwriters will differentiate between good and bad risks at the various works, on the various ships, and in the various stores engaged in the trade, instead of treating all their risks as equal, very little inducement is offered to any individual to lay out money in order to increase his precautions against damage. Underwriters have recently taken steps to bring home to individual shipowners their responsibility for damage done in transit. Were some similar course adopted to fix responsibility for specific damage done in any freezing works or stores and adjust rates accordingly, much good would result to underwriters and to the trade as a whole. With the experience of refrigeration now acquired, there ought to be no serious damage to frozen meats, and but little to chilled, except in cases where an unforeseen accident happens."

Underwriters were restless about this time, and in 1909 they put in force their new clause, which introduced some drastic changes pressing hard on the trade. They stated that the insurance of meat did not pay them, and that the new regulations were brought into the policy to make the business profitable. Negotiations took place between the importers, under the auspices of the Frozen Meat Trade Association, and the Institute of London Underwriters, but the insurance interests were obdurate, and the new clause gradually worked into the Australasian frozen meat industry. Some of the changes were welcomed by all, such as the one insisting that in loading and discharge only trucks, vans, and barges which were provided with efficient insulation were to be used.

The A 1 Clause.

CLAUSES FOR UNITED KINGDOM SHIPMENTS.

FROZEN MUTTON, LAMB, BEEF, VEAL AND PORK.

Clause A 1—(Freezing Works, Voyage and 60 Days).

1. The risk commences from the time the interest is passed into the Cooling and/or Freezing Chambers of the Works at and, unless previously terminated, continues on board the vessel and/or in Refrigerating Stores in the United Kingdom (subject to the conditions hereinafter mentioned) for a period not exceeding 60 days (warranted not more than 30 days on board the vessel) from arrival of vessel at destination as per policy, provided always:

2. That it is warranted by the Assured that the meat is in good condition and properly dressed, cooled, and frozen at the Freezing Works, and that the period between the time the risk commences and shipment on ocean-going vessel shall not exceed 60 days.

3. That where the interest has to be conveyed by rail and/or street vans and/or lighters prior to shipment by oversea vessel, such railway trucks and/or street vans and/or lighters must be insulated, otherwise an additional premium of 10/- per cent. to be paid; and after discharge from the vessel the interest shall be carried in insulated street vans and/or insulated lighters, otherwise an additional premium of 10/- per cent. to be paid.

4. That the cold stores in the United Kingdom shall be approved by the Institute of London Underwriters.

5. That any disposal of the interest at destination other than by storage as above (except with the consent of the Underwriters) or any removal of the interest from the cold stores at destination previous to the expiry of the 60 days above mentioned terminates the insurance on such meat, and no claim for damage shall attach, unless, immediately on the first discovery of any damage to or deterioration of any part of the interest hereby insured, notice shall have been given to the Underwriters, and the amount of depreciation agreed to by them prior to the termination of the insurance.

6. During the period (if any) between assessment of depreciation and termination of the insurance the risks covered hereunder are those of fire and breakdown of machinery only.

7. That in the event of interest being transhipped, or forwarded on to destination in the United Kingdom by rail, no risk to attach hereunder unless notice of such transhipment or rail carriage be given to the London Representative of the Company prior to commencement of such risk, the transhipment or forwarding to be only by steamer fitted with refrigerating machinery or by rail in properly insulated vans. An extra premium at the rate of 20/- per cent. to be paid for such risk.

8. That the value to be made good in the case of meat condemned on or after arrival shall in no case exceed the sound market value, less usual charges.

9. That no adjustment charges shall be incurred unless with the written consent of Underwriters who shall not be liable for survey fees other than those of their own surveyors.

9A. That in the event of any claim for loss before shipment, or for damage in consequence of which the meat is not shipped, the same shall be adjusted on the basis of the actual values at the time and place of such loss or damage (plus any freight payable whether the meat be shipped or not, and charges) irrespective of any other value declared in the policy.

10. The insurance covers loss from defective condition of the meat from every cause (except bone-taint and improper dressing, cooling and freezing), which shall arise during the currency of the insurance, but always subject to the Free of Capture and Seizure Clause.

11. Average payable if amounting to 3 per cent. on each carcass, and/or two half carcasses, and/or four haunches, and/or eight legs mutton or lamb, and/or each package beef, veal or pork, and/or each valuation separately, and/or on the whole.

12. The Underwriters to be credited with any compensation or allowance obtainable from the Shipowner in respect of average attaching hereto.

13. It is hereby agreed that, unless expressly otherwise stated herein, carcasses or pieces comprised in any one mark and valuation, or carcasses or pieces of various marks comprising one valuation, shall, for purposes of average adjustment, be deemed of the same weight and insured value.

14. Including all liberties as per Bills of Lading or Contract of Affreightment (subject to additional premiums, if any, as per tariff) but it is warranted that the obligation of the Shipowner to provide a seaworthy vessel fit to carry the cargo shall be fully preserved in such Bills of Lading or Contract of Affreightment.

15. General Average and Salvage Charges payable as per foreign statement if so claimed, or per York-Antwerp Rules if in accordance with the Contract of Affreightment.

16. In the event of the vessel making any deviation or change of voyage, it is mutually agreed that such deviation or change shall be held covered at a premium to be arranged, provided due notice be given by the Assured on receipt of advice of such deviation or change of voyage.

17. In the event of damage notice to be immediately given to The Salvage Association, London.

This clause is now the basis of all the insurance of frozen meat where the risk starts in the freezing works. There are other clauses where the risk starts from time of shipment, but that merely alters paragraph 1 of the clause. In other clauses in force the risk terminates at ship's side on ship's arrival. Paragraph 17 is only found in insurances effected at Lloyd's; where the insurance is effected with a company, notice is given to the company direct. As to how long this clause will remain in force it is difficult to state. Broadly, the clause has really been in force with the insurance companies for many years. Although within the last eighteen months a few alterations have been made by them, these alterations are embodied in the A 1 Clause. Many of the insurances effected at Lloyd's in the past did not contain such stringent clauses as are now in force, and it may be that in order to obtain business the underwriters will revert later on to the more lenient clauses, or, alternatively, traders will more and more adopt f.p.a. insurance (*i.e.*, against the ordinary marine risks) plus breakdown of machinery, which is a much less expensive method of insuring.

Renunciation of the Passive Attitude.

For the twenty-five years during which the question of insurance has been a burning one the underwriters have apparently considered that claims were out of proportion to the damage. Until 1909 they made little effort to enforce practical reforms in the working of the trade, preferring to cover themselves by raising rates when the balance was on the wrong side. But in the new clause of 1909 this passive policy was departed from. They devoted themselves, with a view to alteration, particularly to the consideration of the system under which consignees in London made claims, and whilst, of course, recognizing their right to call for surveys on goods (this is done now practically in all cases, as the thirty days' cover often runs off before goods are taken out of store, and the meat owners must protect themselves), the underwriters desired to cut down expenses of survey, which were often very heavy. Their new scale of survey fees charged from June, 1909, was 10*s.* 6*d.* per 500 carcasses, 21*s.* per 1,000, and 10*s.* 6*d.* for each succeeding 1,000 ; beef quarters double the above.

Some of the new rules introduced in this clause of 1909 appear somewhat one-sided, but underwriters evidently thought the position required a firm hand. The view has often been expressed that underwriters might have departed, with great advantage to insurers and insured, from their *laisser faire* attitude ten or fifteen years earlier by endeavouring to check by their policy clauses any faults in the system of handling frozen meat which affected their interests. By doing so they would have taken their part in the building up of a correct system and helped to place the trade generally on a sound footing. Now, with the new rules, they have brought up the business with a round turn. The underwriters' answer to this is that it is the duty of the merchants to endeavour to discover means whereby their goods can be carried efficiently, and that the underwriter has no control over the handling of goods, and can only gauge his risk by losses and arrange his premium accordingly. But the warranties in the A 1 Clause show that underwriters can exercise, and have exercised, a salutary and even arbitrary "control over the handling of the goods."

The incidence of damage to frozen meat, where occurring, divides itself roughly into three classes: first, that which takes place in the freezing works or during transit to the ship; secondly, that which occurs afloat owing to faulty refrigeration or defective insulation of the ship's holds; and, thirdly, that occurring ashore, whether it be during discharge, in transit from ship, in the cold stores, or in the general process of transport, handling, and marketing. Under the first head, losses of late years have been greatly reduced, and in this connection there may be mentioned a point which the shipowners make, viz., that very frequently in commercial and official statements, etc., the damage occurring to parcels of meat subsequent to discharge is attributed to the ship which carried them. The reason why this is done is because the shipowner is an easier party to locate than one or other of the various interests subsequently handling the meat; but it is hard on the shipowner as a rule, though there are some defective ships now, just as there were ten years ago, and also easy-going captains who take defective meat on board.

Three Different Views of Meat Insurance.

The Shipowner's View.—To quote the shipowner's own words, he says his duty is to carry meat, and that whatever happens *en route* is an insurance risk, assuming that he has done his utmost in appointing skilled officers and engineers, and in having his refrigerating plant and insulation tested and passed by Lloyd's surveyors representing the underwriters. He only charges a carrying rate, not one to include insurance. The shipper pays the insurance premium to the underwriter to cover the risks outside actual carriage of meat. In practice, the latter pays the insurer the loss, in case of damage, and then attacks the shipowner to recover the money, although the latter never received any premium for insurance. The negligence of the shipowner's servants is the human fallible element which cannot be guarded against entirely. The shipowner inserts a negligence clause in his bill of lading, and he says that such a proviso is in force in all other trades. Under such a clause, he maintains, the ship should not be liable for damage to frozen meat, subject to the conditions mentioned above as to care to be taken by the owner. He maintains that the time has now come for a rearrangement—the shipowner cannot stand the strain. He says it is unfair that the underwriter who takes the premium should put this risk on him. The shipowner is willing to do everything reasonably possible to ensure good carriage, but he says that there is some point at which his responsibility should cease. He would like to have a final certificate that the underwriter accepts the steamer as a good insurance risk, assuming its seaworthiness, before the meat is put on board. Thus he wishes to define the point at which the risk of the insurer begins, he and the shipowner taking the meat at carrying rates, and the underwriter taking the insurance risks. The shipowner suggests that the underwriter should pay more attention to the circumstances of the trade and to claims relating to damage after the goods leave the ship. He maintains that with the present insulation and duplicate system of refrigeration there is not much chance of damage to meat whilst on board. He proposes that two policies should be issued, the one operating from the moment the meat is received from the works till delivered ex ship, the other (if necessary) for the risk of transit and store. Theoretically, this suggestion is admirable, but it is not workable, as it involves the survey of all meat as landed, *i.e.*, before passing under the second cover—a practical impossibility.

The Underwriter's View.—No shipowner, says he, has been made liable for any meat which has been damaged *en route* unless it has been shown that the refrigerating machinery has not been in a fit condition to perform its work before the cargo was put on board. Small claims which he has had to meet for carcasses being dirty through his men or stevedores trampling over them are rightly made. The underwriter states that the shipowner's suggestion that the underwriter should pay more attention to the circumstances of the trade, and to claims relating to damage after the goods leave the ship, seems to be entirely outside his purview, as it is quite certain that for any damage that occurs to goods after they leave his ship the shipowner is not responsible. Speaking broadly, it appears to the underwriter that, if the meat is shipped in good hard-frozen condition, and the ship's machinery does its work properly, it will be landed in good order. "Instead of that, what do we find ?" the underwriter argues. "From vessels arriving, especially from Australia, much meat lands here badly misshapen, carcasses are found in some cases quite flattened out, off colour, and otherwise seriously damaged. Probably, this damage is due to the goods being shipped insufficiently frozen. But if the shipowner has given a clean bill of lading for goods that are shipped soft he has only himself to thank if he is held to blame."

The Merchant's View.—As regards the shipowner's liability the shipper and the underwriter have much the same standpoint. The shipowner's case has been stated in his own words, likewise the underwriter's ; now here is the view of a merchant engaged in the frozen meat trade : "In the carriage of goods by land and sea there is an underlying principle that the carrier shall provide a conveyance reasonably fit and proper for the work required of it. Whether the conveyance is really fit and proper can only be judged on completion of the voyage or journey by the condition in which the goods (signed for as in good condition when shipped) are handed over to the receivers. It is, therefore, manifestly unreasonable for a shipowner in the case of frozen meat to seek to free himself from all liability for the condition in which the meat arrives at its destination. If steamers were surveyed by thoroughly competent men and declared to be in a fit and proper state to carry frozen meat, the shipowner need have little fear of being saddled with a claim for damage, provided reasonable care be exercised by the engineers in charge, and even then he is amply protected by the 'negligence' clause in the bills of lading. It is not in the ordinary business of an underwriter of goods to bear the shipowner's obligations as regards seaworthiness. Such risks can be insured, and, if the shipowner feels that these obligations are more than he can stand by, it is for him to take steps to insure against his own liability and not try to force the shipper into doing so."

The Appeal to the Courts.

This radical difference in point of view as between the carrier and the insurer of frozen meat is as sharp to-day as fifteen years ago. The shipowner, having done his best to provide a seaworthy and "meatworthy" vessel, proceeds, where possible, to contract himself out of liability for damage that may occur on board by special clauses in his bill of lading. He cannot do this now with regard to Australia, because in December, 1904, the Commonwealth Sea Carriage of Goods Act was assented to. Under this Act all clauses in bills of lading whereby the ship or owners are relieved of liability for loss or damage to goods occurring on the ship arising from faulty conditions or negligence of employees are rendered null and void. This sweeping measure settled the question for Australian exporters, as the Act applies to ships carrying goods from Australia to places outside Australia. The measure had its origin probably in the celebrated *Nairnshire* case, in which the shipowner contended that his bill of lading covered him against loss arising from any defects that may have existed in the vessel previous to the loading of the meat. This case was carried to the House of Lords and went against the shipowner.

Frozen Meat Insurance Details.

In the early days when insuring meat was a new and unpopular risk (it was from the first, and is now, unpopular on account of the numerous losses made) the premiums charged were very high ; 80*s*., 85*s*., and 90*s*. per cent. were not unknown in the New Zealand trade. Then premiums were lowered to 80*s*. and 70*s*. per cent., at which figure they stood for a long time. About ten years ago rates were lowered to 65*s*. and 60*s*.

Up to that period the business had been done by companies, both in London and New Zealand; when Lloyd's took up insuring frozen meat at the end of the nineties, premiums fell. From then onwards rates have gradually fallen, with 40*s*. as a minimum for mutton. Premium rates have stiffened for some time past owing to underwriters having suffered loss through the secondary risk, the loss of vessels. There is now little elasticity in rates, and but slight differentiating between firms, ships, freezing works, etc. A great deal of meat insurance is done in New Zealand by the local insurance companies, but Lloyd's have the greater part of the Australian business as well as most of the South American connection. The large works in Argentina have floating policies both with Lloyd's and the companies. The Thames and Mersey, Commercial Union, Indemnity Marine, Ocean Marine, and other British and foreign offices handle a great amount of frozen meat insurance.

Consignees of meat often elect to insure f.p.a., but the more general plan—almost invariably so in the c.i.f. trade—is to have the ordinary "all risks" policy. This policy has the special frozen meat clause (already set forth in this chapter) attached to it; this does not cover bone-taint. Usually, however, insurance starts at the freezing works, including freezing risks, all sea risks, and continues until 30 or 60 days after the meat is stored. In the last conference clause bone-taint and improper dressing are specially excluded. Lloyd's A 1 Clause, "Freezing works, voyage, and sixty days," gives a list of stores at English ports where the meat can be discharged; 90 per cent. of Australasian frozen meat is carried under this clause. The premium for this risk has been lately about 45*s*. per cent. on approved covers. F.p.a. cover, including breakdown of machinery, is issued at about 25*s*. per cent.; such a policy would be for the voyage only, the shipper giving a warranty that the meat was in sound condition when put aboard. A risk adopted by one of the leading New Zealand freezing companies is to cover 75 per cent. only of the goods against all risks, for which they pay about 40*s*. per cent. They reinsure the 25 per cent. balance f.p.a., including breakdown of machinery, at 15*s.* per cent. An increasing vogue in the New Zealand meat export trade is to cover total loss of vessel only, which can be done for about 7*s.* 6*d.* per cent.

The mutual clubs formed amongst shipowners for the pooling of losses have included the frozen meat risk. These clubs have, however, considered the suggestion that losses falling upon the owners resulting from frozen meat carriage should be excluded, because of the heavy drafts upon them in connection with law cases—these cases were fought by the clubs.

The suggestion has been put forward by merchants and shippers in Australasia and England that underwriters should charge differential rates for insuring frozen meat according to the records of individual vessels in the "refrigerated fleet" for conveying meat well or imperfectly. But such a proposal is not workable, some merchants argue, and any attempt, they say, to carry it out would introduce all sorts of complexities into the business. But it is already done every day in connection with ordinary cargo insured f.p.a., and does come into operation to that extent in "all risks" cover. As it is, frozen meat rates vary on the different shipping lines.

The Surveyor's Duties.

The needs of the frozen meat trade have called into existence the surveyor, who is required to possess special experience in the meat business and knowledge of refrigeration and cold storage. The surveyor makes his appearance on the scene when the consignee of a parcel of meat calls for a survey on it. This survey takes place in the presence of two surveyors, acting respectively for the importer and the underwriter. They go to the store and inspect a percentage of the meat, usually 10 per cent., sometimes more. Occasionally small parcels are found without blemish, but in a large consignment it is difficult to escape some cases of injury, and an allowance is made, at so much per lb. on so much per cent. of the whole, according to the results of the investigation of the percentage examined. The surveyors aim at doing no more than to compensate the owner for the difference between the selling value of his damaged meat and its sound value. This is the practice with regard to the great bulk of Australasian meat landed in London. In the cases of some large importers handling their own goods, the underwriters accept their claims made upon all the rejected carcasses of any shipment without question ; such claims are based on the condition of the meat and the market prices of the day.

Classes of Damage.

The most serious form of damage to frozen meat in transit for which claims are made on the underwriter is that which arises from the ordinary sea perils, such as vessels stranding, collisions, etc. When such disasters happen, they involve heavy claims, especially if sea water gets into the meat holds : meat which has suffered in this way makes a " bad salvage." Rarely nowadays have claims to be made on account of breakdown of machinery, although claims were not uncommon in the early days of frozen meat transit. All vessels now engaged in carrying frozen and chilled meat have their refrigerating machinery in duplicate. The risk, therefore, from breakdown is very slight.

The classes of minor damage for which the underwriters have to pay under the circumstances just set forth are as follow : (1) " off colour " ; (2) misshapen ; (3) stained, torn, and broken carcasses, brine-stained carcasses, bone taint (as a rule this damage is not included), dirty carcasses, mould. Bone taint is practically confined to beef ; at one time it affected mutton, too. It is found in beef from Australia, New Zealand, and Argentina. Nos. 1 and 2 are caused by the carcasses getting soft at some stage after leaving the freezing works ; dirty carcasses come from bad stevedore work in loading and discharging ; torn and broken carcasses are the result of mishandling ; and brine-stained meat (which is condemned by the sanitary authorities) results when leaking pipes drip on to the carcass. Though the underwriters have taken no action to prevent the use of unduly low temperatures in the holds of meat-carrying ships, their surveyors hold strong views on the subject. From 15° to 17° F. is considered by them to be the ideal temperature on land or sea for holding meat in a frozen state. An exceedingly dry atmosphere accompanies temperatures at or approaching zero F., at which meat has been carried sometimes, and this, surveyors say, draws moisture from the meat, rendering it " off colour " and dull, besides causing needless loss in weight. Many importers believe that more damage of this kind is done by frequent and considerable variations of freezing temperature than by a very low temperature kept fairly constant.

A special risk attending the transit of chilled beef may be glanced at. This is where the rods by which the beef quarters are suspended break, through the straining of the vessel or from some other cause. The meat is, of course, injured owing to its falling in a heap on the floor. It is a moot point, in considering underwriters' liability, whether the said rods can be considered part of the refrigerating machinery of a vessel, and, therefore, what party has to bear the loss caused by such accidents. This chapter may be concluded by drawing attention to the risk of damage occurring through barges conveying frozen meat to the up-town stores colliding with other craft.

CHAPTER XI

COLD STORAGE

One of the most important branches of the frozen produce trade is that connected with the cold store or refrigerating warehouse, which forms one of the main and strongest links in the chain of industries uniting the pastures of the producing countries with the consumer's table, the ultimate destination of the food produce. The chilling or freezing store which forms an adjunct of the freezing works at the point of production is not the subject of this chapter, but rather the storage depôt at the receiving and marketing end, wherein the frozen meat is deposited for a longer or shorter time after the vessels have discharged their burden at the ocean docks. The storekeeper owes his business entirely to the growth of this overseas transport, the beginning of which was seen in the memorable pioneer voyage of the *Strathleven*. The cold storekeeper is a product of the industry. As a rule, the shipowner gives up only part of his space to the carriage of frozen produce, but frozen meat, butter, fruit, etc., are the storekeeper's all in all. The business he does in the preservation of hops, furs, etc., though a considerable one if reckoned separately, shrinks to small proportions if compared with his "legitimate" trade, that of the storage of food produce. The pioneers of the frozen meat trade and their representatives in London and Liverpool had no conception of the gigantic and highly organized institution which the modern cold store was to become. In their tentative way they shipped forward their meat for marketing in London, holding very modest views as to the future, and having no fixed ideas as to the necessity of an intermediate stage between ship and mart. The pioneers who survive and can carry their minds back clearly to 1880 must marvel when they contemplate present day figures of the cold storage industry in the United

Kingdom. In London alone the public cold stores have a total capacity of $2\frac{3}{4}$ million carcasses, or, roughly, about a quarter of the cubic space occupied by a whole year's imports of frozen meat into the metropolis.

The first experiment in cold storage at the London Docks was an installation made for the American trade (which began in 1874). It was a store in which there was a guttering round the top to hold ice. The first mention of public cold stores in London occurred in a paragraph in the *City Press* of June 19, 1880. Therein it was reported that at a Corporation meeting it was agreed to let to Mr. Stevenson Nos. 7 to 12, Market Buildings, Charterhouse Street, for a "refrigerated store" for twenty-one years from Midsummer, 1880, at £1,000 a year. A condition of the tenancy was that the market toll on meat entering Smithfield was to be paid. A Mr. Judd is reported to have opposed the lease, arguing that it was against public morals to do anything which would prevent the getting rid of perishable meat as quickly as possible. "The Court should, therefore, be careful not to start or encourage a new industry for preserving that in which decay has taken place." Probably this movement for a store—which came to naught, at least for the time—was engineered by the market salesmen who were interested in the success of the *Strathleven* venture and by the interests in the City concerned in the enterprise.

The Dock Stores.

No architectural skill was lavished on the first cold store in London, that at A Jetty, Victoria Dock, which was then the property of the London and St. Katherine Docks Co. (later amalgamated with the East and West India Docks Co. as the London and India Docks Co.). The company was pressed in 1881 by its Australian friends to receive and store frozen meat, and in the following year the first cold store opened in London was working. It was considered that an underground vault was most suitable for holding the temperature, and such an apartment was fitted up. A small machine was supplied by Messrs. J. and E.

Hall, and the capacity of this store was only equal to 500 sheep. It was, however, soon considerably increased.

A very interesting description of the first provision of cold storage accommodation by the London and St. Katharine Docks Co. was given by Colonel B. H. Martindale, C.B., R.E., general manager of that company, in the course of the discussion of a paper "On Refrigerating and Ice-making Machinery and Appliances," read by Mr. T. B. Lightfoot before the Institution of Mechanical Engineers in 1886. This paper, by the way, is one of the only two papers on refrigeration ever read before that Institution, both being contributed by Mr. Lightfoot. Colonel Martindale on that occasion said he had been connected for the past five years

"with the practical working of arrangements for preserving refrigerated meat from the Colonies on perhaps the largest scale that had ever been carried out in England. In 1881 the dock company were pressed by some of their Australian friends to make arrangements for receiving frozen meat and storing and distributing it. They began necessarily on a very small scale. Happening to have under one of their warehouses a large vault about 500 feet long divided longitudinally into four arches, each 16 feet wide, they made use of part of it for their storage. They began with a small engine obtained from Messrs. Hall, of Dartford, delivering 10,000 cubic feet of cold air per hour; and that engine did good work until 1884, when it was removed to make room for a larger one. They gradually increased the number of their cold storage chambers, until they had now got 56 chambers in two vaults. The smallest chamber had a cubic content of 2,273 feet, and the largest of 9,280 feet, the total content of the 56 chambers being something over 183,000 cubic feet. It was found that the carcasses averaged in weight 56, 60, and 72 lbs., and if the chambers could be completely filled they would hold about 59,000 sheep of the first weight, 56,000 of the second, and 44,000 of the third. But in practice a space had to be left for gangways, and for separating different marks, especially in consignments from New Zealand; and a proportionate deduction had, therefore, to be made from what could otherwise be stored. Still the chambers could always store about 44,000 sheep, taking the shipments as they chanced to arrive. The construction of the chambers had varied a little in detail. The last chambers that had been built had been constructed according to the recommendation of Mr. Haslam, of Derby. On the original concrete floor of the vault there was placed a longitudinal layer of 1¼-inch rough boards, on which were laid transverse bearers, 4½ inches deep by 3 inches wide and 21 inches apart. On these bearers was laid a 2½-inch batten floor, grooved and tongued. The sides and roof were constructed with 5½ by 3-inch uprights, fixed on the floor bearers. On these uprights there was an outer skin of 2-inch boards, and an inner skin formed by two thicknesses of 1¼-inch boards, between which was a thickness of specially prepared brown paper. The 5½-inch space between the inner and outer skins of the sides and roof and the 4½-inch space between the floor and rough boards were filled with carefully dried wood-charcoal. All the boards were finished to a uniform thickness and were grooved and tongued. Cold air was conveyed by wooden trunks from the refrigerating machines into the chambers, entering them at one end close to the roof, and being drawn off at the other end, also close to the roof, by return air trunks leading back to the refrigerating machines. From the quantity of snow made by the engines, the air trunks had to be cleared out every twenty-four hours, and the engine snow-boxes about every four hours."

Subsequent to the establishment of the stores mentioned, the East and West India Docks Co. provided stores upon two floating hulks, the *Seawitch* and the *Robert Morrison*; these craft together stored 15,000 sheep, and they were moved from dock to dock as required, one of them also making a trip with a cargo of meat from London to a French port. These floating stores were followed by stores situated on land in the South West India Dock, with a capacity of 14,000 sheep, and it is said the company then felt that all the requirements of the trade had been met. Docks policy, however, has expanded with the trade, and the development of the docks stores has kept pace with its growth; the 500 carcasses capacity of 1882 have become 822,000 in 1911, including the store at Smithfield. The dock companies and their clients always emphasized the advantage of storing frozen meat at the point of discharge, and for a long time after the start of the trade held a large part of the storing business. The development of cold storage as a public service was largely due to the companies, and their schedules of charges and regulations form the basis of business to this day to a great extent. The dock stores were taken over by the Port of London Authority in 1909.

The Smithfield Market Store.

About 1883 a London cold storage company was promoted and duly registered, the names of Messrs. James Anning, E. Montague Nelson, Alfred Seale Haslam, Ebenezer Cayford, and Alfred Towers being associated with the venture. It was styled the Dead Meat Storage Co., Ltd., and had a capital of £100,000. Messrs. Cayford and Towers had been granted a lease of the vaults under the Poultry Market, and these it was proposed to turn into a cold store equipped with Haslam's cold air machinery. The Dead Meat Storage Co., however, did not proceed to operations, and was dissolved by notice in the *London Gazette* in 1890.

The concession, however, was taken up by the Central Markets Cold Air Stores, Ltd., a company formed in 1884 with a nominal capital of £30,000. Among the subscribers were Messrs. E. S. Houlder, E. Cayford, Thomas L. Devitt, Joseph Moore, C. E. Green, John Bell, and Alfred Towers (managing director). Mr. E. Penman was also associated with Mr. Towers in the management of the store. The secretary was Mr. H. E. Kaye, who retired in 1890 and is now the manager of the Blackfriars Cold Storage Co., Ltd. The appearance of the names of so many important shipowners on the subscribers' list of this cold storage company is interesting as marking early recognition by the Australian shipowning interest of the commercial importance of the coming trade. The company was wound up in 1901. This store, that is, in the vaults under the poultry section of Smithfield, was not a success, its failure, however, being by no means due to the Haslam refrigerating machinery installed, which worked well, but to defective insulation. Adapting the peculiar conditions of these underground spaces to refrigerating work was probably too grave a problem for the refrigerating engineering knowledge available twenty years ago.

In 1898 a prominent group of market tenants subscribed the necessary capital, and the Smithfield Market Cold Storage Co. was formed to take over this property. The title of this company was in the following year altered to the London Central Markets Cold Storage Co., Ltd., and its premises at that time comprised the original section under the Poultry Market and the imposing building erected in King Street, adjacent to the market buildings. A few years later this company acquired a portion of the Midland Railway Co.'s depôt at Poplar, and completed the building as a cold store with an entirely new refrigerating installation, thus adding to the importance of the company in its relation to the service for the Central Markets by having a dock and receiving depôt upon the river entirely at its command. From this depôt at Poplar specially constructed insulated motor vans reach the market stores in one hour. The new building in Charterhouse Street facing the poultry section of the markets was acquired in 1905. These efficiently equipped stores together form an important part of the London cold storage system, more particularly in relation to the business of the London Central Markets, as they are actually a part of the market structure. The same company has recently acquired depôts at Liverpool (two), Nottingham, and Chesterfield. Its stores have a capacity of about 600,000 carcasses. Mr. J. H. Geddes was the leading spirit in the organization of these stores, and he remained managing director of the company till his death in 1908, Mr. William Hawkins succeeding him in that position.

Nelson Brothers' Stores and Depôt.

Messrs. Nelson Brothers, who had been using the dock stores in London for their importations of frozen meat up to 1885, in that year opened what might be called the third of the three pioneer cold stores, in the arches under Cannon Street Railway Station, and the effect of this was to lower charges for storage of meat. The Tyser Line from New Zealand to London started its refrigerated traffic at about this time, and the competition set up reduced freight charges. Little experience in cold store construction was available in those days, and the efforts of architects and builders in this connection were accordingly crude. The following description of Nelson Brothers' early cold store is interesting. Under Cannon Street Station there was a central arch extending from Thames Street to the river, and from it other arches ran at right angles. From the central arch as a corridor the other arches were closed in and insulated. The door in the centre of each arch had a trapdoor through which carcasses were passed into the chamber, and through which they were delivered, the main door hardly ever being opened. At the river end of the central arch was a landing platform, alongside which barges from the docks were unloaded; the carcasses were passed up by hand into trucks and run along a tram line to the chambers. At the Thames Street end were the loading platform, scales, and offices. Trucks filled at the chambers were run up to the weighbridges and loaded into market and railway vans for the country. (This store was in 1898 sold to the Union Cold Storage Co. and re-modelled.)

As Messrs. Nelson Brothers found the Thames Street store inadequate for their expanding business—in one year they had handled about one and a quarter million carcasses and 50,000 quarters of beef—the firm opened additional premises, a splendid cold store in Commercial Road, Lambeth, in 1892. Sir Frederick Bramwell and Mr. H. Graham Harris were the architects, and the definite plan on which these stores were constructed was that formed by Sir Montague Nelson, whose proposition was that the store should be a "gigantic tank": everything was to go in at the top and go out from the top. This idea was rigidly adhered to. The "tank" was divided into floors, but though it was intended that these floors should extend from side to side without divisions, the building regulations of London forbade this, and a dividing wall had to be built down the centre, cross partitions at intervals dividing each floor into several bays with fireproof doors. Even the upper floor, intended only for receiving and delivering goods, had to be similarly divided, even to the iron doors. The store, exclusive of the cost of the freehold, cost £150,000.

This store was opened for business on March 17, 1892, and was, and is, one of the biggest and best-equipped cold store warehouses in the world, with a capacity of 250,000 carcasses, equal to 750,000 cubic feet. The opening of the establishment was one of the features of progress in the 1890—1900 period, a time of such great expansion in the frozen meat trade. A feature of this riverside store is the system of taking delivery from the barges. The elevator is an endless chain running up by the side of the building and working between the upper floor of the store and the barges down in the bay. The carcasses are placed in the cradles of this elevator, which is lowered so that its under end is in the barge; they are then carried by hydraulic power to the floor and dropped on to the sorting table. When the elevator deposits its frozen carcasses there, they are placed in little iron trucks, all of equal weight; the trucks are lowered to the chambers and the meat packed away. This method is expeditious and cheap, and saves handling the meat. The standing instruction to the men at Nelson's Wharf is to "handle the carcasses as if they are eggs";

this care, combined with as little handling as possible, saves damage and consequent insurance claims. These stores passed to the Colonial Consignment and Distributing Company in 1895.

The store at Nelson's Wharf was designed to facilitate the company's country business, and the spacious floors are used for assembling, packing, and despatching the frozen meats distributed throughout the Provinces. Various labour-saving appliances are used, notably cutting machines (Nelson-Dicks patents), which separate the parts of a sheep swiftly and evenly, ready for packing, where joints are required. With regard to the Smithfield part of the company's business, the meat is conveyed by van to the Central Markets during the night. When required for Smithfield the carcasses are repacked on the trucks, raised to the delivery floor, passed over a weighbridge, then on to a lift, and lowered to the loading platform and carried into the vans. Two De la Vergne ammonia compression refrigerating machines and one Haslam cold air machine are installed, the Haslam machine, the older plant, being used only occasionally. There are nineteen hydraulic lifts and hoists and ten miles of refrigerating pipes in the buildings. There are five storage floors, each divided into three sections, in the arrangement of which the " bay " system is used. An idea of the working facilities may be gathered from the fact that a cargo of 50,000 carcasses of meat has been received and housed in less than four days. To conclude reference to the features of the premises, it may be added that recently some disused cellars were turned into a miniature rifle range, which is very popular with the staff.

The Union Cold Storage Company's Stores.

The group of cold stores owned by the Union Cold Storage Co., Ltd., is world-wide in its ramifications. There are stores at London, Liverpool, Manchester, Hull, Glasgow, as far as the United Kingdom is concerned, and in connection with the Russian-Siberian butter export business, the Union Co. has opened large cold stores at St. Petersburg, Riga, Kosloff, and

Kourgan. The combined capacities of the Union Co.'s cold stores run into millions of carcasses, and, according to the company's statement, would comfortably hold at any time more than a year's total shipments of mutton from New Zealand. Indeed, the Union cold stores grapple with a storing business which is by far the largest of anything of this nature in the world—a notable instance of the pre-eminence of British enterprise. Some thousands of tons of ice per week are also manufactured by this company.

There are now about 6,000 shareholders in the Union Cold Storage Co., Ltd., and the net earnings work out at the rate of £2,000 per annum for every 100,000 carcasses storage capacity (say 350,000 cubic feet) of the undertaking, before taking into account management, head office expenses, and depreciation. The Union cold stores do such a miscellaneous business that all the well-known systems of refrigeration are employed and the machinery of most of the principal makers is used. The capacity of the refrigerating machinery installed at the company's various stores exceeds a total refrigerating power of 3,000 tons per day of twenty-four hours. The company began operations in 1893 in Liverpool; in 1896 their first London store was started, at Blackfriars, and by rapid strides, as the result of the closest study of all the problems connected with the new industry, the Union cold stores grew to their present far-reaching establishment.

Of late years extension of the Union Co.'s enterprise has for the most part been abroad, except in the direction of taking over existing cold stores in England, which had been erected by public authorities and others. Amongst these were, about 1898, the old cold stores of the Colonial Consignment and Distributing Co., under Cannon Street Station, London, E.C., and later the cold stores of the North Eastern Railway Co. at Hull and the cold stores of the Scottish Cold Storage and Ice Co., Ltd., at Glasgow. During 1910 the Union Co. acquired the large cold stores and ice factory on the Albert Dock belonging to the Liverpool Riverside Cold Storage Co., Ltd.; this had been run by a group of local merchants, who after more than ten years had grown weary of contributing to its running expenses without receiving dividends.

Other Cold Stores: London and Country.

To the above account of the growth of public cold storage in London there should, to make complete the reference to London cold stores, be added mention of the large stores owned by the several meat importing companies, *e.g.*, Borthwick's, Eastman's, James Nelson and Sons, the River Plate Fresh Meat Co., and Sansinena's. But this chapter has in detail treated alone of cold storage erected for the use of the public.

It is evident that competition in the public cold store business is running fiercely nowadays, and that in the big centres of population where this competition exists, only those cold storage enterprises which are erected, equipped, and managed on the most modern lines and in the most businesslike way can be successful commercially.

The meat trade has been mainly responsible for the growth of public cold storage in the various towns throughout the Kingdom, though in most cases the public store owner is open to receive other produce than meat. There are now over a hundred towns provided with public cold storage, and about one hundred and seventy public cold stores in these centres, establishments of a capacity from one or two thousand up to two million cubic feet, which is the total capacity of the stores at Southampton of the International Cold Storage and Ice Co., Ltd., the biggest cold store in Europe. Although in the majority of cases the stores are controlled by proprietors or companies trading in this business alone, a considerable number of the stores are owned by individual traders or trading companies carrying on business as merchants and making the provision of refrigerated accommodation for others an auxiliary to their business. Thirteen towns have municipal cold storage.

CHAPTER XII

THE STOREKEEPERS' DUTIES

It will have been seen from the previous chapter that the cold storage industry which has been built upon the modern development of the refrigerated produce trade is a great and complex undertaking. Some explanation of the various features of its operative customs will be of service in the present chapter, as on the working of the cold store depends to a large extent the success of the frozen meat industry.

Periods of Storage.

With regard to the cold stores at the ports of debarkation, for it is with those establishments that this chapter deals, space is usually arranged for at the stores before the arrival of the vessel, and in London the storage charge on frozen meat begins from the time the steamer breaks bulk with her meat cargo. The object of every consignee is to take delivery of his meat within the first month. Speaking roughly, perhaps 75 per cent. of the frozen meat in the London cold stores is marketed within the month. At any rate, it changes hands within that period, but this question is solved according to the state of the market ; at times meat is held as long as six months, and even more, and nine months' storage is not unknown. Though frozen meat loses its bloom and also weight through long storage, it may be marketed in fairly sound condition even after such lengthened storage. There was a case fifteen years ago of 10,000 Canterbury lambs being " bottled up " for six months—as a market speculation. The carcasses were locked up in a chamber, the doors of which were not opened. The meat stood the ordeal well, and a profitable deal was made. In April, 1909, a consignment of New Zealand sheep, 23,000, was put into one of the London stores in one lot ; the carcasses were taken out in the following September in first-rate condition. Such procedure, however, is risky, the ordinary experience being that six months' freezing makes frozen carcasses bleach. The general question as to how long frozen meats may be kept in cold store, with regard to loss of quality and appearance, and the structural and other changes which take place, is one for scientific men and trade experts. Experiments, mainly in poultry, have been made in England, Australasia, and America; essays and articles without number have been penned, yet there is still to seek a statement sufficiently clear for the non-scientific mind to grasp and explicit enough to use as a guide for business men.

Reaching the Store.

Much, of course, depends upon the conditions attending the actual transit of meat to the cold stores. In the case of up-town stores, that is, the refrigerating warehouses other than those connected with the docks, transit is made by insulated lighters to the stores situated on the riverside, and by insulated lighters and vans to those situated inland. The principal dock stores of London are situated in the Victoria and Albert docks, and have working capacities equivalent to 552,000 sheep. So far as concerns the stores in the Victoria and Albert docks, some of the meat is conveyed thither by insulated lighters from the Tilbury docks, but the bulk is received from steamers discharged in the docks. The principal railways run alongside these stores, and meat is delivered direct to railway wagons for distribution in the provinces. Delivery is also made at night for the market, but as these stores are situated at a considerable distance from Smithfield, orders must be given some hours beforehand, so as to allow time for conveyance and delivery into the market. The inconvenience experienced in these arrangements led to the construction of the West Smithfield stores belonging to the Port Authority, which have a capacity of 95,000 carcasses (they are to be supplemented by another store in the vicinity of a capacity of 84,000 carcasses), and, being adjacent to the market, permit of delivery being obtained at call. Conveyance to the West Smithfield stores is usually made direct from steamers by insulated vans. The owners of the meat decide on the arrival of the steamers the stores into which they require their shipments placed, and, of course, they are guided to some extent by the probable disposal of the meat. There are also stores in the West India and Surrey Commercial docks.

The large white insulated vans that convey the frozen meat to Smithfield's portals hold about 120 sheep; the ordinary river barges take from 1,000 to 1,200 sheep, but some will take considerably more. Meat is not usually weighed on delivery to the store, but when it is sent out, and there is occasionally an intermediate "weighing over" for the convenience of the customer. On meat being received at the store, the first thing to do is to stow it according to marks, readily accessible for delivery.

Marks.

About sixteen years ago the "multiplicity of marks" trouble became acute; the storekeeper was asked to keep separate the various small lots of carcasses represented by the sub-marks—50 sheep, five sheep, and so on. The separation of these lots involved an enormous waste of storage space, and, to meet the difficulty, a line was drawn at 100 carcasses, or 400 pieces of mutton or lamb, such as haunches, legs, and shoulders, as a minimum. Any parcel under 100 carcasses or 400 pieces was, and is now, charged one-third additional. Say, for example, the charges under the (old) management rate of 20*s.* 9*d.* a ton for twenty-eight days for some rent on 50 sheep amount at the usual rates to £2 5*s.* The parcel would be charged one-third extra, or £3, and so on, and the same practice for the same reason has recently been extended to beef, the line being drawn at a pile of 30 quarters.

If several sub-marks belong to one consignee, he may have them piled together, in order to avoid the one-third extra charge, but the marks would be mixed, and would be delivered as they "rose from the pile."

Frozen beef is piled in store similarly to mutton and lamb. Chilled beef, which is hung on hooks, is sent to the cold stores to a far less extent than frozen beef, but with the greatly increased importation of chilled beef from South America the need for chilled, as opposed to frozen, storage has increased. The chilled beef steamers are used as stores until the balances of the shipments have to be cleared, according to the state of the market. There are chambers for chilled beef at the Victoria Dock stores, which have recently been increased from a capacity of 1,000 to about 4,000 quarters.

Cold Storage Rates.

Like all the other charges in the frozen meat cycle, the cold store rates have fallen considerably from the original $\frac{3}{4}d$. per lb. per 28 days in the very early days of cold storage. According to a schedule issued by the docks store in November, 1883, the charges were as follow :—

Receiving from ship, conveying to stores, rent for one week and delivery was		Rent per week or part of a week after one week.
	Per lb.	Per lb.
When a parcel consisted of less than 1,000 carcasses :—		
For the first 400 or less	$\frac{6}{16}d$.	$\frac{2}{16}d$.
On all between 400 and 1,000	$\frac{5}{16}d$.	$\frac{2}{16}d$.
When a parcel consisted of 1,000 carcasses or more :—		
For the first 1,000	$\frac{4}{16}d$.	$\frac{2}{16}d$.
On all between 1,000 and 2,000	$\frac{3}{16}d$.	$\frac{3}{32}d$. } For any portion of the consignment.
On all in excess of 2,000	$\frac{2}{16}d$.	$\frac{1}{16}d$. }

Several conditions as to minimum charges were set out. In July, 1884, the docks store rate was fixed at 8*d.* per carcass for storage for forty-eight hours from breaking bulk ; "when several marks are imported in the same ship," 9*d.* ; rent after forty-eight hours, $1\frac{1}{2}d$. per cwt. per day. Weighing, "if required," was charged for at $\frac{1}{2}d$. per carcass. In March, 1891, the "management rate," $\frac{1}{9}d$. per lb. (20*s.* 9*d.* per ton), came into force, and practically no change, except the application of the one-third additional charge, as already explained, has taken place in twenty years.[1] After twenty-eight days' storage the rent is now $\frac{3}{4}d$. per cwt. per day, but with the proviso that the rent to be charged never exceeds $\frac{1}{9}d$. per lb.

[1] This chapter was written before the labour troubles, and the rise in wages which followed in 1911 caused an increase in the 20*s.* 9*d.* management rate. It is possible that in 1912 the question of cold storage charges at the docks will be revised.

for any period of twenty-eight days on the gross weight in store on the first day of such period. This charge prevails at all the London public cold stores. The warehouse keeper is bound to cover himself by making a substantial charge for the first month, even though the meat may only stay in store for a day, for he has to pay conveying from ship to store and various other items. The competition which brought about that reduction in 1891 in the London cold storage rates (we have seen the same force at work in the freezing and shipping charges) seems to have ceased—at least so far as the occurrence of further reductions is concerned—with the establishment of the existing "management rate." Up to 1898 demand for cold storage was greater than the supply, but in that year cold storage space got ahead of requirements. A "combine," or working arrangement of some sort, was arranged between the London stores in 1899, and under this the proprietors have, except for a short period in 1906, been able to prevent the $\frac{1}{8}d.$ per lb. being reduced, and in this respect all the cold stores are worked on the same lines. There was a breakdown of the London "combine" at the beginning of 1906, and until a fresh arrangement was made, after three months of civil war, there was a period of severe rate cutting; goods were accepted at as low as 10*s*. a ton. It is obvious that, from a proprietor's point of view, the public cold stores of London lend themselves to combined management to prevent ruinous cutting of rates.

Cold Store Dividends.

The cold storage industry is one of many ups and downs, a feast or famine business. In some seasons the stores are used mainly as depositories for frozen meat for a day or two or a week or two before marketing; in others, great quantities of meat are "bottled up" for months, and the stores become congested. About the beginning of the nineties cold storage was a paying business. During the previous decade the frozen meat trade developed heavily in supplies, with frequent seasons when the market was unable to absorb shipments as they came along. On the whole, however, only modest dividends have been earned in the cold storage business. The average ordinary dividend for 1909 of twenty of the principal cold storage companies in the United Kingdom was slightly over 5 per cent., and would have been appreciably lower than this, did the calculation not include three particularly successful concerns whose ordinary dividend averaged 15 per cent. The business of ice manufacture was also included in most of the companies whose dividends formed the basis of this calculation. The largest stores of all, those of the Port Authority of London, are not, however, included, because no separate returns of this department of its undertaking are issued by the Authority.

In 1888 the cold storage space available in London and Liverpool was only equal to the accommodation of 400,000 carcasses, and in 1894 London's cold stores could only hold 500,000. From that date cold store construction went ahead rapidly. In 1895 the capacity of the cold stores of London was equal 1,000,000 carcasses, in 1900 to 1,648,000 carcasses, in 1905 to 2,631,500 carcasses, in 1911 to 2,840,000 carcasses. Readers consulting Appendix IV. will observe that the available storage space at the chief ports of Great Britain, other than London, is equal to accommodating 5,124,500 carcasses. For the last ten years or so storage space has been readily obtainable in London, with the exception of 1909, a year of low values and heavy storage, when for some weeks the stores were congested; for four or five days in August they were absolutely full.

The chief problem which a cold store proprietor has to solve is how to keep his store as full of produce as possible. (In these remarks only the frozen meat department of the business is dealt with, but, of course, there are many other kinds of perishable merchandise, stored at varying temperatures.) One reason why the cold storage proprietor likes to have his chambers full of frozen produce is because of the assistance rendered by the goods in keeping down temperatures; a half-full store requires more engine power than does a full one.

Responsibilities and Risks.

One might have expected, with the advance of cold storage to a position of considerable importance among the mercantile industries of the United Kingdom, especially in London, that some legislation would have accompanied the movement. A good many nice questions must arise, but apparently all matters causing friction between cold store customer and warehouse keeper are settled in accordance with the customs of the trade that have gradually encrusted round the industry. It is assumed that cold stores come in a general way under the Warehouseman's Acts, and legal proceedings concerning the responsibilities of proprietors would be entered under common law. The custom of "general lien," whereby, in the event of non-payment of cold storage charges, a cold store owner has legal claim on goods held (*i.e.*, all goods held by the same customer, and not merely on the specific goods in respect of which charges are due), was formally established in the cold storage industry in 1902 by the Cold Storage and Ice Association, this society taking the necessary steps by public resolution and advertisement.

What are the responsibilities to his client of the cold store proprietor ? They may possibly be summed up as follow. His stores are inspected by the underwriter's surveyors, and, if passed, are certificated as "approved." All the resources of modern science are drawn upon in the construction of the buildings and the installations of plant and appliances, and a skilled and efficient staff is provided to work the business. The meat while in course of being put into store is externally inspected, while here and there a shirt is cut for making an internal inspection, and any imperfections seen are recorded and the owners informed of them in the "landing account," or earlier if necessary. The cold store does not hold itself responsible for any loss of condition which may take place in meat warehoused. The proprietor provides cold air—at about 18° F. in the case of frozen meat—and he keeps temperature logs day and night. If he delivers the goods for which he has given a clean receipt, he says that by taking reasonable care he has done all that can be expected of him. Any damage to meat in store is the merchant's or his underwriter's, not the warehouse keeper's, concern. Of course, if too high temperatures were proved to be due to neglect, that would be a different matter, but the ordinary gradual depreciation in the condition of meat accompanying storage cannot be insured against after sixty days.

Meat in store is covered by the marine policy current for thirty days, occasionally for sixty, and before this runs out the underwriter's surveyor has inspected the meat, and any allowances for damage have been agreed upon. The only other store risks against which the owner of the meat can then insure are "fire" and "breakdown of machinery." The exact situation is stated in the following paragraph from the docks cold store tariff: "The Port Authority will not hold themselves responsible for the condition of the meat stored with them, nor for any loss which may be sustained through failure of machinery or otherwise. But they will render all assistance in their power in the investigation of any question which may be raised, provided that the meat is not removed from the stores, and the investigation takes place on the same day as that on which the question is raised, or at the latest on the following working day."

To continue this part of the subject in some detail, it may be useful to give the clauses printed on the receipt form of one of the London cold store companies:—

1.—The Company will use every endeavour to keep the goods in sound condition, but will not be responsible for loss or damage to goods stored, through maintaining too high or too low a temperature in the stores, failure of machinery or plant, fire, vermin, or any other cause whatsoever other than theft. In case of fire, storage is payable to date.

2.—Goods are only received subject to a general lien for all charges accrued and accruing against the storer, and if not removed after seven days' notice has been given to the storer, or sent by post to his last known address, may be sold to defray the lien and all expenses incurred.

3.—Transfers are allowed subject to a general lien on the goods transferred for all sums due from the original storer.

4.—Where the Company do cartage, it is understood they are not liable for any loss or damage which can be covered by Insurance, and those interested, in taking out Policy, must effect same without recourse, as the Company do not accept responsibility for insurable risks.

5.—The Company will not be answerable for any delay, loss or damage arising from combinations or strikes of any persons in their employ or in the service of others, nor for any consequences arising therefrom.

6.—The Company has the right to remove from the premises, if necessary without notice, any goods found to be of an offensive nature, or such as will damage other goods in the store, and has the right to remove goods to other cold stores, if for any reason they find it advisable.

7.—Contents of package, and condition, unknown.

Working all Round the Clock.

One of the storekeeper's responsibilities to his clients in the ordinary way of business is the delivery and weighing of goods of specified brand at any hour, day or night. Sometimes notice is sent, but the storekeeper has to work all round the clock and hold himself in readiness to deliver meat on demand. There is no form of agreement between the storer and the storekeeper defining the responsibilities of the latter in this and other respects; custom controls the matter. The public cold-storekeeper is the weighing authority; thousands of sales are made on the storekeeper's weights, which in ordinary business transactions are regarded as final.

In connection with the numerous parcels held in store, surveys are of daily occurrence, and facilities are provided by the proprietors, inspection chamber accommodation in some cases being provided, for which a charge is made. By this inspection is meant the examination which is nowadays almost invariably made under the marine policy. In the noteworthy case of *Kidman* v. *Blofeld and Lisenden*, tried in 1903, the plaintiff sought to show that it was customary in the trade for holders of meat in cold stores, whether as principals or agents, to have the goods "inspected from time to time with the view of ascertaining their condition." In this case the meat in question was seriously depreciating in value while being held in store for a market by the shipper's instructions. The judge found that the alleged practice of periodical inspections was not made out.

The satisfactory condition of frozen meat as delivered at Smithfield or elsewhere from the London cold stores, and, indeed, from the well-managed public cold stores to be met with in all parts of the United Kingdom, speaks well for the system and management of these establishments. The keen competition between the stores and the close observation by importers' and underwriters' surveyors combine to produce excellent results.

CHAPTER XIII

THE GREAT SMITHFIELD MARKET

THE London Central Markets, from that fateful day in February, 1880, when the 40 tons of frozen meat ex *Strathleven* were sold at $5\frac{1}{2}d.$ per lb. to the present time, when 250,000 tons of chilled and frozen meat are handled annually at the salesmen's stalls, have played an all-important part in the rise of the frozen meat industry. Smithfield has been the arbiter whose verdict the farmers, graziers, and estancieros of the lands in the South have awaited with expectant and anxious feelings. But the response of the great London market was never in doubt. Whilst the producers of frozen meat have been able to build up an immense trade with the co-operation of Smithfield, the salesmen were quick to perceive what boundless possibilities were opened up to them with the coming of the refrigerator and the transport of the sheep and cattle in frozen form from the Australian runs. So they welcomed the *Strathleven's* cargo and the meat from New Zealand and Argentina as it came along in 1882 and 1883.

For the last thirty years Smithfield has taken the frozen meat trade under its wing, and, the greatest of markets in the greatest of cities, has impressed the imagination of the Australian and New Zealand meat exporters to a remarkable extent. No institution connected with the realization of merchandise has been so much discussed and keenly criticized by its supporters as has Smithfield. "Smithfield scandals," "Smithfield rings," "Smithfield practices"—these and other topics of like nature have furnished interesting material for the newspapers and public speakers 10,000 miles away full many a time and oft. The relations of the great market and its customers have provided frequently cause for friction, and Smithfield itself often comments, in no gentle tones, upon the methods of her

SMITHFIELD MARKET.

A General View, showing the Approach to the Railway Goods Station, a portion of the West Smithfield Buildings, and the "Japanese Village" at the further end of the Central Market structure.

To face p. 182.

Australasian and American suppliers. But criticism and recrimination have never stayed the sailings of merchant vessels laden with the frozen carcasses, nor checked the return flow of British money into the hands of the sheep and cattle kings of the southern hemisphere; the producer in Australasia and the salesman at Smithfield are necessary to each other.

From the tentative transactions of 1879–1880 Smithfield has moulded, as to marketing methods, the frozen meat trade, limb by limb and feature by feature; and in the process has itself vastly changed and developed. For instance, though for years the frozen meat trade at Smithfield was in the hands of a few salesmen and jobbers, now not less than 200 parties—c.i.f. men, jobbers, agents, etc.—out of the 340 tenants are engaged in this branch of the market's business, and their number is constantly on the increase. Moreover, the large proportion of about 50 per cent. of the meat handled by the great multiple meat shop concerns is frozen, and, as a matter of fact, does not pass through Smithfield at all. First, the few hundred carcasses sent from Australia as an experiment to commission salesmen, then the regular marketing of large quantities from the three great producing countries through the agency of distributing firms and companies, handled in a systematic way, and later the evolution of the c.i.f. trade, enabling the multiple retail shop owner to cover himself for six months ahead—Smithfield has participated in all these eventful developments. All the problems of the trade virtually take their rise in the London Central Markets, in the avenues of which one sees visitors from the Americas and the great lands of the Southern Seas. The New Zealand farmer shipping frozen meat regards Smithfield as his Mecca, and is not always content with one brief pilgrimage. The London Central Markets receive Russian Princes, English Secretaries of State, and Australian Prime Ministers, as occasional visitors, and the "porters' band" wielding knives and cleavers gives them musical honours. The "Tall Hat Brigade" is on duty soon after 9 a.m., gentlemen peregrinating the market avenues, engaged in a quest for business, or for information concerning the frozen meat trade. The more legitimate frequenters of Smithfield are buyers (wholesale dealers, butchers, retailers, restaurant and hotel proprietors), and, of course, the importers and agents.

To some degree Smithfield has lost its importance of late years. In the hope of bettering distribution, the Americans, and some of the distributing firms handling frozen meat, established depôts at Croydon, Kingston, Richmond, Reading, Brighton, Bournemouth, etc., and at these depôts meat is sold that would have been formerly handled at Smithfield. In the process of decentralization which is being applied to the frozen meat trade the London Central Markets suffer severely, and, instead of accompanying this great industry in its rise, Smithfield, as far as the proportion of its pitchings to the total frozen meat trade of the Kingdom is concerned, is not advancing. Importers sell ex store or ex ship considerable quantities of frozen meat which formerly would have been brought into the market, and the purlieus of Smithfield are dotted with the offices of importers' firms whose interests to a great extent are interdependent with that of the Central Markets. Changed methods of business, the development of c.i.f. and ex store buying by the large meat retailers, direct shipping to outports, and sundry other influences, have checked the volume of imported chilled and frozen meat passing through Smithfield from keeping pace with the total trade in frozen and chilled meat. Although the markets' total operations have increased, the percentage of the total imports of these meats marketed at Smithfield has steadily fallen from 65·7 in 1882—1886—the beginning of the frozen meat era—to 41 in 1910. The managers of the markets are sore about this, and regret to find that Smithfield's distributive area is now practically confined to the metropolis. In former days its area reached as far as Birmingham, and salesmen now in Smithfield could tell us that they have supplied customers in Edinburgh. Nowadays much more meat comes to London from the country than goes from Smithfield to the Provinces. Australasia has not been faithful to Smithfield. North and South America have defaulted too. Grouping together the last two sources of

OLD SMITHFIELD (FROM THE GUILDHALL COLLECTION).

To face p. 184.

supply, in 1881 their ratio per cent. of meat marketed at Smithfield to the total importations into the United Kingdom was 68 ; in 1910 it was 40 per cent. The above decreases must not be taken in any way to indicate a smaller amount of meat marketed at Smithfield, but only a smaller ratio to the total imports into Great Britain.

But, speaking of quite recent times, it is doubtful if Smithfield has been losing ground even in ratio to the total imports into Great Britain, when one regards the volume of South American chilled beef brought to the London Central Markets *viâ* Southampton. The shipping development of this chilled beef trade of the Nelson and Royal Mail Company's lines probably has also been instrumental in causing a relative increase in the South American trade done in London during the last few years as well as a decrease in the provincial business.

Old Smithfield.

But before proceeding farther, it is better to revert to the beginnings of things. No part of London has a greater wealth of tradition attaching to it than has Smithfield and its surroundings—" Smoothfield " it was termed in the mediæval days when Bartholomew Fair was held there, when tournaments took place and duels were fought. In the days of Mary and Elizabeth, Catholics and Protestants burnt each other by turn at Smithfield. It was the place for public executions before Tyburn became fashionable. As a market for horses, and live stock for killing, we hear of Smithfield in 1150. Billingsgate was selling fish, by the way, 1,000 years ago. Smithfield market in 1253 was the property of the Corporation, and Edward III. covenanted by charter with the City of London not to grant permission to other parties to set up a market within a radius of seven miles from the City. The erection of markets was the King's prerogative. In those times the market price of food was regulated by the City authorities ; in 1533 it was enacted that butchers should sell their beef not above a halfpenny a pound and mutton three farthings, " which act being devised for the great commodity of the Realm " (as it was then thought), " hath since proved far otherwise, for before that time a fat ox was sold at London for six and twenty shillings and eightpence at the most, a fat wether for three shillings and fourpence, a fat calf the like price, a fat lamb for twelvepence; pieces of beef weighing 2½ lb. at the least, yea, 3 lb. or better, for a penny on every butcher's stall in this City, and of those fat pieces of beef thirteen or fourteen for twelvepence, fat mutton for eightpence the quarter, and one cwt. of beef for four shillings and eightpence at the dearest." There were then 120 butchers in the City and suburbs, and of these every one killed six oxen a week, " which is in forty-six weeks 3,120 oxen, or 720 weekly." The foreign butchers for a long time stood in the High street of Lime street Ward on the north side twice every week—viz., Wednesday and Saturday—" and were some gain to the tenants before whose doors they stood, and into whose houses they set their blocks and stalls; but that advantage being espied they were taken into Leadenhall, there to pay for their standing to the Chamber of London." These references are from Stow's " Survey of London." In 1631 a writer, Howes, gives " Ruffians' Hall " as a cant name for West Smithfield, on account of its being " the usuall place of frayes and common fighting during the time that sword and buckler were in use." The Corporation appear to have claimed market tolls in the fifteenth century. One of the features of the riotous St. Bartholomew's Fair was the enormous sale of roast pork, and beef sausages came into fashion in 1750, at about which time it is noted that the average weight of oxen was 370 lbs., and of sheep 28 lbs. The Fair was closed in 1830; the Corporation bought the Bartholomew Priory rights in 1850. Smithfield was then the market for live stock; and the dead meat mart was at Newgate Market, which was close by.

Smithfield Market in 1853.

An article in the *Quarterly Review*, June, 1854, " The London Commissariat," by Dr. Andrew Wynter, presents the live stock market of Smithfield very vividly, and the following extract is made :—

What they do see in reality, if they have courage to wend their way along any of the tumble-down streets approaching to Smithfield, which the great fire unfortunately spared, is an irregular space bounded by dirty houses and the ragged party walls of demolished habitations, which give it the appearance of the site of a recent conflagration—the whole space comprising just six acres, fifteen perches, roads and public thoroughfares included. . . . Thanks to the common sense which has at length lifted up its potential voice, the days of Smithfield are numbered, and those who wish to see this enormous aggregation of edible quadrupeds before it takes its departure to its spacious new abode at Copenhagen Fields must not delay the visit much longer. The best time is early in the morning—say, one or two o'clock of the "great day," as the last market before Christmas-day is called. On this occasion, not only the space—calculated to hold 4,100 oxen and 30,000 sheep, besides calves and pigs—is crammed, but the approaches around it overflow with live stock for many hundred feet, and sometimes the cattle are seen blocking up the passage as far as St. Sepulchre's church. . . . The meat itself suffers in quality, for anything like fright or passion is well known to affect the blood, and consequently the flesh. Beasts subjected to such disturbances will often turn green within twenty-four hours after death.

The same writer, after careful examination of all the sources of supply, gives the following estimate—in those days there were no exact statistics—of the butchers' meat consumed by the 2,500,000 people who formed the population of the capital in 1853. Dr. Wynter values these marketing stock at £14,000,000.

—	Beasts.	Sheep.	Calves.	Pigs.
Newgate Meat Market	156,000	468,000	31,200	31,200
Leadenhall Meat Market	5,200	41,600	—	—
	161,200	509,600	31,200	31,200
Live stock brought to London	322,188	1,630,793	101,776	127,852
Total supply of live stock and meat to London in 1853	483,388	2,140,393	132,976	159,052

These animals were brought from the neighbourhood of London, the country parts of England, Scotland, Ireland, and the Continent, and were handled partly at the markets and partly by the carcass butchers throughout the metropolis.

Smithfield Market and its Supplies.

The whole district is rich with material for the tourist and antiquarian; the site of St. Bartholomew's Priory on the south and the Carthusian Monastery on the north are hard by the present markets. Shakespeare wrote of Smithfield, and Dickens's Nicholas Nickleby reposed in this neighbourhood, at the Saracen's Head, which hostelry is still in existence. As a popular encyclopædia puts it, Smithfield of old "was available for jousts, tournaments, executions, and burnings." Tremendous congestion prevailed in these districts, which were filthy and criminal beyond description, and in 1851 a Royal Commission was appointed to report as to what should be done. A scheme was adopted under which the live stock markets and slaughter-houses were taken away to Islington and Deptford, and the new London Central Markets, the largest dead mart in the world, were opened in December, 1868, for the sale of meat, poultry, and provisions. The centre of "Old Smithfield" is now laid out with an ornamental garden and fountains.

The first part of the Central Markets was a huge parallelogram covering 3½ acres, with 162 shops. The market was a success from the start, and was soon followed by the erection of the poultry section, opened in December, 1875. In 1879 the fruit and vegetable market was begun, to be followed by the fish market, now termed the Smithfield Market Annexe and used as a meat section. Last, but not least, the extension, which is now entirely given up to frozen and chilled meat, was opened in 1889 for general trade. The westernmost section of Smithfield was nicknamed the "Japanese Village" because of the somewhat Japanese style of the original decorations.

The London Central Markets as a whole now occupy about ten acres; the main building, bounded by Long Lane on the south and Charterhouse Street on the north, stretches 600 feet east and west by 240 feet north and south. The building is in the Italian style with Mansard glass louvre roof; the central avenue is 27 feet wide, and there are six side avenues. Each shop is about 36 feet by 15 feet, and behind every shop

SMITHFIELD MARKET—VIEW OF INTERIOR (CENTRAL AVENUE) AND OF THE NORTH ENTRANCE.

To face p. 188.

is an enclosed counting-house with offices above. The markets are strictly wholesale, except on Saturday evenings, when a few of the employees conduct a retail "People's Market." There are at present 340 tenants holding 344 holdings and giving employment to about 5,000 persons. The toll (2*s.* 3*d.* per ton) on meat, etc., in 1910 came to £46,616, and the stall rents to £81,448. The total capital expended on the markets by the Corporation to 1909 exceeded £2,000,000.

The growth of Smithfield Market, which means the growth of the imported meat movement, is seen at a glance in the following table, extracted from the annual report of the Superintendent of the Market, Mr. H. W. G. Millman :—

Year.	Weight of Supplies marketed.	Origin or sources of supplies in terms per cent.			
		"English killed" and United Kingdom productions.	Imported productions, Chilled or Frozen.		
			North and South American Chilled and Frozen.	Australasian Chilled and Frozen.	Continental.
	Tons.				
1869 .	127,981	97·7	Nil	Nil	2·3
1877 .	197,631	89·0	7·4	Nil	3·6
1887 .	259,383	77·5	9·5	5·8	7·2
1897 .	393,307	47·9	18·8	20·3	13·0
1907 .	417,057	36·6	24·6	25·7	13·1
1910 .	419,550	29·6	32·6	25·2	12·6

The average daily pitchings of meat at the Central Markets during 1910 were 24,000 mutton and lamb carcasses, 2,700 quarters of beef, and 2,500 pork carcasses.

An exceedingly interesting chapter of Smithfield's history is opened up when we examine the effect produced upon the great Central Markets by the coming of chilled and then frozen meat. For the five years prior to the *Strathleven's* arrival, meat produced in the United Kingdom was 86 per cent. of the whole quantity marketed at Smithfield; the 14 per cent. imported had grown to 70·4 in 1910. This drop of 56 per cent. in the proportion of home supplies in the thirty years has, of course, involved a complete change in the methods of the market. (It must be noted that these percentages of home and imported meat supplies apply to London only. For the country as a whole the figures are, approximately, for beef and mutton: home produced, 63·5 per cent.; imported, 36·5 per cent., equal to 88·7 lbs. per head of the population per annum.) When frozen meat first appeared, the small quantity of meat imported was refrigerated American beef and Dutch and French meat; there were large consignments from France of all sorts, calves, sheep, etc., and the beef was of very good quality. In country districts frozen meat has satisfied a newly-created want, but in London it has—by its excellence and cheapness, and its appeal to the seven million consumers within twenty miles radius of Smithfield—been gradually supplanting other kinds of meat.

By about 1883 frozen meat had become so important on the Central Markets that the Corporation had to make special provision for it, but even then that body failed to grasp how indispensable an auxiliary cold storage would prove to be to the meat trade and the market—hence it missed its finest opportunity of becoming the cold storage authority within the area of its market rights. The market salesmen took the incoming of the Australasian and Argentine meat with great calmness; gradually more discrimination was exercised as to quality, brands, etc., as frozen meat became a force at Smithfield. The more enterprising of the salesmen of the markets opened their arms gladly to the produce of the South; many firms saw the potentialities that lay in the new business, took it up on a proper scale, and did well. At first only a few salesmen went into the business, but by degrees frozen meat penetrated farther and farther, and now it has conquered nearly the whole market and has become absolutely necessary to the majority of the tenants.

The American Invasion.

This may be a fitting place to write of the American invasion of Smithfield and what came of it. Mr. T. C. Eastman was the shipper, from New York in October, 1875, of the first lot of American chilled beef to this country, and he must have due credit for this pioneer enterprise. A baron of that beef was sent to Queen Victoria at Windsor Castle, and Eastmans, Ltd., have the Royal Seal in connection with that transaction. The Queen pronounced the meat to be "very good." The shippers of this early American chilled beef included Toffee Brothers, Gillette (Jersey City), Martin Fuller (Philadelphia), and Sherman (Philadelphia). This beef was stitched up in canvas and was hung in the ships' chambers, which were kept at a reduced temperature by the use of ice. Mr. J. D. Link acted as agent for Mr. Eastman up to the time when Messrs. John Bell and Sons took over the agency.

Prior to the beginning of the chilled beef trade small quantities of beef, hard frozen, arrived at Smithfield from the United States; hindquarters arrived in long boxes, "as hard as a stone," but bright and in good condition. The frozen beef made about 2½*d.* to 3*d.* per lb.—it was mostly hindquarters. English beef in 1874 was making 9*d.* a lb., and probably this price tempted the Americans. Quite inconsiderable in volume, and negligible as a market factor, was this early frozen beef from North America. Of course, it was frozen by ice and salt mixture. The American chilled beef when it first came was very large—from four-year-old cattle. For years the American meat did not affect prices to any degree in the Central Markets—it was a "little outside trade." After 1880 the chilled beef became important, but the wonderful jump forward was in the decade 1888—1897, when both chilled and frozen meat took up a commanding position in Smithfield supplies. The quality of the chilled beef from North America was so good that it forced itself into general use.

When the goods first came to Smithfield, and for many years afterwards, they were handled by salesmen on commission. At length there came a time when the Americans said, "Why should we not sell for ourselves?" So they acquired various stalls in the Grand Avenue, paying heavy sums for the goodwill. It is only fair to say that the American firms themselves state that one reason which led them to market their own beef was that they did not "get a very square deal with some of the Smithfield salesmen." There were difficulties between the American importers and some of their salesmen. At any rate, the former made up their minds to sell their own meat. It is on record that the Morris Beef Co., which has three shops, paid for the goodwill of one, the stall which belonged to Mr. Edward Poole, £16,000, £11,500 for another, the Venables stall, and £12,500 for the third, the one acquired from Messrs. Jennings. The Hammond Beef Co. has two stalls on the market ; Armours hold four stalls. By the "Combinations in the Meat Trade" Commission it was put to Mr. Woodruff that Swifts (who hold six shops at Smithfield) in 1901 paid £12,000 for the goodwill of one of the stalls they acquired, transferred by Mr. Frost. To the same Commission one of the American witnesses said that the Americans hold about 5 per cent. of the Smithfield stalls—that would be seventeen. These stalls just referred to are stalls run in the name of the companies, or partners. But it is constantly said and suggested in letters to the Press that the Americans are interested in shops nominally held and conducted by other parties, and that if the truth were known it would be found that they really control a large number of shops at Smithfield.

Administration of the Market.

The control and management of the Central Markets is vested in a committee of the Corporation consisting of six aldermen and twenty-nine commoners ; no tenant of the market is permitted to belong to the committee. The chairman of the committee in 1911 was Mr. James Rowland Brough. The rule excluding tenants is of comparatively recent origin, and is constantly being attacked on the ground that a committee of management skilled, as to its *personnel*, in every mercantile business except that of handling meat must of necessity be incapable of a thoroughly efficient, just, and sympathetic administration of London's great Central Markets. Considering how vast and highly complex the operations of the Central Markets are, and the number and variety of interests involved, it does appear

SMITHFIELD MARKET—VIEW FROM CHARTERHOUSE STREET.

To face p. 192.

reasonable that the market community itself should be represented on the Markets Committee—the market representatives need, of course, never exceed a voting minority of the committee.

An important part of the administration of the Central Markets is that concerned with the inspection of the meat supplies by the Corporation officers. The Medical Officer of Health for the City of London is charged, among other things, with the duty to the public of ensuring that a pure supply of fresh food passes through the marts of Smithfield. Mr. T. D. Young is chief inspector, and there is a staff of twelve inspectors, constantly on the watch to detect unsound meat. It must be stated that the tenants themselves are of great assistance to the inspectors, for a considerable proportion of the meat condemned represents cases voluntarily brought before the inspectors by the salesmen. In 1910 1,427 tons of meat were dealt with by the market authority as unsound: ("diseased," 119 tons, "putrid" [mostly accidentally damaged], 1,164 tons, and "unwholesome," 144 tons)—a mere fraction of the whole quantity supplied to the market.

The meat condemned by the inspectors is chemically dealt with at the Corporation condemned meat sheds adjacent to the market, so as to prevent its being used as foodstuff, and it is then sold for commercial purposes, the proceeds being retained by the Corporation to cover the alleged expenses, or handed over to the owners, less certain charges.

The Corporation's Claims in 1904.

That the Corporation of the City of London keep and have kept a keen eye upon the developments of the frozen meat trade is evidenced by their having raised the question of tolls in a special way in 1904. In that year some excitement was caused in Smithfield frozen meat circles by the Markets' Committee's attempt to revise the Central Markets' constitution by introducing new by-laws. The Corporation of London claimed the right to collect tolls on meat sold at Smithfield although delivered direct from store without going on the market. Such a claim would practically involve payment of toll on every carcass imported into London and would be a tax on the food supplies of the metropolis. The Corporation also desired to make by-laws to control the operations of wholesale dealers and agents who transacted business with Smithfield salesmen. This point is dealt with in the following extract from one of the suggested by-laws :—" Any person not being a tenant of the markets who shall, in the markets or their approaches, hawk or offer for sale any goods, or solicit or tout for or take any order for any provisions or marketable commodities, shall, for every offence, forfeit and pay a sum not exceeding £5." What was meant by " approaches " is understood when it is mentioned that, under the old charter, the Corporation took the term to include a radius of seven miles from the Central Meat Markets. These claims were really a protest against the system which had grown up of delivering frozen meat " ex store " ; such meat, instead of, as formerly, being brought into the market, would be sold there by sample, on brand, or otherwise, and the bulk would be despatched from cold store. Probably, the Corporation in making these claims were actuated to some degree by the knowledge that not a jot of the enormous profits made occasionally in the transfer of Central Avenue stalls—for the " good will "—to American houses came its way. And as to the second part of its new demands, the Corporation objected to the Market being used as an open exchange, for buyers and sellers—not being tenants—to meet there and do business. It was necessary that the Board of Trade should grant permission for the suggested by-laws to be added to the Central Markets' constitution, and at this point the Frozen Meat Trade Association, after having, in conjunction with the Agents-General for Australasia and other trade associations, resisted the proposals, lodged with the Board of Trade a comprehensive statement of objections to certain of the by-laws. The Board of Trade was to have held an enquiry into the suggested by-laws, but the inquiry was postponed at the request of the Corporation in 1905, and has not taken place to the present time. The course adopted by the Association was successful in blocking the proposals of the Corporation in a general way, though some change was from that time made in the form of tenants' agreements to cover the Markets Committee's claims in this matter of tolls. One effect of the Corporation's claims to penalize persons (non-tenants) who attempted to "hawk" or "tout" in the markets was to cause firms who did the sort of business aimed at to establish themselves in offices round the market in West Smithfield. This had been going on for some time, and the possibility of cumulative £5 fines made firms get a stand outside the market. The whole matter has now narrowed down into arrangements between the Corporation and its tenants, and where the agreements (which vary in different parts of the Market) admit of such action, the Corporation exacts toll from tenants on all meat sold at their stalls whether delivered from the stalls or from outside cold store. The episode is interesting as a protest on the part of the Corporation of London against the decentralization of the chilled and frozen meat trade from the Central Markets.

Getting the Meat to Market.

Insulated vans convey the meat from the dock stores to the market. The cold stores begin loading up meat for Smithfield at about 10 p.m., and the vans reach the market about 3 a.m. the following morning. The salesmen and their staff arrive before 4 a.m., at which time the retailers begin to come along, all in a hurry to get their daily supplies for London's flesh-food needs.

There are several toll offices outside the market, and each van-load of meat is weighed, and the toll (2*s.* 3*d.* per ton) paid to the clerk, who gives a pass, which is handed to the policeman at the portals of the great Smithfield market. The meat is borne on the backs of stalwart market porters to the various "shops" or "stalls"; these porters are strenuous persons, and stand not upon the order of their going. To meet in full career a porter laden with a 200-lb. quarter of "hard" beef is to experience a "knock-out." Sample carcasses of the different qualities and weights are unclothed and hung up on the salesman's hooks, the bulk being stripped for examination by buyers when purchasing. In this way from one carcass to 200 carcasses are sold at a time, according to the requirements of retail buyers, who consist of suburban butchers, representatives of the large stores, Government contractors, restaurant keepers, etc. The retail buyers may also purchase joints, pieces or "oddments," and some of the Smithfield stallholders make a speciality of cutting up carcasses to oblige their customers. In the chilled beef trade "rumps" and "loins" are very commonly cut from the hindquarters for the West-end butcher. The buyer negotiates swiftly—there's no time to waste at Smithfield!—and if a bargain is struck, wooden skewers are stuck in the carcasses, which are forthwith again shouldered by the porters and carried to the waiting van of the buyer, or, perchance, to one of the two railways which run underneath the market.

Market Methods at Smithfield.

The keen purchaser frequently buys on the brand; he usually inspects the tag attached to the carcass to satisfy himself as to the "sub-marks" for quality and weight. The suburban butcher often likes to secure Canterbury brands, and sees that the ticket is left on the carcass so that he can, if necessary, show his customers that he is selling genuine Canterbury meat. For all comers the salesman is prepared; he will sell you a shoulder of mutton, a dozen kidneys, or 500 sheep, for he caters for "one and all."

The rule in settling in the Central Markets is a week's credit. Credit is given for longer terms in some rare cases, but the frequency of bad debts in the market of late years has limited this practice, and the salesman now pulls his man up very sharply if he fails to pay promptly.

It is only to be expected that the biggest meat mart in the world has complexities which would prevent the onlooker who cursorily regards its operations from getting to the bottom of its methods. Its salesmen are divided, roughly, into sections, and one division is according to the trade in which they are engaged. Though the tendency of late years has been for the Smithfield tenants to cultivate a mixed business, there are some who still confine themselves to the Scotch trade, some to American beef, and others to the frozen business. But the two major classes into which the salesmen are divided are the commission salesmen and the jobbers; there is a third—the direct importer.

The commission salesman receives consignments of meat from anybody and everybody, and his rate of commission varies. On the authority of an old tenant, "for the last forty years 2½ per cent. has been the standing commission for all meat except pork, which is sold at 1*s*. per pig under 12 stone (of 8 lbs.) and 1*d*. per stone above that weight. Poultry is supposed to be consigned at 5 per cent. commission, but in that trade special arrangements are frequently made." The commission salesman pure and simple is fast becoming a rarity in the markets, largely on account of the low prices of frozen meat; little can be made out of the customary 2 per cent., which rate has extensively prevailed notwithstanding what is mentioned just above.

The so-called commission man often indulges in a little speculation on his own account, and in so doing becomes a jobber. The jobber is a trader who buys for resale. He is also prepared to take goods on commission when it suits him; sometimes it does not. In cases of congested markets salesmen may decline to accept goods for sale; cases of 5 per cent. having been offered are on record, and sometimes the salesman may bargain for 1*d*. per stone commission. The jobbers go down early in the morning, and are at their shops between 3 and 4 a.m., before the butchers go to the market. The purchases of frozen meat from the importers are generally effected between 9 a.m. and noon of the day preceding that on which it is pitched on the market, this being the only time when the salesmen have the opportunity of discussing deals with the importers. The rates at which this business is done go a long way to determining the prices ruling next day, though, of course, the total supply and the total demand of the day overrule to some extent the question of the cost price of the meat. Smithfield Market, with its vastness and multiplicity of interests, is large enough to make a study of the influences that determine the market price of the day a subject as elusive and as deep as it is attractive to follow. As a rule, a jobber does not pay toll on any meat bought inside the market, the importer having paid it; but on meat, etc., bought outside the market the jobber has to pay toll, though it is necessary to qualify this by saying that the practice depends on the terms of purchase.

The Americans are doing a lot of harm to the jobber. Their operations make it difficult for him to exist, and the day may come when the jobber will be almost eliminated. He has to be a very smart man indeed, and has to watch the market with lynx-like eyes to secure his "turn." He is essentially a "spot" operator.

The term "direct trader" may serve to describe the position of the third class. The man who buys live stock (the "carcass man") and brings the carcasses to Smithfield for sale on his own account, and the c.i.f. buyer of frozen meat, fall into this category, as they get very near the producer. But the Smithfield tenants who form the majority of this division are the large firms of importers who distribute from Smithfield in a wholesale manner, or sell to their co-stallholders. The American houses and the companies in the Australasian and South American trade, although they may do at times commission and jobbing business, are a very distinctive class of Smithfield tenants, and one likely to increase.

Prices for the Day.

The meat prices vary from day to day and, in times of market excitement, from hour to hour. It is not easy to understand and explain how these rates come to be fixed in and accepted by a market of such enormous proportions and divergent interests. Retailers going from stall to stall could tell us that wonderfully little variation exists in the ideas as to the day's prices on the part of the salesmen. It used to be imagined in

New Zealand that prices for frozen meat were settled day by day by the Smithfield "ring"! The slightest knowledge of Smithfield's business shows that a "ring" cannot exist—it could were frozen meat a market by itself and in the hands of a few. But New Zealand frozen meat is only a small percentage of Smithfield's pitchings, and has to take its place alongside English, Scotch, American, Dutch, Australian, Argentine, and other meats. It is impossible to mention all the factors working automatically towards current meat prices. "We feel it in the air," said one of the salesmen, when asked how the market tenants arrived at the morning's prices. A few of the more obvious influences may be mentioned. First, of course, any change in the public demand for meat, favourable or unfavourable; retailers' buyings would be affected by this, and the weather has a curiously potent (not altogether sentimental) effect in causing the butcher to purchase briskly or cautiously. Then would come the controlling factor in the situation, the supplies of meat—not only frozen meat, but all sorts, for the different descriptions act and react upon one another, and the price movements of superior kinds strongly affect the whole market. The salesman knows roughly the quantities of meat warehoused in London, and the expected imports, and he learns from the carriers the bulk to be pitched on the market on a certain day. Argentine chilled beef, very sensitive as to value, powerfully influences market prices. The state of general trade in the country has its effect. The salesman, with all these currents and influences around him, weighs the general circumstances of the meat market and arrives at a rough idea of the marketing value of his meat. Of course there is much "come and go" in the business. If, for example, he considers that the factors promise well, he will test his buyers with a penny per stone rise on the market rates of the previous day. Finding his sales checked, he drops his price again an hour later, but if sales continue freely, he may try for a further advance. The same general principles, presumably, apply to Smithfield as to all other open markets and exchanges, the fluctuations in price being intensified in the case of meat by reason of its perishable nature.

CHAPTER XIV

REACHING THE CONSUMER

The retail sale of frozen meat puts the whole industry to the touch. The New Zealand farmer, the Argentine estanciero, the Queensland grazier, have in mind the fancies of the English meat-buying public in breeding their animals for freezing. The whole of this business, with its enormous invested capital and widely extended mercantile links, depends as to adequate financial return upon the favours of the British consumer. Engineers burn the midnight oil in designing new and economical machinery and plant for freezing works, ship, and cold store. Shipping and railway systems are changed especially to serve the needs of the transport of frozen meat; marketing and mercantile methods swiftly adapt themselves to its distribution. All these processes await their crucial test at the butcher's shop. Of course, the "butcher" is not now a butcher, but a meat retailer, and the frozen meat trade has brought about the change.

The retail meat trade in England is divided into two classes, the butcher who kills his animals in his own slaughterhouses, and the meat purveyor who purchases his stock-in-trade in the markets. The latter class, with which this chapter is concerned, is again split up into three sections. There is the "family trade" man who sells meat of the highest quality obtainable; he runs accounts, and has often to give long credit. The "mixed trade" man gives some credit, but his business is mostly on a cash basis. Then there is the "Colonial" or "cutting" trade—all cash business. The family trade shop makes its principal show at the beginning of the week, the mixed trade shop at the end; as for the "cutting" shop, there's no show at all. Immediately the frozen meat goes into stock, there it is cut up hard and bright.

In the mixed trade shops home-killed, American, and frozen beef and mutton are sold, as well as small quantities of pork and veal. A small "cutting" shop is generally closed on Monday, as its daily supplies are shut off by the occurrence of Sunday, and its customers seldom require fresh meat on the following day. Frozen mutton and lamb, and forequarter beef, suet, etc., are the goods dealt in: at such a shop about fifteen sheep, five lambs, and three fores of beef will be handled on Saturday. These shops are mostly met with in provincial towns, and the majority are run in the names of, or are "tied" to, the great shop-owning companies. A word in passing here: the family butcher to a great extent has found out that he must keep frozen lamb (New Zealand) in stock, and also, very frequently, New Zealand mutton. If he does not, his customers go to the retailer who does. It is quite a recognized thing for people who would not dream of buying frozen meat in general to purchase Canterbury lamb. Possibly the feelings of these worthy conservatives are soothed by the magic of "Canterbury." There may be recorded the well-authenticated statement that many innocents buy Canterbury (New Zealand) meat sincerely believing that it comes from the Romney Marshes of Old England. The proportion of Canterbury lamb handled in the high-class shops increases when home mutton is dear, and decreases when home mutton is cheap.

The Producer and Retail Business.

It has been the dream of many of the producers in New Zealand and Australia to run their own shops in England, to hold their meat in their own hands right through the whole cycle, and not to surrender the ultimate link to the English meat purveyor. In New Zealand the producer—not the freezing company—has been very keen on this; not altogether because of the retailing profit. He has long been convinced of the excellence of the meat he sends to England, and has naturally resented the good deal of improper substitution that in the past has taken place in the retail vending of frozen meat in various parts of the Old Country. Royal Commissions have showed him that frozen mutton has been sold as home-produced, of which he surely, as well as the English farmer, has a right to complain, and police court prosecutions have revealed the practice of inferior frozen meat being palmed off as "Canterbury" or "New Zealand." The New Zealand grower has desired above all that the article he produces shall not lose its identity when it reaches the shop stage, and he bitterly resents the fraudulent use of the description "New Zealand" in the selling of frozen meat. But the improvement that has taken place of late in the quality of South American and Australian mutton tends to lessen the grounds for the New Zealander's complaint in this connection. According to the oft-stated opinions of all practical men in the meat trade of Great Britain, New Zealand mutton has steadily deteriorated in quality in recent years, chiefly because New Zealand has decided to develop her lamb trade at the expense of mutton. The keen demand for lambs for freezing has tempted farmers to ship many of their most promising young animals instead of keeping them for breeding stock.

Schemes, of which there have been many, have in the past been framed mainly in order to give shippers the benefit of retail profits. To give an instance, there was a very elaborate draft scheme, dated 1887, for selling New Zealand mutton and lamb in twenty shops, to be opened in fifteen of the most important centres of Great Britain. It was proposed to invest £10,000 in the venture, and to turn over 2,000 carcasses a week. A respectable profit was counted upon, but the scheme, intended to conserve the retail selling returns for the New Zealand interests, never got beyond paper.

With regard to proposals in this direction that were carried to completion, a bold move was made in 1899 by the Christchurch Meat Co. in inducing Mr. H. Woodley to open up at Queen Street, Cardiff, a shop for the sale of choicest Canterbury mutton and lamb. The shop was fitted up in an attractive style, with agricultural scenes pictured on the tiled walls. This enterprise was a joint affair between the company and Mr. Woodley, but after a few years the business was taken over entirely by the latter. The venture was not intended in any way to compete with the company's c.i.f. buyers, but, on the contrary, was intended to assist them by advertising the choicest qualities of New Zealand meat. This action of the Christchurch Meat Co. in giving up their interest in the Cardiff shop showed that the directors came to the conclusion that it was not their business to run the retail trade, and no doubt this is the position wisely taken up by the managers of the Australasian meat works in general. Another venture connected with the retail sale of the highest qualities of New Zealand meat and other produce was that of Mr. H. C. Cameron, who opened the New Zealand Produce Stores in Manchester in 1894. He brought New Zealand meat successfully before classes of Manchester consumers who had been accustomed hitherto to see lower grades of frozen meat vended in by no means attractive shops. This business was taken over in 1898 by Messrs. W. and R. Fletcher, Ltd.

An ambitious scheme was propounded in 1903, when the British New Zealand Meat and Produce Co., Ltd., was formed in New Zealand with a share capital of £150,000. Quoting from the prospectus, this company was formed " for the purpose of supplying direct from the producer in New Zealand to the consumer in Great Britain and elsewhere New Zealand meat, butter, cheese, and other produce." Many leading men connected with the meat export industry helped forward the founding of the company, which had the blessing of Mr. Seddon, Prime Minister. A prominent feature of the proposals was the adoption of a defrosting process. The capital was reduced to £50,000, of which sum something under £20,000 was paid up. When business was opened in London, four suburban shops were taken. It was discovered, however, that there were too many butchers in London to allow of the original scheme being carried out with success, and at the present time the company's retail business is confined to only one shop—at Finsbury Park. The company has, however, settled down to ordinary importers' business, and now, with a wholesale stall in Smithfield Market Annexe, conducts a profitable trade. The first dividend, 6 per cent., was paid in 1908, no dividends having been earned on the retailing business.

The idea that money could be made by meat producers taking over the retail selling of their own frozen meat has been very widely held in the past. About eight years ago particulars were forthcoming concerning a bold venture in this direction. A company was to be formed with a capital of £150,000, £100,000 of which was to be devoted to the purchase of 400 retail meat shops, £20,000 being for preliminary expenses, and £20,000 for working capital. It was estimated that 10,000 sheep and 2,220 cwts. of beef would be turned over weekly, say, 431,600 cwts. of meat per annum. The cost price of meat was put at 3*d.* per lb. all round, and selling at 4*d.* per lb. left an annual gross profit of over £200,000; net profit close on £30,000. The author of the prospectus, a practical meat man, pointed out that this sum paid a handsome dividend on capital and left a substantial surplus. The principle of business adopted by the great London stores was to be introduced. Close attention was to be given to the utilization of the inferior parts of the carcass by having them sent to the company's shops in poor districts. The promoter wrote: "Our line of policy will be to transfer the 'goodwill' of local shops from the local shopman to the company." The estimates and the plan of campaign in this case were all worked out most closely. What one had to take for granted, it may be presumed, in order to arrive at the £30,000 a year net profit, was that 400 suitable shops could be acquired and manned, and the whole revolutionary system of management set going as smoothly as a clock is wound up. Nothing came of the scheme, but its details are mentioned here, as they possess features of interest.

Of course, one of the recognized difficulties of running multiple meat shops is the unreliability of the employees, especially in the case of businesses of a casual kind. However, the failure of all attempts on the part of New Zealand and Australian meat producers themselves to carry out the retailing of their meat is due to a deeper cause, and can only be attributed to the fact that a good farmer makes a bad tradesman, even as, of course, a good tradesman would make a bad farmer. English and Scotch farmers for the past fifty years have been possessed by the same idea as the New Zealand producers, and every one of the innumerable attempts on their part to run their own retail shops has ended in dismal failure.

Lamb versus Mutton.

Too much stress cannot be placed on the part which New Zealand lamb has played in attracting a better class of customers; frozen meat in general has been popularized extensively by this means. At times retailers have found it profitable to push frozen lamb against home-bred mutton; it was not a difficult task to convince the public of the superior eating quality of the lamb. Restaurants have taken up this practice, finding it a good investment to put New Zealand lamb joints on their tables in place of the more expensive English mutton. But the lamb often appears as "mutton" on the bill! This policy of the retailers, only profitable when the wholesale prices of frozen lamb do not exceed *5d.* per lb., acts prejudicially on the frozen mutton trade, and the depression which has been felt in this section for years past is, doubtless, accounted for partly by the increasing lamb vogue in the shop and the home.

Attention may be drawn to the statement made so frequently of late by writers in trade papers that mutton is becoming "unpopular" in Great Britain. This is, probably, incorrect. That lamb is displacing home-bred and imported mutton to some degree, as noted above, is evident from an examination of the statistics. In the years 1905, 1906, and 1907, imports of frozen lambs increased splendidly. Taking receipts in the United Kingdom from all quarters, the increases in the years mentioned were, respectively, 592,700, 648,000, and 668,000. In the same years the frozen mutton increases were nothing like so considerable. With Argentina now exporting largely, it is plain that frozen lamb is encroaching upon frozen mutton. If consumers find lamb in a way forced upon them, their demand for mutton must lessen; mutton then becomes "dull" in the wholesale marts. And the heavy supplies of English and Scotch mutton have brought down prices for home-bred. But none of these tendencies need convince one that the public are losing their taste for juicy and tender mutton, which is far more popular in the British Isles than on the Continent of Europe, where, generally, mutton is not fancied.

One reason why frozen mutton has not made its way so thoroughly and successfully into the better-class houses is, probably, because to a great extent it is cooked without sufficient hanging. Frozen meat has no chance at all unless the frost has been entirely removed from the carcass or the joint. Frequently the meat which was in the cold store on Friday morning is in the citizen's oven on Sunday; the joint is bound to be tough. One cannot be surprised at the dislike for frozen meat which would follow the consumption of such an indigestible joint! Many retailers thaw out the meat properly by various means, but some do not. The matter is one well worth taking in hand with a view to the education of the public. Unfortunately, when frozen meat has been well thawed out and has been hung till it has become ripe, it is apt to be unsightly in appearance. On account of its excellent eating qualities, frozen lamb, therefore, commands a special field in the retail vending; neither frozen mutton nor frozen beef approaches it.

It must be confessed that frozen mutton is not as reliable an article as frozen lamb, and one of the reasons why the former has not become as popular with the public as lamb is because of the shipment of secondary and inferior grades, and of ewe mutton sent from Australia and New Zealand and now beginning to arrive from South America. New Zealand is losing her mutton trade in Great Britain to some degree. "Plate" mutton, bright and of excellent quality, and arriving regularly, is getting to be preferred by the retailer, and threatens to command the markets. It is maintained by the English meat merchants and experienced market men that frozen mutton is somewhat lacking in flavour and tenderness, and that shippers should be content with their mutton occupying a secondary place in the markets of Great Britain; they say that the intrinsic merits of the meat are indicated by the price which it fetches on the wholesale markets. They dismiss the idea of "prejudice," pointing out that the word is outworn and ridiculous when Great Britain uses over 10,000,000 frozen carcasses annually and asks for more. But there can be no question that for the first ten to twenty years after the *Strathleven* landed her cargo the force of prejudice acted as a considerable deterrent to frozen meat enjoying the degree of popularity to which its good qualities and cheapness entitled it, prejudice manifested in the servants' hall—at the instigation of the butchers—and the suburban snobbishness because frozen meat was not considered the "proper thing."

The Retailing of Beef.

Speaking of the average frozen meat shop, pure and simple, in England, it is obvious that its style and fittings generally are capable of much improvement. Many visitors from Australasia and other countries remark about this. Something is wanted to make the shops brighter and more attractive.

With regard to frozen beef, the retailer has not found this a very popular class of meat in his trade. That, at any rate, is what he says. Probably, the far better appearance of the chilled beef from North and South America has created some prejudice against the frozen quarter; the dampness in the air condensing upon the exposed surface of 200 lbs. of beef causes the rather unpleasant phenomenon popularly styled "weeping." These things, however, are but externals, and no one questions the intrinsically good quality of the frozen beef from New Zealand and Australia and South America.

Frozen beef is mainly retailed at the cutting shops, though the better parts find their way into the "mixed" trade. The cheap rates at which this beef has been wholesaled must have made it an extremely useful article for the retailer, and the public who have bought the enormous quantities imported have also obtained the maximum of nourishment at minimum prices.

With the coming of the Argentine chilled beef about the beginning of the nineties, the working of the frozen meat shops became immediately easier. Never was any development in the food import trade more welcome! Though the North American supplies of refrigerated beef were a necessity to the meat trade in this country, the suppliers themselves were unpopular, and it was with great joy that wholesale and retail meat men in Great Britain saw plentiful quantities of chilled beef arriving from South America. Chilled beef from the United States of America is too high in price for the cheaper shop to cut, but the retailer is able to sell good joints of Argentine chilled at moderate prices and work this meat conveniently with his frozen mutton and lamb trade. It is stated that frozen beef has come more into favour since the River Plate has been such a large shipper; this is probably owing to the regularity of shipments and steady annual increase which have marked the exports.

Counting up the Retailers.

A useful conclusion to this chapter will be some figures to give an idea of the numbers of shops in Great Britain—we may safely exclude Ireland—at which the meat produced at home, and on farm, station, and estancia overseas is sold. There are 24,000 retail butchers in the United Kingdom. Then there are the stores—there are about 1,500 in the United Kingdom—and the provision and grocers' shops which also sell frozen meat, as well as the stalls in streets and markets and in country towns on market days, and also the humble coster, who must not be forgotten. Nothing in the shape of an accurate calculation can be made, but one may suggest that there are not less than 100,000 establishments in the United Kingdom at which fresh meat is vended. One has it on the authority of Mr. Heap, president of the National Federation of Meat Traders' Associations, that 80 per cent. of imported meat is sold by firms who sell nothing else. A list of the multiple shop concerns handling frozen and chilled meat only, with the number of shops, was prepared for the purposes of this book, but it was found impossible to ensure anything like accuracy. Brief allusion to this branch of the subject may, however, be made in stating that Messrs. James Nelson and Sons, Ltd., run about 1,500 shops in the United Kingdom;

Eastmans, Ltd., about 1,400; the River Plate Fresh Meat Co., Ltd., over 400; W. and R. Fletcher, Ltd., 417; and the London Central Meat Co., Ltd., over 500; these being the largest concerns.

Considering that every town of about 5,000 inhabitants and upwards has its frozen meat shop or shops, and that there are the mixed and family shops to account for which handle the remaining 20 per cent. of imported meat, it cannot be unreasonable to suggest that there must be at least 20,000 shops, stores, etc., cutting frozen meat.

To understand how frozen meat has captured the retail meat trade of England, one has but to study the advertisements in the *Meat Trades Journal*. The great majority of the company shops, as above, are in the populous centres of the north; but Eastmans' are well distributed over the country; Messrs. James Nelson and Sons have about 250 shops in London and suburbs, and Messrs. Fletchers' are mostly in the Midlands and South and West of England. It is interesting to note the turnover of one of the large shop companies' business; the figures appear colossal. One learns from particulars recently published that Messrs. W. and R. Fletcher, Ltd., turned over £1,482,000 in 1910. Eastmans, Ltd., have cold stores in the United Kingdom capable of holding 350,000 carcasses of mutton. The big turnover of the multiple shop owner is done on a very small margin, sometimes as low as ½ per cent., and the business, to admit of success, demands a keen expert knowledge of markets and men. Most of the London suburban retailers sell frozen meat. The restaurants catering for the clerks supply it to their customers freely.

As the multiple shop system has been, and is still more likely to be, such an important factor in the retailing of frozen meat, it is interesting before closing this chapter to take a glance at the development of a concern which was a pioneer in this branch of the trade. This is Eastmans, Ltd., which was formed in January, 1889, with a capital of £900,000, to acquire the cattle and fresh meat business of Messrs. T. C. and Joseph Eastman, of New York, and Messrs. John Bell and Sons, Ltd., of London and Glasgow, the latter concern being the multiple frozen meat shop pioneer in Great Britain. The first directors of Eastmans, Ltd., were Lord Greville, Messrs. George Scheibler, H. Scott Ritchie, Russell Monro, Henry Bell, James Bell, and James John Thomson, managing director. The business of Messrs. Bell was started in 1827, and registered in 1888 as a limited liability company, the whole of the shares being held by members of the firm and their managers—their turnover from 1878 to 1888 was over £17,000,000. It was in 1879 that Messrs. Bell began to open up meat shops in Great Britain, and at the time of the amalgamation they had 330 shops in the British Isles. In 1900 the whole of the American business was discontinued, and the property sold, and from that time onward Eastmans, Ltd., have devoted themselves to their shop trade, in supplement to which they have a wholesale Smithfield connection. The 330 shops in 1889 have now increased to over 1,400; the company have cold store depôts at London, Glasgow, Dublin, Liverpool (two), Manchester, Leeds, Newcastle, Bristol, Chatham, and Sheerness, capable of holding 350,000 carcasses of mutton. Eastmans, Ltd., are very extensive buyers of New Zealand mutton and lamb, and they handle very considerable quantities of frozen and chilled beef from all parts of the world; they have a staff of 4,000 employees, many of whom are shareholders. The issued capital is now £1,126,490. Mr. J. J. Thomson is chairman and managing director.

References to James Nelson and Sons, the retail meat shops of which house, scattered up and down Great Britain, form the great rivals of Eastmans, Ltd., appear on p. 80.

Mention, too, may be made of the part which Messrs. John Rose and Co. took in the retail distribution of Canterbury mutton and lamb. In this department of the trade they ranked as early pioneers (see p. 382).

CHAPTER XV

PROVINCIAL DISTRIBUTION

As the successful marketing of frozen meat depends largely upon careful handling and efficient and rapid distribution, it will be well understood that the comparative merits of ports of destination for frozen meat shipped to Great Britain have always been the subject of considerable controversy. London, with its huge population, its pre-eminent railway facilities, and, generally, its overwhelming importance as compared with other centres, has always occupied first place as a centre of distribution for frozen meat. Twenty-one years ago out of a total import into the United Kingdom of 3,358,823 carcasses of frozen mutton and lamb, London received 2,389,129 ; eleven years ago out of 7,094,782 it took 4,770,801 ; and in 1910 it took 8,572,788 out of a total import for the United Kingdom of 12,981,044 carcasses. So in 1891 London was credited with 71 per cent. of the total importations, in 1901 with 67 per cent., and in 1910 with 66 per cent.

If minimum handling were the only consideration in getting frozen meat to the consumer, it is probable that direct shipment to the various chief ports round our coast nearest to the districts of population would be a method difficult to argue against, and with certain of our large provincial ports this trade has developed to a very large extent in the last few years.

Quite early in the frozen meat campaign, in 1886, vessels from the River Plate were directed to Liverpool, which port quickly became the chief distributing centre for Argentine mutton. In 1902 over 2,000,000 carcasses were landed there, against 412,000 at London, but of late South America has largely increased her London landings. Argentine meat is landed at Cardiff, Manchester, Bristol, Glasgow, Southampton, Hull, and Newcastle, as well as at Liverpool, but the Mersey port has been the great distributive headquarters for the north of England of the importers.

Had Australia and New Zealand—especially Australia—been able to arrange for the regular despatch of vessels to Liverpool and other ports in the early days of the trade, no doubt importers of Australasian meat would have put up a fight with the Plate companies in those markets. But the Australasian shipping services made London their destination, and refrigerated and other produce from those Colonies has, in consequence, been despatched to, and largely handled in, London. In 1892, however, arrangements were made for Australian and New Zealand mutton and lamb to be consigned direct to Liverpool, and from that year to the present time the "outports" have been used by Australasian shippers. The volume of business was, however, quite small till 1904, when the "West Coast" steamer service was started from New Zealand to Liverpool, Cardiff, Avonmouth, Glasgow, and Manchester. In 1907 and 1908 about a million carcasses yearly were shipped direct to those ports from Australia and New Zealand, and by 1910 the totals had grown to over 2,000,000. According to the statistics published by the London Central Markets Committee, during the twenty-nine years, 1880 to 1908, 2,123,839 tons of frozen meat have been imported into Great Britain from Australasia, and of this total 1,533,777 tons, 72·2 per cent., have passed through Smithfield market. This percentage is now, however, lessening year by year. Probably, increasing quantities of Australian meat will be shipped to the provincial ports, but London will always remain the most favourable centre for selling New Zealand meat and the higher grades of frozen mutton and lambs from South America and Australia.

The question of the comparative advantages in shipping to the various ports could be discussed at great length, but as this would be to little purpose here, the discussion need not be entered upon further than to say that it would seem that London and Liverpool are fated long to remain the chief ports for frozen meat; they possess the population and means of distributing to an unequalled degree. There is noted in another part of this book the tendency of River Plate meat since 1905 to go in increasing proportion to London. Whilst shippers in New Zealand and Australia have been hankering after direct shipments to country ports, the Argentine people have quietly directed more and more of their chilled and frozen meats to London. London is the market for the highest quality, especially chilled beef; so South America ships most of her chilled beef to Southampton and London, and mutton tends to follow. The weakness of country markets lies in the fact that they can be so easily overstocked; over and over again importers of frozen meat have found this out to their cost.

It remains to give some idea of the facilities for frozen meat possessed by the various leading ports that compete for this trade. The improved organization of the West Coast service bringing refrigerated produce from Australasia during the last few years has put the Liverpool, Cardiff, Manchester, Glasgow, and Bristol (Avonmouth Docks) markets into more direct touch with the farmer in New Zealand and Australia. This service has now been organized in a thorough manner. Route No. 1, via Suez Canal, is Australia to London, Liverpool, and Avonmouth; No. 2, via the Cape, is New Zealand to Avonmouth, Liverpool, and Glasgow; and No. 3, via Torres Straits and Suez Canal, is Queensland ports to Liverpool and/or London. There are twelve sailings in the year in each route on the homeward journey. The West Coast ports have many conveniences, and shippers are taking advantage of the facilities they offer—non-lightering, handling and railage saved, cheap harbour and landing charges, etc. But the merchants in the cities named have failed to grasp that the frozen meat trade is being, and will in the future increasingly be, done on the selling forward basis. The days of chance or open consignment are waning, and if the provision brokers and merchants at these "outports" desire to compete with London, they will probably find out ere long that they must take up the business in this way.

Liverpool.

The claims of Liverpool as a meat centre require no support: it is the second port of the United Kingdom, and its shipping, cold stores, cattle lairages, markets, and railways comprise a vast and excellent receiving and distributing system. The leading lines of foodstuffs brought to Liverpool in twelve months total about £45,000,000 in value. In 1910, 822,025 quarters of frozen beef from South America were landed there, while in 1890 only about 5,000 quarters reached the port. In 1910 also, 545,642 chilled quarters from South America were imported, and 246,728 lamb carcasses from the same source. The lambs imported into Liverpool from Australia and New Zealand in 1910 numbered 411,132 and 334,341 carcasses respectively, and mutton carcasses 1,166,174 and 37,003 respectively, while frozen beef from the Antipodes was imported to the extent of 186,224 quarters from Australia and 40,683 from New Zealand.

Liverpool is well provided with cold storage accommodation, its seventeen stores having a total capacity of more than two million carcasses of mutton. A glance at the map will show that eight of these cold stores are situated either at, or in handy access to, the line of docks for which the port is famous.

As regards the actual landing facilities for frozen meat at Liverpool, most of the steamers carrying frozen meat discharge at the North end of the docks, and this is most suitable for the consignees of the meat, owing to the close promixity of the principal cold stores to the Canada, Brocklebank, Langton, and Alexandra Docks, where unloading usually takes place. The meat is discharged from the steamer's hold by means of large canvas slings on to the quay, under a covered shed, where it is sorted and delivered according to mark. During the last year or eighteen months a considerable number of insulated vans have been built by cartage contractors for the purpose of conveying the meat from the ship's side to cold stores or railway depôt, a distance of only from about half a mile to a mile. The handling of the meat on the quay is undertaken by the master porter. He, in many instances, is only a nominee of the steamship companies, which arrange their own discharge and employ their own men on the quay to deliver to consignees.

At several—but not all—of the discharging berths allocated to frozen meat steamers, railway lines have been laid down by the Mersey Docks and Harbour Board to enable refrigerator cars to be run alongside for the loading of the meat direct from the steamer. The meat is discharged on to trucks, and these are run across the Dock Board's shed to the railway line, where the meat is then loaded into refrigerator cars. If meat, however, intended for despatch to provincial markets cannot be loaded in railway vans alongside the steamer, it is usually carted in street wagons to the nearest railway depôt or cold store, and loaded up there in ordinary or insulated railway vans, according to destination.

A few railway rates in force from Liverpool to various consuming centres will illustrate the Mersey Port's position in this connection. There is a 2-ton rate of 22*s.* 6*d.* per ton, delivered, to Bradford, Leeds, and Sheffield, and 11*s.* 8*d.* (station to station) to Manchester. The 4-ton (owner's risk rate) to Glasgow is £1 per ton, while the 5-ton rate, delivered, to such towns as Newcastle-on-Tyne, North and South Shields, and Sunderland, is 32*s.* 6*d.* per ton. The steamer rate to Belfast is 15*s.* and to Dublin 22*s.* 11*d.* per ton. The railage to Birmingham is 25*s.* per ton for 3-ton lots.

Manchester.

With the completion and opening in 1894 of the Manchester Ship Canal, and the erection of cold stores in Manchester and on the Canal, enthusiastic efforts were made to establish a direct trade between Australia and that city. A beginning was made in 1895, when the sailer *Timaru* took 16,000 carcasses from Geelong, Australia, direct to Manchester. The meat was sold c.i.f. at $2\frac{5}{8}$*d.* per lb. to local buyers, and opened out in splendid condition. However, the River Plate importers, with the avowed purpose of discouraging further direct imports from Australia, lowered their price for mutton to $2\frac{1}{2}d.$ and under—thus causing the Manchester buyers of the Australian meat to make a heavy loss. Though the claims of Manchester as a suitable port for the direct import of Australasian frozen meat have been advanced with great persistency, the prospect has not been inviting enough to induce the shipper, the shipping companies, and the mercantile interests in England, to land Australian or New Zealand meat on the Ship Canal wharves, except to an insignificant extent.

The Port of Manchester, for port it must be called, with its Canal facilities, vies with London as a centre of population, as within a twenty mile radius of "Cottonopolis" there is a population of nearly eight millions. This centre, too, has special cold storage equipment which is worthy of note. On the Ship Canal is a large refrigerated transit chamber, for the sorting of frozen meat and produce before delivery to railway wagons or carts. This chamber has a capacity of 85,500 cubic feet, and is certainly an excellent provision, constituting that feature of cold storage equipment which has been advocated far and wide by those who study the ideal method of handling big lines of frozen produce imports.

There are also at Manchester good facilities for storing frozen produce. At Weaste, on the Ship Canal, there is the large store of the Colonial Consignment and Distributing Co., now leased and managed by the Union Cold Storage Co. Its capacity is 175,000 carcasses, and alongside large steamers can be berthed and discharged. The Union Co. also has cold stores in Miller Street, Manchester, with a capacity of 80,000 carcasses; the Manchester Corporation has within easy cartage distance from the docks cold storage accommodation for 150,000 carcasses of sheep alongside its extensive market and abattoirs. The Corporation abattoirs and meat market, together with their cold store, are second only to Smithfield, London, and the daily hanging of meat is greater than in any market in the Kingdom excepting Smithfield. The municipal cold store is situated conveniently for the various markets of the city. At the Foreign Animals Wharf and Lairages at the Manchester Docks there is accommodation for 1,850 cattle and 1,500 sheep, with room for extension. The weekly supply of cattle at these lairages from Canadian and North American ports brings butchers from the suburbs of Manchester, who in addition to purchasing fresh beef also obtain their supplies of frozen mutton and lambs, and thus avoid going to Liverpool and Birkenhead.

The Ship Canal Co.'s efficient system of railway lines all round its quays is in direct communication with all the principal railways, and the Canal Co. takes charge of and forwards produce at inclusive through rates.

Cardiff.

This leading port and distributing centre on the Bristol Channel has been closely identified with the frozen meat trade for many years, and although direct shipments from Australasia by the West Coast service are not as numerous as the advantages offered by the port and district seem to suggest they might be, the Argentine shippers take care in distributing their meat to send a large quantity to Cardiff. The first shipment of frozen meat landed in the Bristol Channel reached Cardiff about 1889. It was shipped by the River Plate Fresh Meat Company.

In 1910 Cardiff came next in importance to Liverpool and London amongst the nine ports in the United Kingdom to which South American meat was despatched. As to the various facilities for receiving and handling frozen meat, Cardiff is well to the fore. Its docks are excellent, and railways run alongside. Cold storage space is ample, Cardiff itself possessing, as is shown in Appendix V., about half a million carcasses capacity, an accommodation about to be added to by the Cardiff Railway Co., which has decided to spend £40,000 on another store at the docks. Cardiff, the chief of the three fine ports on the seaboard of the Bristol Channel—Cardiff, Swansea, and Newport—is likely to have a brilliant future as a centre for the distribution of frozen meat and other refrigerated produce, especially if it can hold its own against Avonmouth—its aggressive corporation-owned rival farther up the Channel. A considerable business has been done in the city and district of Cardiff in frozen meat. In Cardiff alone there are about 200 shops dealing exclusively with this article, and one firm of importers, Messrs. H. Woodley and Co., who run a number of retail meat establishments, may be mentioned as being identified with the trade from the eighties to the present time. There is a call for an immense quantity of frozen meat, as well as rabbits and butter, in the Cardiff district, and regular direct shipments would expand the trade.

The Cardiff district includes the densely populated coalfields of South Wales, which are, as far as provisions are concerned, absolutely non-producing. Barry is included in the port of Cardiff, and has large up-to-date cold stores alongside the dock; Newport, twelve miles off, also has large cold stores close to the railway stations. In both of these towns the public cold stores are owned by the Cardiff Pure Ice and Cold Storage Co., Ltd., which has very extensive refrigerated accommodation, consisting of 394,000 carcasses in Cardiff, 100,000 in Barry, and 41,482 in Newport. The managers, Messrs. Neale and West, have been associated with the development of the direct imports of Argentine and Australasian frozen meat ever since these began, over twenty years ago. Cardiff is a fine centre for the distribution of refrigerated provisions of all sorts, and Australasian exports of this character might with advantage be sent direct to that port to a greater extent than is now the case.

Bristol.

Bristol, at the Avonmouth Docks, has provided considerable up-to-date facilities for the import of frozen meat. Its first supply of cold storage was, as a matter of fact, to accommodate the Canadian provision trade, for which a small store was erected in 1896. It was in 1904 that the Federal-Houlder-Shire boats began to call at Bristol and land some of their New Zealand frozen mutton and lamb at Avonmouth; and as the trade proved too large for the existing store another of similar capacity was provided by insulating a portion of the upper floor of a transit shed on the west side of the Dock. In 1907 the importation of frozen meat into Avonmouth began from Australia, and direct shipments from the River Plate in 1911 have helped to make the Bristol Docks Committee decide upon doubling its cold storage accommodation.

The method of discharge of frozen meat at Avonmouth is as follows. The meat is lifted direct from the hold of the vessel and delivered into the reception air-locked chambers, six of which serve the cold store. These air locks are not much used for sorting purposes; the meat in the majority of cases as it comes from the steamer is sorted as it passes through the small door in the end of the store. The air locks are useful, however, for sorting purposes if the stores are crowded, and in addition it is possible in cold weather to make use of the air locks as a storage chamber. The cold store itself is alongside the steamer's berth, so that exposure of the meat between the ship's hold and the cold store is reduced to a minimum by the method of working referred to. Deliveries from cold store are made by shutes direct into refrigerator railway cars, which are loaded under cover.

Glasgow.

The imports of frozen meat at the port of Glasgow during 1911 showed considerable development, and the prospects at this "outport" for shippers in the Southern Hemisphere are quite favourable. It is true that quantities of frozen meat landed at Glasgow prior to and during 1910 were inconsiderable, but the shipping facilities were limited. Glasgow is the centre of the Scotch meat trade; it has a convenient wholesale dead meat market, and a large quantity of Irish meat is handled there. Cold stores and landing and discharging arrangements are available for carrying on the imported frozen and chilled meat trade, a considerable increase in which —especially from Australasia—may be anticipated. (The discharging facilities, by the way, would bear improvement.) Appendix V. shows Glasgow to possess cold storage accommodation equal to 605,000 carcasses, of which more than half is at the stores of Messrs. William Milne, Ltd. (350,000), and

150,000 at the stores of the Union Cold Storage Co. Scotland has, until recent years, been very backward in taking to frozen meats; considering the production of the excellent beef and mutton which the graziers of North Britain supply to the markets, this is not surprising. But the Scotch people in the cities and towns are now becoming customers for frozen meat. The Glasgow market dealt with New Zealand boneless beef for collops until this trade was practically stopped by the Public Health Acts referred to at page 122, but a promising factor in the development of direct shipments to this port is the fact that Glasgow has excellent facilities for absorbing large quantities of frozen beef of good secondary grade to take the place of the said boneless beef, which, in large quantities, was formerly placed—mainly by the North American shippers—on the Glasgow market.

Hull.

Hull as a produce centre has aspirations far beyond the position it holds as a receiving port for the European shipments of eggs, poultry, game, butter, etc., coming to England. It is anxious for more frozen meat imports, stating that owing to its railway distributing facilities it is one of the cheapest centres of distribution for one quarter of the population of the United Kingdom inhabiting about a fifth of its area. This is a big claim, and as yet frozen meat does not bulk largely among Hull's imports. The Humber port possesses four cold stores, one, of 54,000 carcasses capacity, leased by the Compañia Sansinena, and situated at the Alexandra Dock; a town cold store, of 40,000 carcasses capacity, owned by the Union Cold Storage Co., Ltd.; another store run by the same company is in the Sir William Wright Dock, of 200,000 carcasses capacity, and one at the Alexandra Dock, of 50,000 carcasses capacity, also belongs to the Union Cold Storage Co. Frozen and chilled meat is brought from the Argentine for the Sansinena Co. by the Houlder liners, and for the River Plate Fresh Meat Co. by the Royal Mail Steam Packet Co.'s vessels. The Federal-Houlder-Shire Line now books refrigerated cargo from Australia to Hull under direct bill of lading, certain boats in that service making Hull a port of call. The completion of the Hull Joint Dock, which will be owned shortly by the Hull and Barnsley and North Eastern Railway Cos., will give Hull the largest dock in the world, as its water area is to be 86 acres.

The Country Trade: General Observations.

In addition to the frozen meat shipments going direct to the "outports," considerable quantities landed by steamer in London are railed direct to the great marketing centres, where there are ample storage and sale facilities. Meat so railed from London docks or store goes, as a rule, right to the retailer, the goods which are stored in the provincial ports usually being meat shipped direct to those ports. All classes of frozen meats are more or less saleable in the Provinces, although particular descriptions are required for particular districts. Generally speaking, the country trade calls for lighter weight carcasses of mutton and lamb and quarters of beef than that required by London buyers. But fat carcasses find favour in agricultural districts.

Practically the whole of England is carefully worked by commercial travellers on behalf of the great distributing houses. If any Australians or New Zealanders think, because the great bulk of their meat is landed at London, that the forty millions of people outside London are neglected, they fall into a deplorable error. The whole of England, much of Scotland, and all Wales—Ireland has yet to be converted—have been opened up to frozen meat (see the information given in the preceding chapter). The country trade was first tapped by the distributing firms handling Australasian meat—Nelson Brothers, Borthwick's, etc.—before the Argentine meat was good enough for the country trade. Depôts were opened at places like Plymouth and Middlesbrough, which experience has since shown were not good distributing centres. The retailers, then, in the country are supplied by the distributing firms in London, per railway, or they get their frozen meat from Liverpool, Bristol, Cardiff, Glasgow, Hull, or other ports to which it has been conveyed by steamer. Provincial operators are closely in touch with the metropolitan market, and wholesale values in London and country are practically the same. Railway rates for frozen meat from London to the various centres depend to some extent upon the quantities forwarded. The rates on 3-ton lots are as follow: to Cardiff, 25*s.*; Bristol, 20*s.*; Liverpool, 25*s.*; Glasgow, 53*s.* 4*d.* under 20 tons, and 45*s.* over 20 tons. For England ⅛*d.* per lb. covers this item, roughly speaking; for Scotland, ¼*d.* per lb.

By firms owning numerous retail shops very large quantities of frozen meat are distributed without passing through the open markets, and this may be said generally of all kinds of frozen meat handled in English country parts. The Argentine people, who exploited the provinces from Liverpool as their headquarters twenty-five years ago, were until recently much more at home there than the firms distributing Australasian meat. The former have their offices, wholesale and retail, on the spot, their arrangements are cut and dried, and there is a regular all-the-year-round demand. Country markets are very easily over-supplied and cannot absorb (as can Smithfield) unlimited quantities of meat, even at low prices. All classes and qualities of meat are distributed throughout the country; London and the South Coast take the highest quality. Price lists are issued weekly by the Colonial Consignment and Distributing Co., Ltd., Thomas Borthwick and Sons, Ltd., W. and R. Fletcher, Ltd., Towers and Co., Ltd., and others. These houses issue pamphlets, leaflets, and illustrated circulars, got up in a very attractive style.

Turning to the question of direct shipments from the Southern Hemisphere, this notice of the country trade may be concluded with a few lines giving a sketchy idea of the characteristics of the various centres, from the point of view of the importer of Australasian frozen meat, more on general lines than the specific information contained in the articles above. Liverpool, by far the most important of the "outports," the only port and trade centre which offers any serious rivalry to London, has largely increased its importance as a frozen meat port and centre during the last few years. It distributes throughout the Midlands, well across to the east coast, and as far North-East as Newcastle. The sphere of influence of Liverpool, however, in the direction of the East Coast is being checked by Hull, at which fine port large cold stores have been erected, and to which city South American direct shipments of meat are sent. When the new dock referred to above is completed, Liverpool's claim to be the distributing centre for the North of England may be further disputed. In the Liverpool trade lean, small carcasses are preferred, distinctly inferior—from a Londoner's point of view—to the quality handled in the London market. This may arise from the fact that a large proportion of the population are operatives in ironworks, cotton mills, and other factories, where the temperature is high, and, as usual in hot climates, the consumer has no appetite for fat meat. Manchester has been a bit of a disappointment to frozen meat people. The Ship Canal was to have been a highway for argosies laden with the refrigerated produce of New Zealand and Australia, but some error was made in these sanguine anticipations, for Cottonopolis imported direct in 1910 only Australian lambs; a bubble burst, indeed, when one remembers the efforts made years ago to attract direct shipments to Manchester. But although Manchester receives a certain quantity of meat shipped to that port, the shipowner largely takes advantage of the option in his bill of lading, and elects to rail to Manchester a portion of his cargo of meat from Liverpool in preference to the expense of sending his vessel up the Canal. Cardiff is a large consumer of frozen produce, and a very satisfactory and growing trade is done in that district. The Argentine companies have been shipping to Cardiff for many years, and do a considerable business there. Owing to the West Coast service calling at Avonmouth (Bristol), few direct shipments of frozen meat from Australia or New Zealand are now being sent to the Welsh port, which is a pity, as Cardiff possesses all the points which suggest an admirable port and centre for the reception and distribution of Australasian frozen meat and is the centre of a big meat consuming population, amongst whom retail distribution has become well organized. Bristol has not in itself been a very satisfactory market for direct importations. The trade done there is of a healthy description as far as it goes, but demand has not responded to the direct West Coast service shipments on the scale that might have been expected from a city and district with such a large population. Much of the meat landed at Bristol is forwarded by rail to Cardiff for distribution—an unbusinesslike course forced upon shippers by the lack of direct freight to the Welsh port. Glasgow is a useful market for some special grades of beef and for New Zealand lambs of good quality, especially in the earlier part of the season before Scotch lambs have matured fit for the market; and a steadily increasing demand may be hoped for from the Clyde port as frozen meat becomes more popular in the cities of Scotland. Glasgow has not had a fair chance of developing into such a frozen meat centre as the importance of the city and its situation entitle it to, owing to the service of refrigerated vessels being very limited. For some reason no South American frozen meat was despatched direct to Glasgow in 1910. Glasgow is the most likely "outport" to increase during the next few years in the handling of direct-despatched shipments of frozen meat.

What has been the effect of these "outport" markets upon general values? An answer is ready, covering part of the subject, in stating that the premium which, before the days of direct imports, frozen meat—railed from London—enjoyed in the provincial centres has disappeared. Prior to the establishment of the West Coast service, the London importing houses in supplying their country customers were careful to make a good selection of meat that would stand the railway journey and which would meet their clients' requirements in every way, and for frozen meat thus placed upon provincial markets a premium of about 1*d*. per stone of 8 lbs., especially for New Zealand mutton and lamb, over the London ruling rates, plus carriage, was usually obtainable. As competition became pronounced, say, about 1903, 1904, and 1905, and vessels landed meat at the ports of the West Coast, supplies of frozen meats became obtainable at these ports at the London rates, and now mutton and lamb from South America and Australasia in

London and provincial markets practically have the same value.

As direct despatch of frozen meat from Australia and New Zealand to the ports of the United Kingdom, following the example set by the Argentine shippers in dividing their consignments between London and country, became possible in 1903, the "opening up" of the "outports" in the Australasian interest was a matter of course. There was a great quantity of meat to be despatched to Great Britain, and the farmers and meat works managers, in taking advantage of the opportunity to divide their goods between London and provincial ports, adhered to the principle of wide distribution throughout their marketing area. It is probable that the aggregate output of Australasian frozen meat has during the last eight years expanded with less falling of values to the producers by reason of the development of direct supplies to the "outports" than would have been the case had all the meat been consigned to London—which would have staggered had it been made to bear the whole burden of imports, notwithstanding its splendid distributing facilities. There are, however, well recognized limits of usefulness to oversea shippers in the "outport" markets; these country centres tend to become competitive, and their capacity for dealing with large quantities of meat has nothing of the expansiveness and elasticity which constitute London's great attractiveness. There is also to be considered that there is a tendency at the "outports" for the demand to be more or less for only one grade of any kind of meat. For instance, at Bristol light lambs are preferred, at Liverpool secondary lambs, while at Glasgow the market is chiefly for medium weight lambs. Bristol and Liverpool both ask for very light mutton, while Glasgow, on the contrary, requires heavy mutton. Of late years there have been occasions when these markets in the provinces have become congested, with the result that mutton and lamb—South American particularly—sold under London values.

CHAPTER XVI

CUSTOMERS IN OTHER LANDS

WHEN Mr. Mort in his Lithgow Valley speech heralded the era of frozen meat exports, he had in view the markets of England, and those only. For the first few years of the business there were no efforts made by the exporters of Australia and New Zealand to ship frozen meat to other parts than the Mother Country, and they centred their energies upon popularizing their mutton and beef with the English people. Indeed, to the present day New Zealand does practically no "foreign" trade in frozen meat. It was not so, however, with the Argentine meat exporters, who, owing to the greater suitability of their produce for Continental needs, early devoted themselves to pushing their goods in European countries. In many ways Argentina took up and prosecuted the frozen meat trade on broader lines than did Australasia. Her grasp of the general position and prospects of the enterprise was thorough and intelligent, and quite early in the trade's developments the Argentine companies perceived that frozen meat was an article whose possibilities of distribution were world-wide. Whilst Argentina adhered to the business principle "not to put all your eggs into one basket," Australasia was doggedly pounding away at her one market, London, which, of course, showed a marked preference for New Zealand and Australian meat as compared with Argentine in the early days of the trade.

Attempts on the Continent.

Belgium.—The first signs of departure from this principle were, perhaps, to be noted in the despatch to Antwerp in 1885 by Messrs. Nelson Brothers of the hulk (alluded to on p. 378) which they had sent to Plymouth with 10,000 carcasses. The craft was loaded with meat for a Belgian syndicate which had taken up very thoroughly the idea of importing frozen meat. This Belgian company had gone into the question deeply, they had shops in Antwerp, Brussels, Liège, and other Belgian towns, appointed agents, and advertised the meat well. For a time the sales were satisfactory, but after a while the demand died away. The people did not take kindly to the meat, and finally the hulk had to bring back the unsold carcasses. The Belgians insisted on the carcasses bearing the lungs, which were frozen in New Zealand. To inaugurate this departure, Nelson Brothers entertained the Belgian Burgomasters on frozen mutton at Cannon Street Hotel when Mr. de Keyser was Lord Mayor. The freezing hulk did not cease her connection with the frozen meat trade with her double failure at Plymouth and Antwerp. She was sent to New Zealand to act as temporary freezing works, and is now, after fifty years of stout service, still afloat.

The "battlefield of Europe" has always looked a likely customer for frozen meat, as she is a heavy importer of meat. Belgium, as mentioned above, first attracted shipments of New Zealand meat, and Argentina followed up this trade in a determined fashion. A small duty, available storage, and energetic local butchering co-operation, were more than merchants handling frozen meat could resist. A Brussels butcher made £40,000 in five years dealing in foreign meat in his two shops. Messrs. Jules Renard and Co., of Antwerp and Melbourne, acting for Australian meat houses, endeavoured to get frozen meat into Belgium, but the "Lung Law," under which no meat could be imported into Belgium without the lung adhering to the carcass, proved a stumbling block. This statute was repealed in 1894 as to sheep, but it is still in force with cattle. In 1904 the Sansinena Co. made another effort to open the Continent to refrigerated meat supplies. In connection with the company's tallow business in Belgium, the idea of forming a separate company for the importation of frozen meat was entertained. Two shipments of meat were made to Antwerp, the beef having to be sent in sides. The result of the shipments was unsatisfactory, and no established business followed.

France.—In 1886 the New Zealand Loan Co. tried Paris with one or two small parcels ; however, we hear of no satisfactory business following. Probably the expenses of the venture checked further trials—the costs from discharge in London to selling in Paris came to $2\frac{1}{2}$*d.* per lb. Nothing more as regards Australasia seems to have been done till early in the nineties, when the shippers of Queensland beef—the graziers were then taking the market risks themselves—began discussing with their London agents the advisability of looking for "extra outlets."

The River Plate Fresh Meat Co. had a depôt of its own in Paris nearly twenty years ago, starting its business there early in 1892, but giving up its cold store there in 1893. The company, in the search for a fresh outlet, went to the expense of fitting up a special store and sent two boats into Havre. They found it very unremunerative, for after they had spent their money and made all arrangements, the French Government put a lot of restrictions in the way, and they were compelled to give it up. One of the principal restrictions was that the lungs, heart, etc., had to remain in the carcass, which had to be cut up into four separate quarters—that prevails now, and, of course, is prohibitive. This regulation, which was added to the tariff law of 1892 at the request of the agrarians, is applied to all imported meat except Algerian. But a trade is regularly done in legs of mutton and rumps of beef with Parisian salesmen—which meat, by special regulation, eludes the Act, which expressly states that no frozen meat is to be imported into France without the animal organs. The River Plate Co. chartered an old hulk, the *Robert Morrison*, when they put up their store in Paris, discharging into the hulk from their ocean steamers. The hulk, in fact, was a bonded store for the time being, and the company had to send over butchers to cut up the carcasses.

The position of France differs from that of Germany in this important way : the latter country has a deficiency in food supplies for the nation, whereas France grows food enough. But notwithstanding this, France at an early date attracted frozen meat imports. The efforts of the Sansinena Co. are referred to on p. 82; and this firm, which opened depôts in three French cities, did a considerable trade in connection with them until the agriculturists, taking fright, brought pressure to bear upon the Government to raise the import duty from 3 to 12 centimes per kilo., and in 1891 to 33 centimes, which, with the octroi, 12 centimes, brought up the total import duty to about 2*d.* a lb. Still the Argentine-Franco meat trade continued, and in 1891 100,000 Argentine carcasses of mutton were imported, while in 1896 over that number were sent to France. But by 1900 the record was a sorry one, the trade having dwindled to one of 4,000 carcasses. The regulations as to frozen meat imported into France having to be accompanied by the organs of the animals were really protection in disguise.

Why should the Parisians have to eat over 50,000 horses a year? They could have, if they liked, at less than the cost of this unnutritious meat, good beef and mutton from Australasia and South America. But for harassing regulations and biting tariffs frozen meat could well hold its own when shipped direct from the country of production. Meat can be produced in South America and Australasia at a lower cost than in any country of the world. In this connection there may be given the following quotation from an article in *Cold Storage* of March 19, 1908, by Mr. P. B. Proctor: "A rough average of the c.i.f. value of Argentine or Australian meat would be about 3*d.* per lb., and it is sufficient to point out that on this figure the Italian duty amounts to about 36 per cent., the French duty to about 43 per cent., and the German duty to about 62 per cent. *ad valorem.*"

Germany.—The early nineties were a period of seething activity in the frozen meat trade. Meat came along in great bulk, beyond the capacity of London to deal with it, and the frozen meat firms put out feelers in the countries of Europe to see what could be done there. Germany was the goal striven for; the industrial movement there, at that time taking definite form, encouraged America and Australasia in the idea that the standard of living in Germany would be raised in the direction of an increased consumption of animal food.

But the great Agrarian Party became alarmed, and immediately set up the agitation which resulted in the passing in 1903 of the meat "inspection" law, more correctly termed the "meat exclusion" law. German imports of fresh beef from neighbouring countries (Denmark and Holland) in 1899 showed a 400 per cent. increase as compared with 1897; cattle of all kinds, on the other hand, fell off from 307,000 to 268,000. The agrarians in Germany possessed enormous power and thus they succeeded in securing the "protection" they clamoured for. The United States of America meat packers, who shipped 350,000 cwts. of preserved meats to Germany in 1899, protested vigorously against the measure that prohibited the entry of their goods, but without success. So the conditions were not propitious in 1892—1893 for introducing frozen beef into Germany.

Still, something had to be done, for Queensland beef was arriving in London at the rate of 1,500 tons a month. Hamburg was attacked at the end of 1892 by Messrs. W. Weddel and Co., with the object of distributing the Queensland beef through North Germany, the import duty being ¾*d.* per lb. Satisfactory arrangements as to storage in Hamburg having been made, a bold step was taken in 1893 in despatching direct to that port from Townsville, Queensland, the sailing freezer *Turakina* with 300 tons of frozen beef. The venture began auspiciously, but in the end proved a disaster to the German owners of this meat, owing to the opposition raised by the butchers throughout the country, though the meat was of excellent quality when landed and the whole enterprise well managed. As a matter of fact, the opening of separate shops for the sale of frozen meat was one great mistake commonly made in foreign countries. The sale of the meat in existing establishments would have been far less likely to arouse the antagonism of the Continental butchers. These might have been encouraged to secure profits out of the trade from the outset in their own shops, though, of course, import duties were always calculated to afford frozen meat a smaller margin of profit as compared with home-killed than was the case at the start of the trade in England.

Before this the General Steam Navigation Co. insulated one of their steamers for Messrs. Weddel, in order to convey frozen beef from London to Hamburg. The market price for the beef in that city was 3½*d.* per lb. for prime quality meat. Mr. H. Kirsten was responsible for this enterprise, and shops were opened at Hamburg specially for the sale of frozen beef,

Austral. Fleisch-Import.

PREIS-COURANT.

Beefsteak-Fleisch	per Pfd.	100 ₰
do.	„ „	80 „
Bratenstücke	„ „	70 „
do.	„ „	60 „
do.	„ „	50 „
Rollfleisch	„ „	70 „
Suppenfleisch	„ „	40 „
Stückenfleisch	„ „	25 „

Hohe Bleichen 36/37

geöffnet von 8—1 u 4—8 Uhr

Fernsprecher Amt I, 282

unter **Johs Trope**

and the public at first thronged them, the police being actually employed to regulate the crowds. But the trade flagged, both wholesale and retail, the meat hung on hand, and this, coupled with the poor financial returns, forced the relinquishment of a promising enterprise, which had meant the shipment of about 1,500 tons of beef. The above facsimile reproduction of the price list of one of the frozen meat shops opened at Hamburg in 1893 may stand as a reminder of this premature effort to place frozen meat before the Germans.

No country seems in a better position statistically for frozen meat exploitation than Germany. Of late years her meat supplies have been heavily depleted, and the working classes in 1905 raised considerable outcry about high prices for food, all the sections of the industrial life of the nation supporting the demand for modification of the regulations dealing with meat imports. The answer on the part of the Government—impelled by the agrarians — was a new tariff wall higher by 50 to 70 per cent.! The temper of the German Government affords no hope of relaxation, but the General Election of 1912 may ultimately result favourably for those Germans who desire to see a more generous diet placed before the people. The meat consumption of the German working man is small compared with that of his rival in Britain, and in order that the splendid country of Germany may benefit fully by the output of her sons engaged in her vast industrial activities it will be necessary to permit of the import of fresh meats under reasonable conditions.

Austria.—Part of the *Turakina's* previously mentioned meat cargo was despatched to Vienna by Messrs. Weddel, and there again the opposition of the butchers had to be reckoned with. Considerable pomp attended the first arrival of Queensland beef in the Austrian capital, as will be seen from the following extract from the *Wiener Tageblatt* of June 24, 1893.

A large consignment of meat has just come to hand at a time when the supply of meat from Vienna threatens to fall short on account of the scarcity of cattle in the monarchy. The shipment comprises 1,600 Australian oxen preserved in ice from the effects of temperature. To-day the first wagon is due at Vienna, and others will follow on. The meat will be sold next Tuesday in the places of the Society of Victuallers at the Central Market, and also in the district markets. The price is 60 kreuzers for first quality of hinds and 50 kreuzers for first quality of fores, but it is anticipated that prices will fall whenever the contractors organize the sale in the market places on their own account. The Court Counsellor Oser said that so that there might be no doubt about the good condition of the Australian meat, he had ordered the Court Counsellor Professor and Dr. M. Grübe to inspect the meat and report as to its nourishing properties. Afterwards these gentlemen visited the Mayor, at the Town Hall, to obtain his consent and support with regard to the sale of the Australian meat in Vienna. The Mayor replied that he had some knowledge of Australian meat, having eaten it and found it very good and appetising. He also stated that he would do his utmost to introduce the meat to the public.

Both at Hamburg and Vienna municipal assistance was forthcoming. But the opposition of local interests and improving prices in London checked the business. The Argentine shippers took advantage of the situation to despatch a train-load of frozen meat from Antwerp, but its entry into Austria was prohibited, and the country remained closed to frozen meat until the recent shipments from Argentina which followed the inquiry set on foot in 1910 and the visit of an Austrian delegation to Great Britain.

Other European Countries.

Before going farther afield, let us glance at the remaining European States to which frozen meat has been sent. Race feeling naturally tempted Argentina to exploit the land of her fathers, though, sad to say, proud Spain did not for a long time respond in proper parental style. A company was formed in Madrid some years ago to sell Argentine meat in the city, and a shipment was despatched. As the business collapsed, it is fair to suppose that either the demand was disappointing or that other difficulties prevailed. It does not appear that Australian meat was ever imported into Spain, though the formation of a butchers' syndicate at Barcelona for the purpose of such a trade would make it seem that Spain is at last ripe for a frozen meat venture of some kind.

Italy welcomed refrigerated meat in 1905 with small samples from North and South America, sent to the Magazzini Frigoriferi Genovesi's modern stores at Genoa. In 1906 the purchases and total imports were 1,300 tons, of which 300 tons of Queensland beef and mutton arrived by the *Indraghiri*, and 200 tons of Argentine beef by the *Zero* from the River Plate Co. Signor Fausto Scerno, the founder of the stores, wrote as follows: "We commenced better than England, where, in 1875, the total import was about 800 tons. In 1907 we imported about 1,000 tons; low figures, if you compare them with the grand total of 517,000 tons imported into England, but quite enough to prove that business in Italy has started much better and earlier than in France and Germany." Two experimental parcels were landed at Naples in 1907 from Australia. In 1910 the imports from all sources aggregated nearly 3,000 tons.

A market for frozen meat appeared four years ago in Sweden. A Stockholm merchant imported, and still imports, from

London small quantities of Australian mutton and lamb regularly. Switzerland, each of whose Cantons, by the way, has its own import regulations, is pretty bare of meat in the tourist seasons, and some of the frozen meat sent to Genoa finds its way to Zurich. Mr. Coghlan, Agent-General for New South Wales in London, worked hard in 1908–1909 to secure entry for Australian meat into the Swiss Republic, and got the British Foreign Office to intervene. The Central Government was impressed, but from the following information from the Argentine Consul, indicative of the present position, it would seem that the impression did not sink deeply. "The importation of meat preserved by an artificial freezing process will only be allowed on the special authority of the Federal Home Office, request for which will be necessary to be made by the Cantonal Government. This meat will only be able to be sold on a clear understanding as to its nature and origin and complying with the conditions stipulated in the Home Office Authority with the view of protecting the health of the consumers." Nevertheless, small parcels of Australian and Argentine beef are now going into Switzerland regularly through London.

How great the prize of new custom must be when it is won is patent to everyone, but how great the need is for the diversion of heavy supplies from existing markets, only those in touch with the frozen meat markets can tell. The situation in Europe was never so promising as at present, but much has yet to be done before a trade of big dimensions is opened out.

Were it possible to get a footing for frozen meat in the fair land of France, or in Germany, what a godsend it would be for Argentina and Australia, with their surpluses of stock! But the main outstanding fact that these countries manage to supply themselves from internal resources is a sufficient reason for anticipating the greatest possible difficulty in introducing frozen meats into these markets, because it is certain that if, by reason of lower prices, shippers could compete with the local supplies, duties would be promptly raised so as to stop such importation. We are face to face there with the radical and permanent difficulty which it is as well to recognize,

Although France and Germany are coupled in this paragraph, a marked distinction must be made between them. Germany is really the less able to feed itself now, and imports considerably live cattle from surrounding countries, including France, at times. France is the more self-reliant country.

When diplomacy induces these European protectionist countries to refrain from directly prohibiting frozen meats from Australasia or South America, or advancing the duties to a point which would stop imports, they can easily pass regulations, nominally to protect the public, which achieve the end that the interests opposed to meat imports have in view quite as effectually as if the goods were barred by statute. On this point, there may be quoted the opinion of the late Mr. William Cook, who had very considerable experience in trying to persuade the Continentals to take Argentine meat. "Going to France was a mistake," said he. "So inveterate is the protective policy of the Continentals that if the import of frozen mutton and beef cannot be checked by existing arrangements and ordinary means, other and extraordinary ones will be resorted to."

An important point to take into consideration is that before frozen meat can be imported in commercial quantities into the important European cities, well-equipped cold stores will have to be built; there are only enough stores now to accommodate a hand-to-mouth trade.

Customers in the East.

From its geographical position the continent of Australia enjoys a peculiar advantage over its two rivals, New Zealand and South America, in undertaking to supply the East with frozen meat. Directly it dawned upon merchants and freezing works managers that London prices would certainly fall to very unremunerative levels unless other markets could be found, they sent commercial travellers in various directions to get orders. "Farther Ind" was explored in 1893 by a commissioner, who reported unfavourably. Mr. John Cooke ten years later had special inquiries made about

Eastern countries generally, and, as nothing further followed, presumably no good tidings were brought back. There has been much vague talk of getting frozen meat into India, China, and Japan, in all of which countries only the white population might be hoped for as customers, and their numbers are limited. The natives, where they are not entirely vegetarian in diet, are too poor to buy imported meat. It is suggested that the Indian Government might with advantage import frozen beef for the Mussulmans so as to save the friction set up between them and the Hindoos when the former slaughter cattle for food. The only definite commitment in Eastern parts was that of the Queensland Meat Export Co. in investing capital in a local meat and storage company in Singapore, with premises at the Tanjong Pagar Dock. Though the enterprise, as far as the Queensland Co. is concerned, has not brought many shekels to the shareholders, the Singapore white population must have benefited. "If we get good, honest, prime-fed Australian beef and mutton instead of the stringy Calcutta and Bangkok stuff we have to put up with, the better for everybody in Singapore," wrote the Singapore *Free Press* many years ago.

Three years ago Australia's "customers in other lands" were receiving 40 per cent. of her meat exports, chiefly to the Philippines, the Cape, and Malta, and this may well account for the fact that Australia has been at times since the beginning of the century unable to keep pace with New Zealand and Argentina in regular shipments to the United Kingdom.

Wars Stimulate the Trade.

The Australian meat exports to the Philippines, Malta, Egypt, Gibraltar, Hong Kong, and Singapore, are largely for the supply of military garrisons. Some proportion of the total exports go to great maritime junctions, such as Colombo and various Japanese ports, for the supply of the mail steamers, and hotels entertaining European visitors. For troops, either in the field or in garrison, frozen meat has come to be a necessity, and in no direction has the refrigerator been so fruitful of benefit as to masses of soldiery. Supplied in pre-refrigeration days with the "cattle of the country" in which the military operations were being conducted, or with preserved foods, moving armies are now provisioned from a convenient base with fresh frozen beef and mutton, a dietary system enormously to the benefit of the soldiers, and one much more easy for the commissariat to organize by means of contracts made ahead. The British-Boer war in 1901—1903 was a sensational epoch in the history of the frozen meat trade.

Cold Storage in South Africa.

Cold storage in South Africa had, in connection with the supply of meat to the troops in the Boer war, and the events that followed, the main impetus towards its present stage of development. A concise statement has been made in the *Pastoralists' Review* by Mr. R. H. Harrowell as to the growth of public cold storage in South Africa. Cold storage, this writer says, was first introduced into South Africa by the old Dutch firm of Combrinck & Co., who established plants at Cape Town, Port Elizabeth, Durban, Pietermaritzburg, Pretoria, and Bloemfontein. This firm was floated about the year 1898, into the South African Supply and Cold Storage Co., Ltd., which, in the capacity of meat contractor to the British Army during the early stages of the late Boer War, accumulated a reserve fund of £1,000,000, after paying big dividends. The late Cecil Rhodes, with the assistance of the Johannesburg mining houses, formed the Imperial Cold Storage Co., Ltd., with plants at Cape Town, Kimberley, and Johannesburg. The new company was successful in taking away the military contract from the company, but worked it at a loss of about £350,000 per annum. The main source of profit being thus lost to the South African Supply and Cold Storage Co., it was determined to float another company and transfer to the new concern a portion of the assets of the old one.

In 1901 the South African and Australasian Supply and Cold Storage Co. was formed, taking over the business of the South African Supply and Cold Storage Co. at Cape Town, Pretoria, and Bloemfontein, and the remainder of the assets, comprising shares in the companies at Port Elizabeth, Durban, and Pietermaritzburg, were taken over by the Cold Storage Trust.

The Cold Storage Trust, Ltd., has its registered office in London, but is practically controlled and directed by interests in Cape Town. This company controls the following subsidiary undertakings :—Port Elizabeth Cold Storage and Supply Co., Ltd., Pietermaritzburg Cold Storage and Supply Co., Ltd., Sparks and Young, Durban and Johannesburg, and has some interest in the Rand Cold Storage and Supply Co., Ltd.

On the conclusion of the war in 1902 the Imperial Cold Storage Co., Ltd., amalgamated with the South African and Australasian Supply and Cold Storage Co., and the business was floated under the name of the Imperial Cold Storage and Supply Co., Ltd., and a new plant was erected at Durban, and subsidiary companies were formed at East London and Johannesburg for the same purpose. The capital of the reorganized company was £1,750,000, in £1 shares, in addition to a 6 per cent. debenture issue of £500,000. The share capital was cut down by one-half in 1906, and was further reduced in 1911 to £201,000, at which figure it now stands.

The Imperial Cold Storage and Supply Co., Ltd., with its head office in Cape Town, has cold storage plants at Cape Town, Durban, Kimberley, Bloemfontein, and Pretoria, and controls subsidiary undertakings operating at East London and Johannesburg, trading as the East London Cold Storage and Supply Co. and the Rand Cold Storage and Supply Co., Ltd., respectively.

Professor Wallace's Statement.—Professor Robert Wallace, who has written a great deal on stock and agricultural questions relating to British settlement in South and East Africa, writes to the authors as follows :—

" South and East Africa are essentially stock, and not agricultural, food producing countries, and when people begin to realize that European cattle are not suitable for the natural conditions which prevail in those quarters without such a liberal supplement to the natural food which would be prohibitive in cost, they will breed animals with constitutions that will withstand the climatic conditions, and produce large quantities of beef of second-grade quality and suitable to the local demand. South and Central Africa is a black man's country, and the white population is not likely under prospective conditions to increase at such a rapid rate as would require a large amount of prime beef. All I can say is, if South Africa is unable to produce as much meat as the people require, it will be the farmers' own fault, by allowing the country to remain unfenced so that epizootic diseases may continue, as in the past, to ravage the flocks and herds, or by persisting in the impossible attempt to introduce highly-bred European cattle under climatic conditions which are inimical to them."

Manila and Vladivostock.

Australia's export of frozen meat—mainly beef—to Manila dates from the placing of the American garrison in the Philippines in 1898. Queensland, whence the meat is shipped, commands this trade on account of her geographical position. A yearly contract is made between the American military authorities and the Queensland Meat Export Co., and the other freezing companies in Queensland take a share in it; 7 cents per lb. gives a rough idea of the price. The 1910 contract was valued at £12,000; 3,000 carcasses of lamb, 5,000 carcasses of mutton, and 250,000 lbs. of beef, plus pork and rabbits, being supplied. It is understood that the business is cut rather fine, but the Manila contract has proved useful to the Australian meat men in providing an avenue for the disposal of a respectable quantity of frozen beef during dull years.

The market at Vladivostock came with a rush in 1906, and in that and the following year over 9,000 tons of Australian beef found its way to the Siberian port. The negotiations were first opened in 1898, when a cablegram was received by the New South Wales Minister of Agriculture asking if 700 tons of frozen beef could be sent to Vladivostock. The business was mostly negotiated in London by agents of the Russian Government, and considerable competition between suppliers marked its developments. In 1908 the shipments fell off, owing to the port being supplied from North China with live cattle, which were slaughtered at Vladivostock and chilled on board three refrigerator steamers. These Vladivostock contracts were also connected with the supply of meat for the troops.

The Mediterranean.

An important section of Australian exports of frozen meat, representing meat supplies for the British garrisons in Egypt, Malta, and Gibraltar, remains to be touched on.

Messrs. Wills and Co., Ltd., started refrigerating stores at Port Said early in the nineties, capable of holding 200 tons of produce ; from the depôt the requirements in the way of meat of the passing mail boats and the small European population were supplied. At that time the Army contracts at Cairo and Alexandria were for native mutton at about 4*d.* per lb. The thing to do in Egypt was to secure this Army custom for frozen meat, and this has now been done. Practically all of the trade falls to the Australian shippers, and since 1905 Egypt has been an important buyer.

About 1894 the British Government made up its mind to feed the garrisons at Gibraltar and Malta upon frozen meat. They had the necessary cold stores erected in 1895 at those Mediterranean strategic points, and ever since the military forces there have had the benefit of the nutritious diet furnished by beef and mutton grazed on the plains of Australia. Mr. C. A. Lichtenburg, head of Messrs. Wills and Co., Ltd., took an important part in getting frozen meat into Malta, where it is partly used for the civil population, and his firm has held the contract since the trade was introduced. Allusion may be made to the grievance of which frozen meat importers in Malta rightly complain, namely, the differential import duties on live cattle and dead meat. These pressed unfairly on the latter, and the Australian Agents-General have had to appeal to the Imperial Government on several occasions to get the grievance redressed. In 1908 an ordinance was introduced into the Legislative Council of Malta to reduce the duty on frozen meat from 10*s.* per 175 lbs. to 7*s.* 3*d.*, but the elected members unanimously opposed it, and as the Imperial Government did not feel inclined to overrule the elected members, the matter dropped. The Gibraltar contract for frozen meat concerns the troops only. It was first held by Messrs. Wills and Co., and since 1906 it has been in the hands of Messrs. Thomas Borthwick and Sons, Ltd. The contract is for about 500 to 600 tons a year.

There are certain harassing conditions attached to the Gibraltar contract. The vessel with the meat must go to the Mole, and the contractor has to take the risk of conveying the goods up to the store in the Rock. It is well to mention that these and other British Government frozen meat contracts are open to American shippers, the preference to British suppliers, alluded to before, having vanished—in fact, the River Plate houses have shared in the business to a small extent. Australia has to secure the business by competing in the open market, but having a preponderating advantage in being able to despatch meat by the vessels sailing to England through the Mediterranean, she is in a position to secure the whole of the Mediterranean trade.

To Sum Up.

Exports of frozen meat, present and potential, to "Other Lands" divide themselves into four sections, as detailed above.

First, there is the European Continental market, so far almost virgin soil, about which in the past anticipation and realization have widely diverged. Is it too much to hope that when a less conservative spirit pervades German official circles, the much needed supplies of frozen meat may be permitted to be imported? The opinion of practical men who have studied the question is that, in the cases of manufacturing countries unable to grow all their own foodstuffs—like Germany—barriers of the unreasonable kind which now prevail will be removed. Only now are European countries reaching that economic stage attained by England a generation ago, namely, that the absolute limit of the productiveness of the land in relation to the population has been passed, and the only alternatives presented to these nations are reduced feeding or importation. The powers that be may fight as they please against the new demand, but the cry for plentiful food is bound to be listened to in the end, which is not far off in European countries, and is approaching even in the United States of America. In view of England's experience, the Continent has no excuse for prejudice against frozen meat based on inexperience. But prejudice against it is, undoubtedly, cultivated by the land-owning and ruling classes. Unless scientists find unexpected grounds for discarding frozen meat on its merits, this barrier must fall.

Secondly, there is the demand from the peoples of the Far East. About this the obvious suggestion to make is that the present moderate demand will grow according to the development of industrial enterprise in Japan, China, and India. This demand has been disappointing so far, but when the new industries of China and Japan get into their swing, the importation of refrigerated meat supplies should follow; the plains of Queensland, grazing excellent beef and mutton, are but a few weeks away from Eastern centres, and the future will, doubtless, see a large volume of trade passing.

Then, thirdly, there are the Government contracts for military stations, British, American, etc. The dimensions of this avenue can pretty well be measured, and frozen meat supplies passing along it are likely to remain fairly constant whilst the nations are at peace. Wars of the future will cause a great stir in the trade, and the subject constitutes a supplementary division in this list of "Customers."

Fourthly, there remain South and East Africa, and about these the observation has to be made that Cape Colony and Natal are rapidly becoming self-supporting through the development of agricultural and pastoral industries. Except for casual and speculative shipments to Cape Town and Durban, such as took place in 1910 and 1911, and the arising of abnormal conditions, the demand for Australian refrigerated produce is likely to die down. Old residents in South Africa say that they are not at all certain that the States named will not have to draw upon the Commonwealth of Australia for foodstuffs in the future, but others lay emphasis upon the view that South Africa will one day take a place among meat exporters. It may here be suggested that Madagascar, where there are cattle of a secondary grade available for export, may be drawn upon for the needs of the white population of East and South Africa, should these districts require meat imports in the future. A freezing works, fitted with Haslam machinery, has lately been erected on the island.

The holding back from the market of supplies of fat cattle in the United States of America, and the consequent heavy reduction of beef products, a tendency which first became noticeable in 1907, enabled New Zealand meat to be experimentally exported from Liverpool to New York in 1910. Frozen mutton and lambs were despatched, and although an import duty of 1½ cents per lb. had to be paid, the meat was sold at a profit at 2 cents per lb. under the price of native mutton. This small trade was stopped by the authorities demanding veterinary certificates in a form which, not being necessary for the British market, had not been provided.

The Beginning of the End.

The position on the Continent at the latter part of 1910 and during 1911 with regard to meat supplies for the people became acute owing to pronounced shortage of locally produced meat. Serious disturbances took place in Austria and France, with riots causing loss of life. Throughout the more important and thickly populated countries of the Continent the communities rebelled against the monstrously high prices for butchers' meat they were called upon to pay, and the mob in Vienna shouted "Give us frozen meat." The war against the agrarian parties was officially declared in October, 1910, when the Austrian Government gave permission for 25 tons of Argentine frozen beef to be brought into the country: this declaration of war is a modern day crossing of the Rubicon. The signs of the times cannot be mistaken. Notwithstanding the difficulties which were placed in the way of South American and Australasian exporters of frozen meat during 1911, it is plain that the agrarian obstructionists are weakening, and that with a little more pressure from the people the Government will have to relax the prohibitive regulations in force.

Summarizing what has occurred since the historic 25 tons of beef, referred to above, made entry into Vienna, it appears that after a fair run for the Argentine beef, permits of entry into Austria for which were given up to June by the Government for 3,000 tons, the agrarian party in that country again got the upper hand, and in July, 1911, the Austrian Minister for Agriculture announced that no further imports of frozen meat would be allowed to enter the country. Some months later the populace of Vienna were rioting on account of "dear food," and in September some speculative shipments of Argentine frozen beef were made—in anticipation of Austrian official relaxation of the prohibition—to Trieste. That prohibition was not withdrawn, and the meat was mostly sold in Italy and Switzerland.

From Germany little can be expected until the leaven of the 1912 general election has worked. Switzerland in 1911 was a customer for a limited quantity of Australian frozen meat; there are some difficulties of transport of frozen meat to Switzerland to be overcome, but during the summer—tourist—season it is certain that the Swiss market will call for a certain quantity of frozen meat.

One Continental country, Italy, presents most favourable prospects for the South American and Australasian shippers of refrigerated meat and dairy produce. There is no disposition on the part of the authorities to block imports, and the duty can be easily met. A great deal of frozen beef has reached Genoa during the last eighteen months, and the principal exporting houses and their London agents have been making the closest inquiries in Rome, Naples, Milan, and the other principal Italian cities, with a view to the securing of contracts.

In the autumn of 1911 several importing houses in London began to open up business on the Continent, France especially, with small trial shipments of frozen mutton, prepared in accordance with the regulations prevailing in the countries to which the sheep were dispatched, and both in Argentina and Australasia, and London, arrangements were being quietly made in the closing months of the year for the coming Continental campaign. The Compañia Sansinena de Carnes Congeladas fixed up a contract for monthly supplies of mutton with an importer in Havre, and the quartering of the carcasses, required by the French law, takes place in the bonded cold store of that city. It is possible that in France the restrictions upon the importation of frozen meat will be abolished, or modified, but that the duties will remain.

Readers are referred to Appendix III., which gives in tabular form the estimated output of the world's freezing works in 1910, and shows the exact value of "Customers" outside the United Kingdom to the frozen meat trade as a whole.

It only remains to be said that at no stage in the history of the frozen meat trade have its prospects as regards new outlets shown such a tendency to widen as at the present moment, and even between the times of the writing and publishing of the present volume new business may appear for refrigerated meat shippers. This is as it should be in the case of a trade which is still in its early youth and full of vigour, and the reader will, therefore, rest content with deriving from this chapter a more or less coherent idea of how this world-wide expansion of the trade came into being.

CHAPTER XVII

THE CHILLED BEEF TRADE

MANY allusions to the chilled beef industry are scattered throughout this book, but it is in every way desirable that a chapter should be devoted to this branch of the refrigerated meat trade. The Americans, who started their export dressed beef trade to England in 1875, have always avoided complete congealment of the meat, depending at that early date upon the cooling power of freezing mixture in tanks aboard ship to maintain their cargoes at a chilling temperature. In 1874 a few parcels of hard frozen beef were shipped, as mentioned on p. 13. M. Tellier in his *Frigorifique* experiment in 1876 and 1877 held the meat at a temperature of 32° F.

That chilled beef as a marketable article is superior to frozen beef is unquestionable, and the only reason why the South American frigorificos did not adopt this way of preparing their beef for the British market prior to the beginning of the twentieth century lay in their assumption that their chilled beef could not be placed on the London or Liverpool market in sound condition. It was thought that the "life" of chilled beef was strictly limited to about fifteen to twenty days, and that the limit was prohibitive to an industry being carried on where thirty or forty days elapsed between slaughter of the beeves and the marketing of the beef. The same lion in the path blocked anything more than the experiments which are referred to later on in this chapter with regard to Australia and New Zealand. With the improvement of refrigerating machinery, rendering practicable the holding of beef in ships' chambers at an unvarying temperature, the chances of the chilled beef trade between Argentina and the United Kingdom became more promising; and with the adoption later on of a scientific system of sterilization it was proved that the dreaded mould spot could be prevented, and, further, that beef could be brought to London even from far distant Australia in a chilled condition and placed on the market in good order.

Frozen beef has certain advantages as to ease of marketing ; it can be stored without suffering deterioration, and it bears handling better than chilled beef. But there end the points that can be urged in its favour in comparing it with chilled. The chilled beef industry stimulates the bringing to a high degree of excellence the cattle of the country which makes the shipments of the article, for the primest beef is required for chilled exports. Chilled beef enters a high-class retail trade, where the demand is more regular and prices higher than for frozen beef. Chilled beef, to be sure, suffers terribly in its wholesale value when the market is glutted, and its prices on such occasions are apt to fall below those of frozen beef. But the chilling process does far more justice to good meat than does freezing, and it is difficult to avoid holding the view that the great bulk of supplies of refrigerated beef for Great Britain in the future will ultimately be carried at a chilling temperature. Chilling mutton, by the way, has never proved successful. The few experiments that have been made in this direction have been far from encouraging in their results : for one thing the fat of chilled mutton always becomes discoloured.

In the nineties, a period of unrest and development in the frozen meat trade, the first attempts were made to bring meat from Australia and New Zealand at a chilling temperature. The earliest experiment was that initiated by the late Mr. J. H. Geddes, whose friends shipped 1,000 quarters of beef in the s.s. *Port Pirie* from Sydney on August 21, 1894. Dr. Shiels' " thermostat " was used to regulate the temperature of this cargo, a Linde refrigerating machine supplying the cooling power. The " thermostat " consisted of a system of tubing filled with spirit and hermetically sealed. By contraction or expansion of the agent, under the influence of temperature, the " thermostat " controlled the action of the refrigerating machinery and regulated the supply of cold air in the meat chambers, thereby maintaining a uniform temperature. The meat in the *Port Pirie* had to be frozen down during the voyage, so that experiment failed. The second, third, and fourth trials to bring chilled beef to England from Australasia were made in 1895. The *Gothic* brought in two successive voyages from Wellington 2,000 beef quarters consigned to Messrs. Thomas Borthwick and Sons, the "thermostat" being used also in these trials.

These shipments were the first serious attempts, showing an appreciation of the advisability of introducing the chilling process into the Australasian meat export trade, made by English capitalists on a commercial scale. The beef which was thus treated was carried at 28½° and 29° F., and was landed so lightly frozen as to be fairly described as chilled. A certain degree of success was achieved, but the financial results were not sufficiently favourable to induce further shipments from New Zealand, because, as Messrs. Borthwick reported, the quality of the beef was not up to chilling standard. The s.s. *Rakaia* later on brought 500 hindquarters from Brisbane, but the experiment was a failure, as the beef had to be frozen after being forty-nine days on board.

In the annual frozen meat review of Messrs. W. Weddel and Co. for 1895, that firm suggested that in connection with the bringing of chilled beef from Australasia "it will probably be found that only by sterilizing the atmosphere of the refrigerating chambers can the meat be kept sound for a sufficiently long period"—a theory which was not practically tested till 1909. A fifth attempt was the shipment of a few quarters in 1896 in the s.s. *Urmston Grange* from Bowen, Queensland. This meat had been dipped in oil prior to shipment to prevent the formation of mould; in no sense could this beef compare with American chilled on arrival, so this trial was also a failure.

Nine years passed, and in July, 1905, 1,200 quarters of beef were shipped in the s.s. *Tokomaru* from Dunedin, New Zealand, by the New Zealand Refrigerating Co. The shipment should have been made by the s.s. *Matatua*, which had a specially suitable system of refrigeration, but the *Tokomaru* was substituted, and various causes made the voyage a long one.

and it ended disastrously in the seizure of the meat by the Port Sanitary Authority in London. When the cargo was opened up, it was found that black mould had set in. This experiment was based upon the fact that beef had been kept in sound condition in the works for six or seven weeks at chilling temperature, but like many other experiments it failed because it ignored the crucial question of how to prevent the development of mould spores deposited upon the beef at the time of shipment. As no insurance could be effected, the venture resulted in total loss, not even the boiling down value of the beef being returned by the Authority!

Argentina's entry into the chilled beef trade was in 1901, in which year 40,000 cwts. were exported to Great Britain. The River Plate Fresh Meat Co., who were its pioneers, had for eighteen months previously been quietly experimenting. They fitted the s.s. *Zuleika* with a chamber to carry 500 quarters, and made three consecutive successful shipments in that boat—the beef selling at 1*d*. per lb. more than frozen, for hinds—before making arrangements with the Royal Mail Co. It was the improvement in Argentine cattle stock which largely contributed in making chilled beef such an acceptable article of food as it is to-day, and it must therefore be recognized that a great factor in the development of the Argentine refrigerated meat trade was this intelligent action on the part of enterprising estancieros, who in fact began this reform nearly half a century ago. The River Plate chilled beef trade has developed rapidly. In 1901, when 40,000 cwts. were imported from Argentina into Great Britain, the imports from the United States amounted to 3,180,291 cwts; in 1910 Great Britain imported 2,710,747 cwts. from Argentina and only 469,444 cwts. from the United States. It will be noted that the total trade fell off in the nine years.

It was in 1908 that the chilled beef industry became undoubtedly a successful trade as far as South American shippers were concerned. In that year the reduction in exports to the United Kingdom from Chicago, etc., of live cattle and dressed beef was about 40 per cent., and this was Argentina's opportunity. Everything suggests that the South American Republic is to be the great beef producing country for Great Britain; and the American "Beef Trust" people evidently are the strongest holders of this opinion.

Very extensive and serious losses from mould have occurred to shippers of chilled beef from the Argentine Republic during the ten years in which the trade has been built up, and for years following the start of the industry. The loss accruing from the seizures by the sanitary authorities at English ports and markets became a pressing problem to shippers and importers. For years it was practically impossible to get full insurance cover for this article in transport, and, where obtainable, underwriters sometimes charged prohibitive rates, five or six guineas per cent. This state of affairs set people thinking. If the fungus germ on the meat surface could be destroyed and the atmosphere of the ship's chamber kept sterilized, the mould trouble might be scotched. It may be well to state here that Dr. Klein, Lecturer on Advanced Bacteriology at St. Bartholomew's Hospital, and one of the highest authorities on these questions, having made an exhaustive examination in June, 1909, of the "black spots" on some Argentine chilled beef, reported that "the fungus not being able to thrive at a temperature of the animal body (98° F.) no pathogenic action could be expected. . . . It is obvious that the material of the black spots is harmless to the animal body." But it should be noted that this avoidable mould defect has now been almost entirely removed, and, in the case of some companies, altogether eliminated. Many men worked at this scientific problem, and Mr. J. A. Linley patented a sterilizing system named after him and set up a small experimental plant in the Southampton Docks cold stores. The first shipment under this system arrived in London in September, 1907, by the s.s. *Guardiana*—1,059 quarters of beef, which were landed in excellent condition, bright and dry. The English rights of the patent were then taken over by the Improved Chilling Co., whilst the South American Improved Chilling Co. exploited the process in Argentina.

The Linley system is thus described. First, the sterilizing

agent is evaporated by heat and driven by means of fans through ducts into the chambers or rooms, wherein the meat is to be hung, and there circulated. This vapour has the effect of destroying any bacteria that may be on the surface of the walls of the chamber. The chamber is again charged when filled with meat, which is, of course, extremely liable to gather the microbes in the course of butchering and preparation for shipment. After the vapour has been allowed to remain in the meat chamber for at least an hour it is driven off. The ship's hold and the beef loaded in it are similarly treated. The second stage consists of an apparatus designed for drying and purifying the atmosphere. The air of the chamber or room containing the produce is passed through two tanks, the first one containing sticks of chloride of calcium, which remove the major portion of the moisture, the atmosphere then being passed through into a secondary tank in which rotate discs of lead kept moist with sulphuric acid. This latter operation gathers the remainder of the moisture from the atmosphere and also removes any organic impurities. Under this second system of drying and cleaning nothing enters the chamber or the ship's hold but air cleaned and dried. It will be seen that this treatment permits of application to the meat after slaughter, to the cold chamber before the meat enters, and to that chamber also when the meat is stored. The apparatus is simple and comparatively inexpensive, and both the vaporizing and drying plant may be combined on a common bed-plate, and both are actuated by a ventilating fan, and can be worked independently. In the No. 1 process commercial formaldehyde is vaporized in the plant at a temperature of about 300° F. The evaporation takes some fifteen to twenty minutes, and one ounce of commercial formaldehyde to every 100 cubic feet of the gross space is the amount of chemical used. In the case of a store the charged gaseous atmosphere is allowed to remain some three or four hours, and it is then replaced by pure, dry air supplied under No. 2 system, that is, air which has been passed in the apparatus over chloride of calcium and lead discs rotating in commercial sulphuric acid. The system thus provides a complete process of sterilization, and is at once scientific and practical, having been taken advantage of by several of the big Argentine shippers. Nearly 400,000 quarters of chilled beef have been brought to the United Kingdom under the process, the machinery for which had been—at October, 1911—fitted in four frigorificos, and on 19 vessels.

Taking advantage of the facilities offered by the Linley process, in which, it should be mentioned, Messrs. W. Weddel and Co. have throughout taken a keen interest, Messrs. John Cooke and Co., of Australia, sent over to London in 1909, 1910, and 1911 five experimental shipments of chilled beef from their Redbank works, Brisbane, in all 6,484 quarters, in the Aberdeen Line s.s. *Marathon*, in which the owners had installed the necessary Linley apparatus, in addition, of course, to the refrigerating machinery. The first of these trial shipments arrived in London in November, 1909, after a passage of sixty-two days from Brisbane, the cargo consisting of 1,181 hindquarters and 150 forequarters. The Queensland Meat Export Co. participated in these shipments, and the Queensland Government lent its aid in the form of a guarantee. Three shipments arrived in London in the following year. The condition of the beef in three of these four trials was excellent, and it sold readily at prices averaging rather more than $\frac{5}{8}$*d.* per lb. above the rates current for Australian frozen beef of similar quality. In one case the condition of the beef was imperfect, owing to the use of unsatisfactory meat wraps. These trials, resulting so favourably, proved that chilled beef can be brought from Australia or New Zealand to the English market, and that under the Linley sterilizing process it can be delivered in sound condition even after a seventy days' passage. But these small trials ceased in 1911, and the special plant was removed from the *Marathon*. The support given to the enterprise of the firms named by the Queensland graziers lacked the thoroughness which was necessary for the continuance of the operations.

In 1911 an experiment on a small scale under the Nelson-Dicks-Tyser process was tried successfully. The s.s. *Muritai* brought from New Zealand 70 quarters of beef which were marketed in excellent condition. But the process is too expensive for further shipments to be made.

It is satisfactory for the Australian and New Zealand cattle graziers to know that they are not excluded from the chilled beef markets of the United Kingdom, and that with direct transit, and larger and more regular supplies of beef of the high-grade quality required for the chilled trade available for shipment, there is no reason why Australasian chilled beef should not cross the ocean and enter into competition in the English market with that from South America. Until Australia and New Zealand take up this trade, one may say that the various interests composing the export meat trade there have fallen short of the full development of the industry.

CHAPTER XVIII

SOME NOTEWORTHY INCIDENTS: 1880—1910

WHILE the frozen meat industry with all its ramifications needs to be studied separately in a number of its phases for the reader to gain a clear idea as to its operation, it is only a running account of the progress of the great trade as seen from the point of sale or consumption which can indicate with any coherency the importance attaching to the many developments in the history of the industry. Hence an attempt is made in the present chapter to show the evolution of the struggle which has arisen on the part of the three competing producers, Australia, New Zealand, and Argentina, to market their frozen meat supplies, each working out the problem on its own system.

The story of the beginnings of the trade has already been told. Australian frozen meat was first imported into Great Britain in 1880, New Zealand in 1882, and Argentine in 1883. Up to 1910 the respective totals of sheep and lambs stood thus: Australia, 27,824,820; New Zealand, 72,464,591, South America, 53,463,982—a grand total of 153,753,393 carcasses.

The following figures show the growth of the frozen meat trade, and the relative position of the suppliers to Great Britain, from 1885, when all three countries were hard at it, to 1910:—

	Australia.	Percentage of total imports.	New Zealand.	Percentage of total imports.	South America.	Percentage of total imports.	Totals.
	Cwts.		Cwts.		Cwts.		Cwts.
1885	61,352	13	292,857	63	113,153	24	467,362
1890	131,252	9	875,817	60	444,017	31	1,451,086
1895	985,771	34	1,187,365	41	738,680	25	2,911,816
1900	860,040	20	1,797,864	43	1,527,057	37	4,184,961
1905	524,438	8	1,670,319	27	4,042,689	65	6,237,446
1910	2,406,094	21	2,637,003	22	6,551,276	57	11,594,373

Australia's Part.—All three of the countries mentioned were well set at their task of export by 1884—1885. But even at that early date Australia—the pioneer—was handicapped by her erratic shipments. Whilst her two competitors forged ahead well, Australia's exports were most disappointing; by 1889 both New Zealand and Argentina had reached their million carcasses of sheep and lambs, but Australia only shipped 86,000 in that year.

However, in the year 1890 a move was made, and by 1895 Australia was sending 1,000,000 carcasses to the English market. After 1896 the great drought, which culminated in 1902, began to tell, and Australian exports gradually fell away, till in 1904 the total amount of mutton and lamb shipped to Great Britain was under half a million carcasses. From 1905 Australia has been again exporting on an increasing scale to England. One reason to account for the falling off just referred to in Australian shipments of mutton and lamb to England was the demand springing up at South African ports, Manila, and the Mediterranean, and, to a less degree, in the East. Australia is favourably placed geographically for supplying these countries, and, indeed, has customers for her frozen meat in many lands.

For many years after Australia started exporting Great Britain was her only customer, but war works potent changes in this trade. The war in South Africa, the conquest of the Philippines by the Americans, as well as the demands of the British garrisons at Malta and Gibraltar, drew off a tremendous quantity of Australian meat from the English market. In 1902 Australia exported 19,690 tons of frozen mutton and lamb, of which only 8,510 tons were sent to Great Britain. The pioneer shipment of frozen meat made in 1879 by the s.s. *Strathleven,* and selling on Smithfield at from 4½*d.* to 6*d.* per lb., included beef, and this meat for about ten years was shipped very irregularly from the various States of the Commonwealth. In 1893 shipments rose to 10,350 tons, owing to the beginning of the Queensland Meat Export and Agency Co.'s operations. From that point to 1901 the yearly imports averaged 21,300 tons. The drought then dominated the position, and up to 1906 Australia almost dropped out as a regular and large supplier of frozen beef. But in 1909 Australian beef again became a factor to reckon with. Australian beef, as a matter of fact, has been going everywhere but to England; in 1902, whilst South Africa absorbed 23,905 tons of Australian frozen beef, only 3,290 tons were imported into the United Kingdom. Though Australia took fifteen years to get to her first million of sheep and lambs shipped in the twelvemonth, she quickly went ahead of New Zealand and Argentina in beef shipments when the Queensland works started. From 1893 to 1899 Australia (nine-tenths were from Queensland) exported 160,000 tons of beef to Great Britain, the combined contributions of New Zealand and Argentina only amounting to 41,000 tons for that period.

In 1884 Sydney beef was sold for 6*d.* a lb. at Smithfield. Some bone-taint trouble affected early shipments—a fault, unfortunately, that frozen beef (not Australian alone) is occasionally subject to now, depending as it does partly upon atmospheric conditions at time of slaughter. In 1890 Queensland beef of "magnificent quality" was imported, though a disappointing set-off to this item was the decision of the War Office not to allow frozen beef to be supplied, although frozen mutton was permissible, in Army contracts. The view of the War Office was tersely put by Mr. Lawson, Director of Contracts (in 1893): "We find the beef suffers from freezing, and the soldiers do not care for it"—a *dictum* afterwards found to be greatly wide of the mark. The restriction was removed many years ago.

To proceed with the bare chronological record forming a running account of the trade, in 1883 10,000 carcasses of Australian mutton were thrown overboard at the Straits of Magellan when the s.s. *Sorrento* was beached there. In the following year the average weight of frozen Australian sheep was $62\frac{3}{4}$ lbs. To celebrate the opening of the campaign of the Queensland Meat Export Co. in exporting beef to England, a banquet was given at the Hotel Metropole, London, on January 27, 1893. A distinguished company were present, and Queensland frozen beef, prepared by the hotel chef, formed the *pièce de résistance.*

High Water Mark—and After.—From 1896 to 1899 was

Australia's high water mark for both mutton and beef, for later her frozen exports waned, and there was no question of competition in quantities with the River Plate shippers. Shops at Hamburg were opened in 1894 for the retailing of Queensland beef, which article came along so heavily up to 1899 that holders had to combine to sustain prices. About 1896 Australia enlarged her *clientèle* ; the Cape, Colombo, Port Said, Gibraltar, and Malta became customers. The freight on frozen mutton from Sydney to London was reduced to ½*d.* per lb. in 1896. In 1897 the London cold stores were congested. The question of faulty cargoes became acute in 1897. South Africa was buying heavily at this time from both Australia and the Argentine. Vladivostock bought Australian beef in 1905, and this trade continued for a few years. By 1902 Australian beef had sunk to a shadow of its former bulky proportions in British markets, and by 1907 the South African demand for refrigerated produce had practically ceased, the Cape imposing a duty of 1*d.* per lb. on Australian and 1½*d.* on Argentine frozen meat; the duty on foreign meat is now 1¼*d.* The Transvaal has not imposed any duty.

Australia first took hold of the lamb trade seriously in 1899, and surprised British importers by shipping 233,000 carcasses in that year, a two-thirds increase on the exports of 1898. From that time onwards lamb shipments from Australia increased at the rate of about 100,000 a year until 1903, when the drought caused a 50 per cent. reduction. But in 1905 an enormous expansion set in ; out of 1,368,000 carcasses imported in that year 910,000 were lambs. From then onwards Australian shippers have steadily extended their lamb trade, receiving, however, a check in 1908. But the seasonal import of this article, December to March, is now one of the recognized features of the frozen meat trade, and Australian lambs are constantly becoming a more important feature in the English market. In 1911 1,650,000 carcasses were imported into the United Kingdom.

New Zealand's Progress.—As has already been said, the progress made by New Zealand in her frozen meat exports has from the very early days been fairly uniform and uninterrupted.

Nevertheless, the big jump up to be noted in 1903 was followed in 1904 by a setback almost as considerable, and exactly the same features were noted in the import figures for 1907 and 1908. But the expansion of the New Zealand mutton and lamb imports, as the freezing works in the Colony increased and as the demand in Great Britain developed, has been very satisfactory, though market prices were at times so low that curtailment of shipments seemed inevitable. The importance of the lamb trade was foreshadowed by the fact that the first steamer with frozen meat brought 449 lambs out of a total of 3,970 carcasses, and in the earliest days of the trade frozen lamb made 1*d.* per lb. above mutton prices. But one can have too much of a good thing, and in 1889 the market was glutted with New Zealand lamb. In 1893 New Zealand lamb went through the first of the crises to which it has been subject more than once in the autumn months, after the summer trade is gone. It receded to 3*d.* per lb., a price below the price which was then current for New Zealand mutton. In the following year mutton experienced a slump, a record low market price of 2¼*d.* per lb. being quoted. Prime Canterbury mutton fell to 3*d.* per lb. in 1895, and in 1897 another serious crisis in the lamb trade occurred in September. The whole decade 1890—1900 was one of struggle and trial in the frozen meat trade: problems came up for settlement, and as organization and precedent were lacking in many departments, there was a good deal of trouble before things settled down. Damaged cargoes constituted the most serious question. In 1898 20 per cent. of the cargoes from New Zealand were more or less damaged, and conferences between shipowners and importers were held. Damaged cargoes fortunately are now practically a thing of the past, and in chronicling this feature which disfigured the transit arrangements of the period referred to, it is pleasant to note the excellent record of the vessels conveying chilled and frozen meat from Australasia and South America.

In 1903 the Seddon shop scheme was promulgated and caused much excitement and feeling on Smithfield Market. About the end of the previous year, a splendid one for frozen meat importers, prime Canterbury mutton had touched 5½*d.* per lb. In 1903 New Zealand tried a direct shipment to Cardiff, and in 1906 freight arrangements were definitely made with the Federal Line for regular sailings to the west coast of England. This service is running now, and is carried on by the vessels of the Federal-Shire-Houlder combination. By 1908 the shipments of boned frozen beef in boxes from New Zealand assumed large proportions. This meat was from dairy and other cattle of a light weight, and was mostly shipped to Glasgow to be used for "minced collops" and in other ways. During the year to July, 1908, 3,500 tons of this beef were exported from New Zealand, valued, f.o.b. cost, at £66,000. Under the regulations of the Public Health Act of 1907, port and market inspectors in Great Britain received additional authority from January 1, 1909, and the importation of boned beef in boxes was rendered practically impossible because inspectors found it difficult to examine the separate pieces of meat when frozen hard in a box.

New Zealand, with herds of cattle chiefly devoted to dairying and, therefore, not permitting any great shipment of beef, has been an erratic exporter of this article. About 1899 she raised the scale of exports, and for a few years placed over 100,000 quarters on the English market per annum, the high water mark being in 1910, when 344,000 quarters were handled in Great Britain.

Early Prices for New Zealand Meat.—As already mentioned, the pioneer shipment from New Zealand was that by the sailer *Dunedin*, and the mutton brought by this vessel was delivered in sound condition and sold for 6½*d.* per lb. During the first few years of the trade it was not all plain sailing, as some cargoes were brought in unsatisfactory condition, and two years after the *Dunedin's* arrival prices for New Zealand mutton at Smithfield had dropped to 4½*d.* to 5*d.* per lb. As early as 1883 the practice of the consignees of distributing New Zealand meat throughout Smithfield was inaugurated. The year 1885 was one of severe trial, and the growing trade had to struggle against adverse influences, mainly low prices. In 1886 the Colonial and Indian Exhibition proved helpful, as the opportunity was taken of organizing a prominent exhibit, by which the public were favourably impressed.

The second shipment of mutton brought by the *Dunedin*, in 1883, made 7*d.* to 8*d.* per lb., which, as far as records go, was high water mark in rates. "If unfrozen," the chronicler states, "the meat would have topped the market." In 1884 New Zealand mutton was $4\frac{1}{2}$*d.* to 5*d.* per lb., and beef (hind-quarters) at one time touched $6\frac{1}{4}$*d.* to $6\frac{3}{4}$*d.* per lb., against American town-killed (sides) $5\frac{1}{2}$*d.* to $6\frac{1}{4}$*d.* per lb. The average weight of New Zealand sheep imported in 1884 was $65\frac{1}{2}$ lbs.

Rise of South American Exports.—The first arrival of Argentine frozen meat in London was in 1883. Both the mutton sent and the market prices it fetched were very poor at the start; the carcasses were almost lamb-like in proportions. But by 1886 a considerable improvement had taken place in the type of sheep exported. In 1884 River Plate mutton was making 3*d.* to 4*d.* per lb.; in that year the average weight of the frozen sheep was 48 lbs., 15 lbs. less than the Australian. In 1887 a trade with the Continent of Europe was attempted by the Argentine companies, depôts at Antwerp and Havre were established, and in 1891 100,000 carcasses were sent to France. By 1888 Argentina, in volume of exports and by her excellent trade methods, showed great strength in competing with importers of Australasian frozen meat in Great Britain.

Argentina's Success.—On an average Argentina shipped about 5,000 frozen lambs a year up to the end of the nineties. For the next six years her shipments averaged 150,000, and in 1909 South America became a factor in the lamb trade in Great Britain with an export of 634,000 carcasses. In 1911 the total shipments to England exceeded 1,000,000. Argentina was a long time making headway with beef shipments. For ten years after the establishment of freezing works frozen beef was quite a minor department of her business; the meat works were constructed for dealing with sheep, and the live cattle trade, which began in 1890, rapidly grew to large proportions and proved a useful avenue for the disposal of a part of River Plate surplus stock. This, by the way, was the year in which the export of live sheep also started, with 22,000 head, which grew to 306,000 in 1895 without causing any scarcity of freezing sheep. By 1896 Argentina was, in frozen beef exports, becoming a competitor with Australia, and in 1900 Argentine imports into Great Britain jumped to 20,600 tons, exceeding for the first time the receipts of Australian beef. This sudden increase was brought about by the stoppage of the live cattle trade. Having once secured the lead, Argentina went ahead fast, Australia falling behind as rapidly as her rival advanced. By 1901 River Plate frozen beef formed two-thirds of the total imports of this meat into Great Britain, and soon it was a case of Argentina first, and the rest nowhere. Imports of frozen and chilled beef in 1910 into Great Britain from South America were 252,922 tons, from Australia 44,034 tons, and from New Zealand 27,641 tons. In Chapter XVII. will be found detailed reference to the progress of the Argentine chilled meat trade.

Argentine Advance.—As early as 1893 an experiment in the export of chilled beef was tried; by 1901 this had become a regular trade. It took some time to secure a footing in English markets, but when the initial difficulties were overcome butchers soon took to Argentine chilled beef, and in course of time it proved a severe competitor of North American chilled beef, which has now nearly disappeared from the market. Roughly speaking, North American chilled beef has been marketed at 1*d.* per lb. above Argentine chilled, which in its turn may be put at from ½*d.* to 1*d.* per lb. above Argentine frozen beef.

For some years previous to the beginning of this century the Argentine meat companies had been quietly adapting their plant and general arrangements to increased handling of cattle for freezing. So thoroughly did they take in hand their new enterprise, that from 1900 to 1908 over 75 per cent. of the frozen beef imported into Great Britain was from South America; and the competition of Argentine chilled beef with the North American dressed beef—the choicest dead meat that England imports—became very keen. The rapid development of the beef export business of the Argentine Republic stands out as one of the most sensational features of the meat trade. It should perhaps be mentioned that the year 1902 was an *annus mirabilis* for the River Plate companies and their shareholders ; Sansinenas paid 50 per cent., and James Nelson and Sons paid 50 per cent. The dividend paid by the River Plate Fresh Meat Co. was only 25 per cent., but earlier in the year a distribution of the accumulated reserve fund was made to the shareholders by way of bonus. The bulk of this reserve fund had, of course, been accumulated out of previous years' profits. The shareholders were immediately asked to put this 100 per cent. bonus back again into the company in the form of capital, so that it really amounted to capitalizing the reserve fund. Of course, this splendid year was followed by the Nemesis of competition, and a crop of new freezing concerns sprang up under the stimulating effect of the golden showers of 1902.

Frozen Pork.—From 1903 the imports of fresh pork into Great Britain declined ; in that year the total imports were 705,844 cwts. (Holland 527,269 cwts.), and by 1906 the quantity was reduced to 492,171 cwts. Seizing this opportunity, Australia and New Zealand shipped frozen pigs to help fill the gap. In 1906 20,779 cwts. of frozen pork were received from Australasia, and the goods were welcomed at Smithfield market, and were sold fairly profitably. But it seems that the enterprise is not a paying one for the shippers unless the home market is under supplied, for after 1906, when the Dutch pork was again exported freely, supplies from Australasia fell away to an inconsiderable quantity. In 1910, only 11,000 cwts. of frozen pork were exported from Australasia to Great Britain, but pork is too expensive an article for the Australian and New Zealand freezing works to handle for export unless the shipper can expect a London market price in the neighbourhood of 5*d*. per lb.

Vitality of the Canning Trade.—With the advent of frozen meat in the eighties, canning meat became of secondary importance, and this method of handling stock has been resorted to in Australia during the last thirty years, mostly as a collateral to a freezing works, in dealing with the less prime parts of slaughtered stock. But at some works all the carcass is tinned.

In 1905 Queensland developed her exports of tinned meat very heavily: there was a big glut of these goods in stock in Great Britain, also of extract of beef, and prices went down to a ruinous level. A large proportion of this canned meat was of inferior quality, and in December, 1895, as low a price as 13*s.* per dozen 6-lb. cases ($2\frac{1}{6}$*d.* per lb.) was accepted. Tallow was selling as low as 20*s.* per cwt. at that time. The decennial figures of imports of canned meat into the United Kingdom may be given for the last thirty years:—

	Australia.	New Zealand.	South America.
	Cwts.	Cwts.	Cwts.
1880 . . .	120,472	21,808	3,163
1890 . . .	69,821	50,301	10,783
1900 . . .	232,465	28,422	41,926
1910 . . .	246,604	44,786	*301,527

* Uruguay 191,538 cwts.

Random Jottings: 1890—1908.

Having considered the progress of the trade in connection with the separate developments of New Zealand, Australia, and Argentina, it may now be well to sketch leading historical events that embrace all three sections. In the nineties difficulties and problems presented themselves in England as imports increased; in 1892 France established frozen meat import regulations, requiring that certain organs should be left in the carcass, which practically closed the markets of that country to shippers. Another obstacle to opening up the Continent for frozen meat was raised in 1905 by Germany advancing the import duty on meat by 50 per cent. Early in the nineties selling prices became lessened, and it was plain that to make profits the costs of production and realization of frozen meat must be lowered.

In 1894 the London cold stores were blocked for nine months, and vessels were kept on demurrage as floating stores : Queensland beef was then arriving in large quantity. In that year the working capacity of the London stores was 500,000 carcasses, stowed to marks, and there were 100 vessels engaged in the trade. By 1895 the capacity of the London stores had risen to 1,000,000 carcasses. The British Government built cold stores in the Mediterranean to make the supply of frozen meat for the troops possible. By 1898 English cold stores got well ahead of requirements, and the " refrigerated fleet " numbered 131 vessels. Frozen rabbits by 1899 became a menace to the cheaper kinds of frozen mutton, and in the following year 19,000 tons were imported into Great Britain. In 1899 beef was first shipped from Australasia with jute bags over the calico wrappers. In 1901, the heyday of the Australian exports to South Africa of refrigerated produce, there were twenty-two cold stores in the latter country.

Owing to the scarcity of New Zealand mutton for some months at the end of 1902 and the beginning of 1903, 120,000 carcasses of chilled mutton from North America were imported into England, the mutton being sold from $3\frac{1}{2}d.$ to $4\frac{3}{4}d.$ per lb.

Contrasting Systems.—The main difference between the Australasian and Argentine sale systems in Great Britain is the direct outcome of the fact that the Argentine companies have their head or branch offices in London. This is the case with most of them, though the Smithfield and Argentine Meat Co., the South American Export Syndicate (Rio Seco), Compañia Frigorifica de Patagonia (San Gregorio), and the Venezuela Co., transact their business through agents. But in the historic periods of the Argentine frozen meat trade, 1883—1902, during which the industry was in the making, the whole system was exceedingly compact. The three great companies bought their stock, froze the meat, shipped it in vessels owned or chartered by themselves, and landed and sold it in Great Britain. The Sansinena Co. and the River Plate Fresh Meat Co. distributed and sold their meat through the usual wholesale channels, whilst Messrs. James Nelson and Sons relied mainly upon their shops. The c.i.f. selling system was introduced when the La Plata Co. sent its meats to England, but the sale of its works in 1907 stopped this method of doing business.

The New Zealand and Australian systems were essentially different from those of the Argentine. Up to a comparatively recent time all frozen meat from those countries was either consigned to brokers or agents in London for sale on commission, or was sold outright to English c.i.f. buyers. The opening of London offices by the Christchurch Meat Co. marked a new departure; another system was introduced by English capitalists, Messrs. Thomas Borthwick and Sons, Ltd., establishing offices in New Zealand, and buying and shipping their own meat, and, later again, acquiring their own meat works in New Zealand and Australia. Messrs. W. and R. Fletcher, Ltd., acquired freezing works at Geelong, Victoria, in 1901, and Messrs. Henry S. Fitter and Sons have had an office at Christchurch, New Zealand, for many years.

Producers' Conferences.—A survey of the various movements undertaken for the advancement of trade interests reveals the repeated formulation of schemes of various kinds intended to improve the conditions of the trade, the product of ingenious brains in Australia and New Zealand and elsewhere. In the River Plate trade, where frozen meat interests were concentrated, one does not hear of such things; but so many people are concerned in the Australasian meat export business that divergent views as to the management of a crisis have from time to time necessitated conferences and the submission of special proposals. It may be interesting to record four of these schemes.

In 1887 the first conference was held, and on October 18 of that year the following firms and gentlemen met to consider "combined action amongst consignees of New Zealand mutton to support prices":—Nelson Brothers and Co., Ltd.; Miles Brothers and Co.; John Bell and Sons; Shaw, Savill and Albion Co., Ltd.; New Zealand Shipping Co., Ltd.; Gear Meat Preserving Co., Ltd.; P. Comiskey, T. Russell, F. Larkworthy; New Zealand Loan and Mercantile Agency Co., Ltd. A scheme to regulate supplies, limit prices, and concentrate sales was considered and rejected.

A New Zealand Scheme.—On September 5, 1893, at the request of the agricultural and pastoral societies, a conference was called by the New Zealand Government of delegates representing the sheepowners and freezing companies of the Colony, and this was held at Wellington and numerously attended. The meeting was summoned to discuss certain schemes presented by Messrs. D. J. Nathan and M. C. Orbell, and the following resolutions were put before the conference :—

1. "That the present methods of conducting the frozen meat trade being unsatisfactory and unprofitable to the sheep farmer, it is desirable that a controlling company representing the various interests concerned be formed for the purposes set forth below :—

"(*a*) To arrange for the amalgamation, buying-out, or federation of existing freezing companies.

"(*b*) To negotiate with shipping companies as to freights and kindred matters.

"(*c*) To inspect, report, and act, as to the suitability and class of ships employed in the trade, and in regard to the insulation, etc.

"(*d*) To attend to grading and insurance, watch the unloading in London, arrange for storage accommodation near port of discharge, and thus enable supply to be regulated, also save many handlings and cost of barges and delays caused thereby.

"(*e*) To arrange for the concentration of the trade in Great Britain and elsewhere, and to open up new distributing centres.

"(*f*) To issue debentures for the purchase of existing works, new plant, or establishing other works, and for raising the necessary capital.

"(*g*) To provide a sinking fund for the repayment of debentures."

2. "That, in order to achieve the objects set forth in the foregoing resolution, this conference desires the Government to introduce a Bill empowering such company to levy tax upon all sheep in the Colony. Shares to be allocated to each stockholder in proportion to the total amount of his paid-up sheep tax."

After a long debate the resolutions were withdrawn, and the

following ones, proposed by Mr. A. C. Begg, of the New Zealand Refrigeration Co., were adopted :—

1. " In the opinion of this conference any attempt to establish a monopoly of the frozen meat industry of the Colony would be both undesirable and impracticable."

2. " That, in order to give confidence to buyers and to secure as far as possible good quality of the meat exported, it is desirable that the freezing companies in the Colony should agree to a uniform system of grading, as far as consistent with due regard to local conditions."

3. " That, in order to regulate the supplies to the home market, and to prevent the glut which has been occasioned by excessive shipments during the first half of each year, it would be very advantageous if provision were made for storage accommodation in the Colony, so that supplies sent forward may be regulated and any glut prevented."

This conference marked one of the stages of development of the meat export trade of New Zealand. From 1889 to 1894 the average price realized by prime Canterbury wether mutton at Smithfield Market had fallen to the following rates: $4\frac{7}{8}d.$ per lb., $4\frac{5}{8}d.$, $4\frac{3}{8}d.$, $4\frac{1}{4}d.$, $4\frac{1}{8}d.$, and $4\frac{1}{16}d.$—these were top quotations. The flockowners in the Colony became alarmed about the future, and schemes and plans without end were put forward to effect improvement in the conditions, locally and in England, under which the frozen meat industry was carried on. It is curious that the practical (positive) outcome of a conference called to sanction a revolutionary and visionary scheme should have been the acceptance of two such businesslike methods as standardized grading and local storing, neither of which has yet been effectually carried out!

Two London Conferences.—In September, 1897, a desperate slump occurred in the lamb trade. The chief holders met on September 22, and a declaration of stocks (200,000) held by them was made. In March, 1898, a series of meetings was held in London and attended by these firms for the purpose of considering the lamb position, which had again become acute. These conferences, under the auspices of the Frozen Meat Trade Association, were considered to have checked the " rot " which had set in. Skipping forward a decade, allusion may be made to one more meeting of the trade. That was on May 4, 1909, when Sir Montague Nelson invited all leading importers of frozen mutton, Australasian, South American, and North American, to discuss the position—an exceedingly dismal one of over-supply and under-demand. Nothing came of this conference except a useful exchange of views. In the marketing of Australasian frozen meat it has often been necessary to attempt to secure combined action, either in limiting quantities offered or in fixing minimum prices; but such movements have rarely been completely successful owing to the number of holders and their widely divergent interests.

Mr. Twopeny's Mission.—One of the most authoritative and well-supported movements engaged in by meat exporters was the formation in Sydney in September, 1896, of the Australian Meat Export Association. Mr. R. E. N. Twopeny was appointed delegate to visit London for the purpose of forming a committee of the London representatives of Australian pastoral interests to supervise the disposal of Australian meat in the United Kingdom. Mr. Twopeny succeeded in forming the committee, which sat for some months under the presidency of Mr. E. T. Doxat. The committee recommended that there should be a regulation of supplies of meat by agreement amongst the freezing companies, and also that a limitation of the number of consignees in the United Kingdom was desirable. Mr. John Cooke was somewhat in opposition to the London committee scheme, holding that "our first duty in the Colonies is to set our own houses in order by shipping only first-class meat, and arranging for its transport and delivery in first-class condition." Although Mr. Twopeny carried out his mission in first-rate style, the London committee failed to send prices upward. The *British Australasian* of March 11, 1897, thus alluded to the matter: "It is apparent that what has occurred has been one of those occasional deviations from established usage prompted by the pressure of depressed conditions of trade. Ten Anglo-Australian merchants and bankers have been discussing academic problems, such as the c.i.f. trade, concentration of shipments, etc., at a round table, and still we are no nearer a solution of the one important problem—how to raise prices to a permanently paying level."

Regarding these conferences connected with the frozen meat trade, whether of producers in Australasia or of importers and merchants in London, one cannot gather that much in the shape of practical results attended them. A healthy interchange of views took place, but the clashing of interests, both at producing and selling point, to be expected in a trade where there is such keen competition, has hitherto stood in the way of the acceptance of proposals involving uniformity of action.

Meat Marking.—The Select Committee of the House of Lords appointed on August 25, 1893, with the late Lord Onslow as its chairman, to consider the marking of foreign and colonial produce, issued a voluminous report. The Merchandise Marks and the Sale of Food and Drugs Acts had not been completely satisfactory in stopping misrepresentations and dishonest trading. Bills advocating the marking of imported meat and (or) the licensing of retailers of the same have, of course, been before Parliament ever since the refrigerator became a factor in meat supplies. The Committee's Report has become a classic, and much valuable literature bearing on the retail vending of imported meat is to be found therein. The Committee, whilst reporting that the consumer would benefit by the marking of meat, did not show any enthusiasm in recommending compulsory meat marking; but they were agreed that marking could be done satisfactorily. A metal tag passed through the shank bone of legs and shoulders, and sealed, was suggested in the case of mutton. Aniline dyes were not approved on account of the lack of permanence of the mark. The Committee reported clearly in favour of the registration of the retailers of imported meat and the affixing of a notice to that effect over their shops. The evidence placed before the Committee showed that misrepresentation in the (retail) meat trade existed, chiefly in the substitution of American chilled beef for English and Scotch. The Report said: "It does not appear that retail butchers habitually inform their customers of the source of origin of their meat. The usual practice is to supply such quality of meat as is likely to meet with the approval of the customer without giving any actual guarantee of origin." That "usual practice" still prevails. But the Committee had cases before them of gross fraudulent sales, the most glaring of which was "The Old Established Welsh Mutton House" in the Strand (until ten years ago), where New Zealand mutton, the retail value of which was 7*d.* per lb., was sold as "Welsh" at 10*d.* A witness stated: "I have seen 'Prime Canterbury' stuck above a shop that had nothing but River Plate mutton in it."

False Trade Description.—Allusion may be made here to an episode in the history of the New Zealand meat trade. The Government of that State has always been careful to safeguard the interests of its meat exporters. In 1900 the Agent-General in London determined to institute prosecutions of retailers against whom a clear case could be brought of passing off inferior meat as New Zealand produce. "New Zealand" and "Canterbury" as terms had got to be well known and liked, and so unscrupulous retailers were in the habit of applying them recklessly. The matter was first put to the test by Mr. H. C. Cameron, the New Zealand Government's Produce Commissioner in the United Kingdom, bringing proceedings against a Blackpool butcher for applying a false trade description to a leg of mutton which the defendant sold to him on March 16, 1900. This and similar cases were taken under the Merchandise Marks Act, 1887, under which measure for a prosecution to be successful it is necessary that the "false description" shall be in writing. The magistrates dismissed the case, but the High Court sent it back to them, and the retailer was convicted. Other prosecutions were brought afterwards on similar grounds, and the New Zealand Government went to considerable expense and trouble to put a stop to this fraudulent practice, which, if persisted in, might well have injured the high reputation of the Colony's produce. In 1905 the New Zealand Government in conjunction with the London County Council summoned a meat salesman for applying the labels of the New Zealand Refrigerating Co. and other New Zealand labels to Australian lambs supplied under contract to some of the London County Council asylums as New Zealand lambs. This case was under the same Act and on the same indictment. On the evidence of the prosecutors, there was about 1½*d.* to 2*d.* per lb. difference in value between the two descriptions of lambs. The defendant was fined £20 for having in his possession the goods falsely described. A " hair-pin "—of the kind then used to pin down Australian lambs' tails at the time of freezing—played no small part in the evidence against the loser. Another conviction was secured in 1907 on much the same grounds; in this case Argentine meat was supplied under contract in place of the New Zealand article.

That the substitution of frozen meat for the home-grown article in the retail trade does not prevail as largely as is thought by many people is proved by the relatively low wholesale values for the former. Undoubtedly, some people buying " Canterbury " mutton and lamb think they are getting meat produced in the English Canterbury district, such is the magic ring of the trade description " Prime Canterbury." Salesmen say that the Canterbury brand will sell anything. That improper misrepresentation takes place, and constantly, no one can doubt. Each grade is substituted for the one above; American chilled beef for Scotch, Canterbury meat for home-grown, and Australian and Argentine mutton (to a diminishing extent) for New Zealand. The persons who have a right to complain are the English and Scotch farmers, who feel to some extent in restricted values the effect of the substitution of the imported article for their home-raised meat. The consumer, too, has certainly good reason to find fault with his butcher, though the number of those who will not admit the excellence of the quality of the chilled beef and frozen mutton and lamb is becoming less and less. The admitted and open sale of " town-killed " beef for English, that is to say, at " English " retail values, is an unjustifiable practice, pressing hard upon the home producer in England. One of the reasons to account for the popularity of the live cattle trade with the carcass butchers is the profit which they make in buying American cattle landed at English ports and retailing this " town-killed " beef as home-grown. There was a case in the English High Court in October, 1908 (*British*

Tea Table Co. v. *Gardner*), in which Mr. Justice Ridley made this remark: "I understand English beef to be beef bred and killed in England. One might as well call the English soldiers killed at the battle of Waterloo 'Belgians.'" The judge referred in terms of disapproval to the plea that "town-killed" is recognized by market custom as English meat.

Mr. Edward Lloyd, master butcher, of Chester, gave evidence before the Marking of Foreign Meat Committee on June 27, 1893, and the following paragraphs from the Report are instructive:—

Lord Onslow (Chairman): "You assert that if a purchaser went into a shop and asked where the meat [Birkenhead-killed American cattle] came from he would be told that it came from England?"

Mr. Lloyd: "Yes, in 99 cases out of 100. Scotch bullocks they call them often enough."

Lord Onslow: "That which is bought a penny a pound cheaper wholesale is sold at the same price as that which is bought at a penny a pound dearer?"

Mr. Lloyd: "Exactly."

Lord Onslow: "And therefore the purchaser and the consumer have to pay more, and the butcher makes a great profit?"

Mr. Lloyd: "Exactly."

The Scotch and English graziers are the principal victims, for were the American beef sold as "American beef," the public, to some extent at least, would not buy it, however good it might be.

The above remarks have not the force to-day which they possessed some years ago, for North American exports of both fat stock and chilled beef to Great Britain have very considerably fallen off of late. Also it is well to say that, as Argentine and Australian mutton have so greatly improved and advanced relatively in price, there is not now so much temptation for the retailer to substitute these meats for New Zealand as there was some years back—but there is a readier opportunity!

The most recent appeal to the law to invoke penalties for selling as New Zealand mutton meat which, as alleged by the prosecutor, the New Zealand Government, was of Australian origin, was tried in the Liverpool Police Court in May, 1911. The defendants were shipping merchants, and they sold to the Allan Line a quantity of mutton marked with the "Crescent" brand of the New Zealand Refrigerating Co. The contract was for New Zealand mutton, and the prosecution suggested that the meat was Australian. The "Crescent" brand has not been used since the New Zealand Refrigerating Co. was taken over by the Christchurch Meat Co. in 1906, but the magistrate ignored that, and only applied himself to the point, had Australian mutton been substituted for New Zealand? In conducting a case of this kind the prosecutor is fighting a lone hand, and to provide the necessary array of witnesses is far from easy. The magistrate held that no *primâ facie* case had been made out, and dismissed the charge, but without costs.

Rise and Fall of North American Beef Export Trade.—When beef was first exported from the United States of America (1874), cattle numbered 27,000,000. The stocks rose year by year until 1892, by which time the 1874 figures had increased to 54,000,000—a 100 per cent. increase. The population in 1892 had expanded to about 65,000,000. From 1892 cattle statistics report a falling tendency down to 1900, but in 1901, owing to altered methods of "enumerating," an extraordinary jump upwards from about 44,000,000 head of cattle in 1900 to 62,000,000 occurred. The high-water mark of cattle in the United States of America was in 1907 with 72,500,000. If the appendix dealing with imports of fresh beef is consulted, it will be seen at a glance how important have been supplies of beef under refrigerated conditions from the North American meat works. Up to the end of 1909, from the beginning of imports from all supplying countries, imports from the United States of America of chilled beef exceeded in quantity the supplies from South America and Australasia put together, frozen and chilled beef. In 1901 there were imported into Great Britain over 150,000 tons of chilled beef from North America, and from that point the demands of the increasing population in the exporting country began to tell, and in 1911 only 8,720 tons were despatched to Great Britain, with a tendency to further serious curtailment.

CHAPTER XIX

A MISCELLANY

In the compilation of a record of an industry with so many ramifications as the frozen meat trade, there is necessarily difficulty in assigning a proper position in the story to many side issues which have to be dealt with. A chapter of miscellanea is almost inevitable, and the following items of interest are, without apology, given, regardless of their heterogeneous character.

Congress of Refrigeration.

It is generally recognized that the campaign to secure markets for frozen meat among the countries of the Continent of Europe, a fight only now proceeding—slowly—to its more hopeful stages, has had a valiant protagonist in the International Congress of Refrigeration. The Premier Congrès International du Froid was held in Paris in October, 1908, and nearly 4,000 delegates, from forty-three countries, attended this successful gathering, which was organized by a French engineer, the late M. J. de Loverdo. The international movement has done much to stimulate an industry which, because of its special character, had previously lacked the stimulus of combined action within its ranks, and for this, as well as for the wide publicity gained for commercial refrigeration, M. de Loverdo, with his initiative and organizing genius, is principally to be thanked. His death, which took place on January 12, 1912, was felt to be due in some measure to his tremendous labours for this cause. At the Paris Congress 174 papers were read, and among the numerous resolutions carried were several bearing upon the subjects treated in this book. Two of them may be given. Mr. T. A. Coghlan, Agent-General in London for New South Wales, had devoted his attention at the Congress to the modification or abolition of regulations hindering the import of refrigerated produce into any country, in the interests of cheaper food, and his resolution, as accepted, was as follows :—

"That the Congress expresses its opinion that in order to reduce the cost of living to the working-classes, and to promote international trade, regulations which hamper the introduction into any country of frozen or chilled produce, and the storage, distribution, and sale of such produce in any such country, should be modified or abolished."

The other resolution (which was carried) was one proposed by Mr. Gilbert Anderson :—

"That, in view of the large expansion of the trade in refrigerated products, it is desirable that an international uniform standard of meat inspection be established and agreed to by the various countries exporting and importing animal foods so as to ensure the healthy condition of the meat."

Although no immediate action was taken with regard to this resolution, M. de Loverdo rather more than a year before his death proceeded to organize an International Meat Inspection Conference, to which it is still hoped the Powers may be officially invited by the French Government to send delegates in 1912. This is a great step in the right direction, and it is felt that the discussion of the subject among expert delegates of both meat producing and consuming countries may do much to convince the latter that the high standard of inspection ruling in the British exporting countries is a strong argument for the removal of the barriers that are now raised against this trade. It will, doubtless, be the policy of British producers to ask European delegates their highest demands as to an inspection standard, and then to satisfy those demands as far as their exports are concerned.

The veteran refrigerating engineer and inventor, Charles Tellier, whose pioneer work is recorded elsewhere, was present at the Paris Congress, and was accorded a great ovation. As an outcome of the Congress the Association Internationale du Froid was formed, besides which it was decided to hold a second Congress in Vienna two years later and further Congresses triennially. The British organization of these Congresses has been conducted by the Cold Storage and Ice

Association, the scientific body in the United Kingdom representative of the refrigerating industry.

The Second Congrès International du Froid was held in Vienna in October, 1910, Governmental representatives of the meat exporting countries were well to the fore, and Sir William Hall-Jones, High Commissioner for New Zealand, moved the following resolution, which was carried unanimously in full Congress :—

"That, subject to every reasonable regulation to ensure sound and perfect condition, restrictions operating to prevent the introduction of refrigerated meats and other food products into countries whose inhabitants would benefit by their addition to their food supplies, should be abolished or modified."

At the Vienna Congress it was decided to hold the Third Congress in America in 1913, and Chicago has been fixed upon as the centre for these meetings on September 15 to 20 of that year.

Lord Bacon's Frigorific Experiment.

It is reported that as early as in 1816 three Esquimaux were the forerunners of the commercial pioneers of sixty-four years later. They brought frozen ptarmigan and other game to Harwich packed in air-tight cases. They had to pay £50 duty and £10 for carriage, but this produce from the frozen North sold well. Back farther, to 1626, we come to the incident, which no conscientious chronicler can neglect, in which Lord Bacon fell a victim to his praiseworthy endeavours to open up the frozen meat trade. Here is the biographer's statement :—

King James died in 1625. His unfortunate and ill-requited Chancellor (Bacon) survived him for little more than a year. Always in feeble health from his youth, his life was finally sacrificed to an experiment. He believed that decomposition might be prevented by freezing (then an original idea), and he determined to ascertain, experimentally, if he was right. Therefore, one cold spring morning he drove to Highgate, alighted, bought a fowl at a neighbouring cottage, and stuffed it with snow which lay on the ground around him. By the time this operation was finished he felt greatly chilled, and sought warmth and shelter at Lord Arundel's house, which was near at hand. Here he was gladly welcomed by the household, given warm cordials, etc., and was put into a damp bed (!). From this fatal hospitality he never recovered ; and he seems to have been aware that he was in great danger, for he wrote to his absent host, comparing himself to the elder Pliny, who lost his life by too near an approach to Vesuvius when watching a terrible eruption, but adding that his own experiment had ended "excellently well." A fever and cold on the lungs closed the career of one of the greatest Englishmen one week afterwards. He died on Easter morn, 9th April, 1626, at the age of 66.

Let the reader note that it was not refrigeration but a damp bed which ended the career of Lord Bacon. Many people think his frigorific experiment brought the fatality.

Nor must the mammoth be lost sight of! During the last fifty years mammoths have been dug up in Siberia and their flesh found in excellent condition, after preservation for who knows how many centuries. Only the other day came the story from Russia of certain men of science who cooked and ate part of a mammoth found encased in ice in Siberia. Curiously enough, the Life of Bishop Ridding, which has lately been published, tells of a similar banquet. It was at the Westminster Deanery, whither, in his undergraduate days, Frank Buckland sometimes carried off young Ridding to partake of the trying hospitality of his father, the Dean. "His sideboard bristled with fossils, and his tables groaned under meats of which his guests ate sparingly. 'That was mammoth soup, made from the bones of a mammoth encased in Siberian ice from prehistoric times!' the host triumphantly informed his guests one day after they had eaten it."

South Africa as a Possible Meat Exporter.

The export of frozen meat of late years has been adopted in an experimental style in various countries which are not in the ordinary way on the list of suppliers to the British markets. One or two small lots of beef and lamb have arrived at Smithfield, London, from time to time from South Africa, Natal particularly. The venture probably has been undertaken more to test the quality of the Natalians' stock than with the idea of establishing a regular business. However, it tends to show that British South Africa hopes to supply herself with beef and mutton. Mr. Francis Harrison, the Acting Trades Commissioner for the Union of South Africa, communicated the following account of a tiny consignment of beef from Cape Colony that arrived in London in the summer of 1910 :—" The South African beef in question, consisting of a carcass weighing 126 stones, realized 3½*d*. per lb. The quality of this South African consignment was reported by Smithfield to be, if anything, superior to that of the beef from other countries, although some few improvements were suggested in the matter of dressing, and the recommendation was made to ship younger cattle in the future, and, if possible, in a cooled state, instead of being frozen hard."

Frozen Beef from St. Helena.

The unexpected happened when the import of a parcel of frozen beef (8 sides) from this island into London occurred in 1909. The Imperial Government are adamant, as a rule, to applications for subsidies or assistance for commercial purposes, but the authors learn that on behalf of the struggling farmers of this Crown Colony, the Government of Great Britain extended a helping hand in the way of supervising and assisting the small shipment of beef, which, unfortunately, realized an unremunerative price for the consignors.

Argentine Meat at 18d. an Ounce.

A startling incident occurred in 1910 at the Palermo International Fat Stock Show, held on July 15. Nothing more remarkable is chronicled in this book than the contest between the Las Palmas frigorifico and La Plata frigorifico for the possession of a bunch of five Hereford steers owned by Messrs. Duggan Brothers. A world's record was established when the La Plata people secured the animals with the fabulous bid of $11,500 m/n (£1,004) per beast, the cost of the meat working out at 1*s*. 6*d*. per ounce. The La Plata company took in all 177 head of cattle, at a cost of $310,600 m/n (£27,116). These cattle were all chilled and exported to the London market. The beef was of splendid quality, but suited to the winter rather than the summer trade, and the price at which it sold was not above that ruling for chilled beef of primest quality. When asked, after the Show, for some explanation of his action in buying the Herefords at over £1,000 a head, Mr. Pryor, the La Plata manager, stated:—"We want Argentine breeders to be assured that if they will only produce the right sort of animals, prices will be forthcoming to remunerate them amply."

Canterbury Mutton at the Lord Mayor's Show.

In the early days of the frozen meat trade one of the principal problems which confronted the importer was the bringing of the meat prominently under the notice of the masses of the population. One of the most successful efforts in this direction was made with carcasses of Canterbury mutton, selected from parcels consigned to the New Zealand Loan and Mercantile Agency Co., Ltd. By the combined ingenuity of the consignees, their Smithfield salesmen, Messrs. Ward and Stimpson, and a leading firm of carriers, a lorry load of these carcasses formed one of the exhibits of the Lord Mayor's Show in the year 1885. The mystery as to how this exhibit came to form part of the famous procession has never been satisfactorily cleared up by the City authorities, but the fact remains that the lorry with these carcasses hanging from a specially prepared framework of wooden beams traversed the streets of the City and formed an object of curiosity and interest to the thousands of cheering sightseers who were informed by placard that the carcasses in question were "New Zealand Frozen Mutton—the Meat of the Future." As a matter of fact, the lorry was driven by Mr. Fardell, principal of the firm of van proprietors referred to above, into the unformed procession behind the Guildhall. Excellent as was this advertisement, it did not end with the mere exhibition of the meat in the streets of the City, for the morning papers of the following day all commented upon this novel feature of the Lord Mayor's Show, some treating it as part of the authorized procession, and others indignantly inquiring how it was the lorry with its burden found a place in the time-honoured pageant. It is also recorded that though the carcasses lost some of their bloom by exposure to the murky November atmosphere, they were sold at Smithfield on the following morning at prices above the market value of the day for prime carcasses, as the buyers were anxious to exhibit them in their shops as frozen mutton which had formed part of the Lord Mayor's Show. A very smart advertisement!

Kosher Frozen Meat.

One of the modern distribution developments of the trade in Great Britain is indicated in the following advertisement which appeared in *The Jewish World* of August 19, 1910 :—

NOTICE TO THE JEWISH PUBLIC :—We hereby give notice that the Ecclesiastical Authorities and the London Board of Shechita have granted facilities for the importation of Chilled and Frozen Kosher Meat from Argentine. This meat is prepared by officials authorised by the Ecclesiastical Authorities, and appointed by the Board of Shechita. The meat is porged and Koshered prior to being shipped, and is therefore ready for use. The first consignment has arrived and will be followed by consignments at regular intervals. It is now on sale in the shops of retail butchers holding the Licence of the Board. Horwitz & Abrahams, Limited, Importers, 56 Aldgate High Street, E.C.

The first lot of kosher frozen fores of beef—from the Campana works—arrived in London in August, 1910. A five years' contract was entered into with the River Plate Fresh Meat Co. for the supply of regular consignments. The whole scheme was pushed forward by the Jewish authorities in the interest of the poorer sections of the Jewish community, owing to the high prices of home and European supplies. At the time of the arrival of the above-mentioned shipment, prime English Kosher forequarters sold at $6\frac{1}{4}d.$ per lb. ; the Argentine kosher fores made $3d.$ to $3\frac{1}{2}d.$ Over ten years ago Nelson Brothers made a bold bid for this business, but the scruples then of the Shechita Board on religious grounds could not be got over, and it is unfortunate that six months after the launching of the later enterprise recounted above the official announcement had to be made that the trade had not proved a success.

When the Jewish retail butchers expressed their failure to create a demand for the meat, the Jewish authorities arranged for its sale from a shop opened especially for this purpose, but even this enterprise was not rewarded with success, thanks to prejudice which, presumably, is to be found among Jews as well as Gentiles.

Doubts are expressed as to the likelihood of a trade in kosher frozen meat ever being built up, unless the authorities are prepared to alter their regulations to permit of salting after delivery in Great Britain. For it is stated that the salt used in koshering renders the meat unfit for the application of freezing.

Some description of koshering meat may here be given in the words of a Jewish Rabbi:—"With one swoop of the knife we have to cut two pipes in the throat—the œsophagus and the thorax. If at least half of each of the pipes are cut with one swoop, the animal is considered to be properly killed. It does not live more than a minute after the cut, and there is not the slightest doubt that within ten seconds after the knife has passed across the throat there is no consciousness of pain, and in that respect, at any rate, it has a great advantage over the ordinary method of pole-axing. When an animal has ceased to live, the slaughterer makes an examination of the lungs, and the slightest anatomical derangement or defect may cause the animal to be discarded as unfit for human food. If, on manual examination, everything is found to be satisfactory, then the lungs are removed and examined optically, and if then there is nothing unsatisfactory, the shochet requests one of the attendants to distend the lungs by blowing, and soon they reach the size and become just as they were in the body before death. If he puts a little moisture on any suspected part during the time of distention the air will escape. If any air escapes, it will be seen by the bubbles rising from the moisture placed there. If there are bubbles, the animal is considered to be 'trifah,' or unfit. If it is all right, the shochet places a little leaden stamp upon the various parts of the carcass, on which is written the word 'kosher' in Hebrew on one side and the day of the week on the other, and thus the meat is guaranteed to be fit for food. If our contention is correct, and we can trace most diseases by examination of the lungs, it is evident we do, at least, something from the sanitary standpoint in discarding such animals as are unfit. Even after the meat has been purchased and brought home, it is not done with. The blood is removed, so far as possible, by soaking the meat in water for half an hour, and then it is covered with salt, the salt being again removed by rinsing. That process is called making the meat 'kosher.' The flesh of beasts dying from any other cause than death by the shochet is forbidden. The veins and arteries are, before eating, removed by a peculiar process. Diseased or dying animals must not be killed. The 'kosher' Jew does not eat the hindquarters of the animal."

An Early Welcome to Frozen Meat.

In the account of the efforts of the pioneers of the frozen meat trade, the work of Mr. John Grigg, of Longbeach, New Zealand, must be noted. The following account of the reception in this country of meat shipped by Mr. Grigg on the *Dunedin* in 1881 is given by his son, Mr. J. C. N. Grigg :—

"When the sailing ship *Dunedin* left Port Chalmers, part of the cargo were some half-bred Shropshire wethers and lambs railed from Longbeach by my father, John Grigg, and consigned to the New Zealand Loan and Mercantile Agency Co. in London. Three lambs and two wethers were consigned direct to me at Cambridge, they were delivered to me on a Monday, and directly they arrived I had them put in a fishmonger's ice room and wrapped in a blanket, where they remained thawing until Thursday. Kettle was the name of the Jesus College cook who agreed to cook the three lambs and sheep all at once. On Thursday afternoon I went to the kitchen at 4 p.m., and saw them all being spitted, the joints being turned by the old smoke jack. I told Kettle the sheep were from New Zealand, but asked him not to tell anyone. Next day the men in the Jesus first boat (then head of the river) were lunching with me, and I had two cold joints of Canterbury lamb on the table. It was only natural that men in training should have two helpings, but when one said, 'You will think me damned greedy, Grigg, but the lamb is so good, I must ask for a third helping,' and another followed suit, I was delighted, and then I told them the lamb was from New Zealand, much to their surprise.

"Stephen Fairbairn, of Melbourne, was one of the men, and he was very excited, as his father was chairman of the first freezing company formed in Victoria. I wrote at once to my father to say how excellent the mutton and lamb were, especially the latter, which was perfectly sweet and good several days after leaving the freezing chamber.

"After this first shipment I received letters from my father expressing great confidence and hope in the future of the frozen meat trade. My father with others in Canterbury at once started to build freezing works at Belfast, after forming a small farmers' company."

Flock Maintenance and Exports.

It is interesting to note that the export of frozen sheep and lambs from New Zealand, increasing steadily (the period 1903—1904 was the only exception) as it has from 1882 to the present time, has not depleted the flocks of the Dominion; indeed, the frozen meat export has been accompanied by a steady growth, from about 13,000,000 head of sheep in 1881 to about 20,000,000 in 1901, at which date New Zealand was shipping 16 per cent. of her flocks to Great Britain in the form of frozen sheep and lambs, a percentage which rose in 1903 to 24 per cent. In 1903—1904 sheep fell in numbers, but an upward movement set in again in 1905, and has continued to the date of the last official enumeration. In the United Kingdom, with an area not very much more extensive than that of New Zealand, there are 32,000,000 sheep, and the annual killing of sheep and lambs is about 40 per cent., and of cattle 25 per cent.

So New Zealand's sheep stock will, no doubt, still further increase. The following figures illustrate this paragraph :—

—	Sheep in New Zealand.	Frozen Mutton and Lambs Exported.
		Carcasses.
1886 . . .	16,565,000	656,000
1891 . . .	18,128,000	1,894,000
1901 . . .	20,233,000	3,234,000
1909 . . .	23,481,000	5,035,000
1910 . . .	24,269,620	5,407,000

The wonderful position of New Zealand to-day in the export of frozen meat is due to the suitability of the land for sheep breeding and the evenness and mildness of her climate. That the Dominion can keep on exporting such a very large proportion of its sheep, and still go on increasing the number of its permanent flocks, is due to the high percentage of lambs reared. Roughly speaking, the percentage of lambs reared year after year is about 90 per cent. Many farmers in Canterbury rear 120 per cent., and in some cases still higher percentages are recorded.

Enter Mr. Hooley.

An incident that may be mentioned here was the temporary appearance of Mr. Ernest T. Hooley in the frozen meat trade arena in 1897. On May 21 of that year, his solicitors, Messrs. Ashwell and Tutin, issued a circular letter in which it was stated that he was prepared to "form a combination," if supported by the industry, with the object of placing the frozen meat trade upon a satisfactory footing. His plan was to amalgamate the freezing companies in Australasia, and he was prepared to purchase the companies on the basis of the net assets as per balance sheet plus a bonus equal to the aggregate amount of dividends paid during the previous seven years. A new company would be formed in London with a larger capital than the aggregate values of the concerns taken over. Mr. Hooley's idea was that the trade on this side lacked capital and organization, and he thought a combine would improve market prices. Mr. Marshall Stevens, the late managing director of the Manchester Ship Canal, worked out the details of this proposal, which came to nought. The complexities of the frozen meat business did not yield, as other problems had done, to Mr. Hooley's magic touch.

A Cargo in Coffins.

A rather amusing but disastrous incident connected with the early shipments of frozen meat in the New Zealand trade, is told by Mr. Frank Coxon, the well-known Australian consulting engineer. The incident may be told in Mr. Coxon's own words: "It occurred in connection with the sailing ship *Mataura*, I think on her third voyage. She was sent to Auckland to freeze and load a cargo of mutton at that port, and the charterers of the frozen space were anxious to encourage every local industry in their power, with the result that it was decided that each carcass of mutton should be placed in a Kauri box, or 'coffin,' as these receptacles were popularly called on board, in lieu of the usual cotton bags. This was carried out, and after considerable freezing had been done, the captain, who was not satisfied with the appearance of things, communicated with his owners, who at once sent me to Auckland to look into matters. On examination of the cargo, I condemned the system as being quite unsuitable, the boxes entirely obstructing the circulation of the air round the meat; this I explained to the charterers, but they refused to act on my advice, and the ship sailed for London. Needless to say, the whole cargo had to be jettisoned at sea before its destination was reached."

A Frozen Meat Cooking Recipe.

The following "directions for cooking New Zealand and other frozen mutton," issued by the New Zealand Loan and Mercantile Agency Co. in the early eighties, shows that more care was then taken in this important department of the trade than is the case now. "Frozen meat, like English, improves by hanging. The hindquarters will keep a week in cool weather, the forequarters may be cooked sooner. As there is a tendency for the juice to run from the mutton while thawing, it should be hung in such a way as to check this. The hind-quarter, haunch, and leg, should be hung by the flaps, the knuckles hanging down, the loins and saddles also by the flaps, giving them a horizontal position. This meat should not be soaked in water for the purpose of thawing (as some suppose), but hung in the larder or other dry, draughty place, and wiped occasionally with a dry cloth in damp weather. Flour should not be used, as it is apt to turn sour. When put down for cooking, the chump part of the leg or loin should be exposed to the fire, or hottest part of the oven, for a few minutes, to toast the part cut and so seal it up, thus keeping the gravy in the joint."

A Frigid Message.

The housewife who, about 1888, found pushed up alongside the bone in a leg of mutton (which she had purchased as English at 9*d*. a lb.), a piece of paper bearing this legend :—"Where did you buy this leg, and what price did you pay ? inform J. C., Ashburton, Christchurch, N.Z.," must have been surprised.

Tallerman Enterprises.

Some interest attaches to the work done by Mr. Daniel Tallerman in introducing Australian preserved meats at cheap price during the times of scarcity that preceded the *Strathleven* voyage. Mr Tallerman arrived from Melbourne to settle in London in 1868, bringing some packages of meat in tallow with him. This meat was mild-cured, boned mutton, enclosed in a linen envelope, and was rolled like sides of bacon. It was packed in casks, and tallow was run round the meat. This mutton carried well, and when chopped into mincemeat and cooked with potatoes it made an appetizing, nutritious, and economical dish, the potatoes absorbing the fat. At one time there were 100 tons of the meat on hand in London. Mr. Tallerman and his "Penny Dinners" at Norton Folgate were famous from 1868 to 1872. The movement was initiated by a gigantic banquet to 1,400 London working men and women at the Lambeth Baths on December 1, 1869, at which many Australians were present. The following is from *Punch*, January 8, 1870 :—

"The French Emperor having expressed a desire to test some of the Australian meat, which furnishes the penny dinners in Norton Folgate, Mr. Tallerman, manager of the Australian Meat Agency, at once submitted samples to the Tuileries. The Emperor, on the principle of *fiat experimentum in corpore vili*, caused some of the meat to be cooked for the soldiers on guard. Finding that they survived it, and even, like Oliver Twist, asked for more, he ordered the same dish to be set before the principal officers of the Imperial Household. The officers, unlike the privates, shuddered, but ate, and to their own amazement, relished ; and then the Emperor tried it himself, pronounced it good, and expressed his gracious intention of causing a more extended trial to be made—we presume on the Empress and the *entourage*."

The British public were very much prejudiced against preserved meat in any form owing to the disastrous outturn of some shipments from South America of "charqui" (1866 to 1868). The *Observer* of October 31, 1868, spoke of "the macerated caoutchouc-like lumps of charqui which earned the execrations of the populace"; Australian meat was popularly termed "charqui" for some time.

Mr. Tallerman formed a company for each class of meat he handled. The Norton Folgate campaign was conducted by him at his own risk and expense. Later on he turned his attention to tinned meats, which he showed at the Vienna Exhibition of 1873. The Emperor of Austria tasted Australian meat, and was so pleased with it that he made Mr. Tallerman a "Ritter Kreitz" of the Order of Francis Joseph. Mr. Tallerman acted in London for the French company which financed M. Tellier's *Frigorifique* venture, and he imported into Great Britain about 10 tons of the meat. The newspapers from 1868 for ten years or so were full of Mr. Tallerman's enterprises for introducing meat preserved in all sorts of ways.

In 1874 he took premises in Upper Thames Street, and fitted up a patent refrigerator. In 1879 he opened up a retail department for the selling of American dressed beef in one of the arches of the railway ; Mr. Tallerman and the salesman in charge (Mr. Burket) worked in co-operation. This is mentioned on account of the enormous interest excited. Mr. Tallerman declares that 50,000 people a week visited the premises, many of whom purchased meat, and that £100 a week was paid in tolls (½*d.* per lb.). A merry season of fourteen weeks was experienced, in which £30,000 was taken. This enterprise, however, was short lived, the salesman-partner left, the boom collapsed, and the place was soon closed.

Frozen Meat Squibs.

An amusing hoax was perpetrated in the seventies on the Press and public of Australia. It proceeded from the pen of a well-known gentleman in Brisbane, who gravely described how experiments had been successfully carried on at freezing works in Sydney Harbour on a live lamb, which after having been frozen hard under the influence of drugs had been unfrozen again at the end of six months and thawed out alive (!) and, with the exception of the tail, absolutely sound. (The tail had been snapped off in the process.) This was cabled home and created great commotion in interested circles, and a leading London journal, carried away with the excitement, devoted a leader to the subject. Allied to this yarn there was current at the same period, say about 1875, at the squatters' clubs in Queensland, the story that steps had been taken to analyse the poison which the "mason wasp"—to temporarily poison its victims—injects into the bodies of the spiders placed in the nest in which the eggs of the wasp are sealed up. (These clever insects, well known to all residents in Queensland, stuff the bag charged with eggs and spiders into keyholes, gun-barrels, etc., and then close up the apertures. When the young wasps emerge from the eggs they break up the sac and admit the air, whereupon the unfortunate spiders, which had been in a condition of suspended life, revive, only to be devoured by the young wasps.) This suggested a lively squib, directed at Mort and his work. When the mason wasp's poison had been fixed, it was to be artificially manufactured and injected into the bodies of cattle and sheep intended for the English market. After the animals had been treated, they were to be placed in ships' holds and hermetically sealed up. On docking at London, the inrush of air, following upon the opening up of the chambers, would reanimate the stock, like the mason wasp's spider victims, and, hey presto! fat beeves and plump woollies from Australia's plains would be soliciting the custom of the English butcher.

CHAPTER XX

THE DIETETICS OF FROZEN MEAT

THE dietetic value of frozen meat is really proved by the fact that it is consumed and found perfectly acceptable by all classes in Great Britain. But, possibly, its food-value may not be completely represented by its popularity. Public experience and scientific tests have long ago vindicated frozen meat, so that when, for instance, in September, 1909, Mr. Rowland Hunt in the House of Commons asked ironically if the War Minister was aware that, according to a pamphlet issued, experts consider that one pound of freshly-killed beef is worth a stone of frozen meat, nothing more than an outrage on common sense was perpetrated. Mr. (now Lord) Haldane neatly countered by saying that "he was sure that the hon. member had, without knowing it, flourished on frozen meat." The Minister realized that to reply directly to the query would be an insult to the intelligence of the House of Commons! It may be well, however, to bring forward in this chapter conclusive evidence that frozen meat, as to primeness of quality and excellence of condition, is sound and sterling food; that the processes through which it has passed have caused no injury and deprived the meat of none of its nourishing value. Further, there may be put on record statements of practical men, and the results of scientists' researches.

To begin at the beginning, on p. 98 are given some particulars of the enormous sums paid by breeders in the Argentine for pure-bred sheep and cattle imported from Great Britain. The progeny of the rams and bulls purchased from the pedigree herds and flocks of the British breeder have been sent to England in the form of frozen meat. The object of the South American buyer of these expensive animals was to improve the native stock in the Republic so that the dead meat exported might be of unexceptionable quality. Without forcing any comparison between the frozen meat imported into Great Britain and the home stock slaughtered for food, there can be no doubt that, if it were possible to compare the frozen and chilled meats marketed throughout England and Scotland on a given day with home-fed meats of all grades, and strike averages for quality, the imported meat would not be the inferior. As in Argentina and Uruguay, so in Australia and New Zealand, where the whole of the flocks and herds are of British blood. For many years past the stockmen in those countries have been steadily buying pedigree sheep and, to a less extent, cattle, from English and Scotch breeders. The Kent or Romney Marsh, Shropshire, Southdown, Hampshire and Oxford Downs, Dorset Horn—numbers of rams from these meat breeds are exported to Australasia yearly. What is the meaning of all this ? Simply that in eating frozen meat we are eating—to a great extent—just the same class and grades of meat as we should order from English and Scotch grazing districts. It would be a good thing if the Government of New Zealand were to cause a circular to be sent to every householder in England announcing the fact that the Kent or Romney Marsh sheep exhibited by Mr. Ernest Short, of Parorangi, New Zealand, at the Argentine 1910 Centenary Show beat the British sheep of the same class and took championship honours. This is the stock from which New Zealand meat is bred !

Scientific Tests.

In previous chapters have been described the processes through which the frozen meat passes from the time it leaves the freezing works until it arrives at the English butcher's shop. The experience of thirty years in conducting the trade has eliminated all faulty methods and has established such a smooth procedure that accidents and mischances in the conveyance of the meat from point to point which would interfere with good condition are relatively rare, and if the treatment to which reference has several times been made, viz., sufficient thawing out, is observed frozen meat ranks with the best!

To revert a while and meet the objection of the critic who says that freezing injures meat, the point might almost be answered in general terms. Did the Beefsteak Club find that the freezing process had spoiled the flavour of the Canterbury wethers that pleased them so mightily? Do the millions who enjoy the tender frozen lambs find the pleasure to their palate lessened owing to the carcasses having been subject to King Frost for a couple of months or so? But let us see what scientific men say.

Back in 1896 the London journal the *Hospital* printed some articles giving the results of certain quantitative cooking tests made by Mr. Samuel Rideal, D.Sc. (Lond.), F.I.C., which tests showed that frozen meat was digestible, nutritious, palatable, and economical. This examination certainly has historical interest, in view of its being undertaken for the purpose of publicly dissipating current prejudices against frozen meat. Under the direction of a clever practical cook, two legs of mutton of nearly equal weight were baked for an equal length of time in ovens equally heated. No. 1 was English grown and killed and cost 10*d.* per lb.; No. 2 was the best New Zealand frozen mutton, thawed at the butcher's, and sent in ready for use. Two and a quarter hours were given to the baking, and the following are the resultant figures:

	English.		New Zealand.	
	lbs.	ozs.	lbs.	ozs.
Weight when delivered by the butcher	8	6	7	14
Weight when taken from the oven	5	15	5	13

Seven sub-tests were made as to weight of slices suitable for hospital diet, bone and waste, gravy, etc., in all of which the frozen joint held its own. "It is clear," wrote the *Hospital*, "even from one experiment that the assertion that New Zealand meat is essentially and invariably more wasteful than English cannot be supported." Then the *Hospital* commissioned Dr. Rideal to undertake tests concerning the nutritive and digestible properties of frozen meat. The three kinds of mutton selected for comparison were London-killed Scotch, at 10*d*. per lb.; frozen Canterbury, at 8*d*. per lb.; and frozen Australian, from merino sheep, at 4*d*. per lb. In these experiments the Australian and New Zealand mutton well held their own with the Scotch meat.

In 1897 Dr. Rideal made an exhaustive examination of Queensland beef, comparing it with chilled American and English ox beef. In view of the oft-repeated statement as to the effect of freezing upon meat, it is of importance to place on record this sentence from the report: "In the microscopical examination, the samples when thawed showed that the meat fibres had not been ruptured or altered in any way by the hard freezing process adopted by Queensland shippers." Dr. Rideal continued :—

I can confidently assert that both with regard to digestibility and for the preparation of soups or beef tea the hard frozen meat is of intrinsically the same value as that which has been chilled or freshly killed.

The above quotation of a scientist's statement that freezing has no injurious effect on the structure of the meat is necessary in view of the fact that critics constantly assert that thawing frozen meat acts on the fibre as does melting ice in water pipes in a sudden thaw.

In 1907 Dr. Rideal made further analytical experiments to determine the nutritive value of frozen and chilled Argentine beef and of Australian lamb and mutton as compared with English meat. Analysis of lean from samples of beef tested gave the following percentages :—

—	Beef Shins.			Beef Steaks.		
	Argentine Chilled.	Argentine Frozen.	English.	Argentine Chilled.	Argentine Frozen.	English.
Water	74·32	74·84	75·66	66·28	74·84	64·47
Fat	1·27	0·88	1·11	9·18	2·90	13·04
Meat fibre extractives and associated mineral matter	24·41	24·28	23·23	24·54	22·26	22·49
Total nitrogen ...	3·21	3·34	3·13	2·88	2·56	2·73

In testing for the comparative digestibility of steak the following results were obtained :—

Argentine Chilled.	Argentine Frozen.	English.
61·99	69·89	63·78

the percentages of lean portions of meat available for digestive purposes turning out in the three cases as follow :—

Argentine Chilled.	Argentine Frozen.	English.
84·4	85·3	72·2

The housewife, who in the past has had preached to her such tales as to the wasting qualities of imported meat on cooking, might with profit study the following results of baking tests made in the course of Dr. Rideal's investigations :—

—	Welsh Lamb.		Australian Lamb.		English Mutton.		Australian Mutton.	
	lbs.	ozs.	lbs.	ozs.	lbs.	ozs.	lbs.	ozs.
Weight when delivered ...	4	15	5	7½	8	10	8	9
Weight when taken from oven	4	2½	4	15	6	13½	6	7¼
Weight of slices suitable for hospital diet ...	2	13½	3	8¾	4	4½	4	6
Weight of bone and waste	1	0½	1	1½	2	4	1	11
Pure bone		9		8¾	1	2		13
Dripping		4¾		4¾		11¾	1	0½
Gravy in dish after carving		1½		2		¾		3½
Gravy under dripping ...		½		¼		¼		½

Then as to its digestibility, the same authority arrived at the following figures of percentages :—

	Welsh Lamb.	Australian Lamb.	English Mutton.	Australian Mutton.
Water	58·55	57·65	57·79	57·32
Fat	4·11	6·62	7·53	9·68
Organic matter and associated mineral matter...	37·34	35·73	34·68	33·00
Percentage of nitrogenous organic matter digested	19·91	15·25	18·20	12·48

The concluding remark of Dr. Rideal in his report as to this close scrutiny of the value of refrigerated meat is that "It is satisfactory to find the general opinion confirmed that no incipient decomposition or hydrolysis takes place under cold storage; while this further series of tests also fully supports the favourable conclusions arrived at as the result of previous experiments, by further illustrating the satisfactory food values of frozen mutton and lamb." To sum up, the analyst, in a paper read before the first International Refrigerating Congress in Paris in 1908 said: "In a series of quantitative cooking trials I found that the food value of frozen was not less than that of fresh meat."

There is also solid material at hand in favour of frozen meat in the highly technical paper of Mr. W. D. Richardson, chairman of the Chicago Section of the American Chemical Society, delivered at the same Congress, on the subject of "The Cold Storage of Beef and Poultry." This essay is illustrated by a wealth of diagrams from photographs of meat substances under freezing and thawing conditions. Changes in flesh foods under cold storage were the problem, and the chief change investigated was bacterial action. Experiments were made on fresh and frozen beef knuckle; the latter was observed up to 554 days, and the conclusion the scientist came to was that "frozen meat from 93 to 554 days old is in the same condition bacterially as meat from freshly slaughtered animals." This is a high testimony to the preserving powers of refrigeration, though, of course, the cost involved in prolonged cold storage would of itself serve to ensure that the public meat supplies are never submitted to such lengthy warehousing.

Two paragraphs may be quoted from this valuable paper:—

> There are no facts known at present which would militate against the possibility of flesh preservation for an indefinite length of time under proper conditions of storage. Ostertag-Wilcox (*Handbook of Meat Inspection*, p. 824) says: "Cold is unquestionably the best method of preserving meat. It causes no alteration in the meat, either with regard to taste or nutritive value."
>
> In conclusion, cold storage appears to be the best method of preservation of flesh foods at present known to man, inasmuch as it modifies to a less extent the appearance and quality of the product than do other methods. That improvements may be and will be introduced into cold storage practice must be admitted, but that in principle cold storage prevents or inhibits to a large extent the forces of deterioration, chemical and biochemical, cannot be denied. It is a satisfactory, efficient and safe means of preservation of flesh foods for long periods of time.

Another endorsement of the fact that the nutritive matter is identical in home grown and frozen mutton has been made by Professor James Long in the record of the results of a test which he undertook a few years ago at the request of the War Office. To make his test Professor Long took thirty legs of wether mutton, Scotch, English, New Zealand, Argentine, and Australian, and the following extract may be made from his report :—

If we make a general comparison between the home-grown and the imported mutton we obtain the following results, which represent the average of the fifteen joints :—

	Average Weight per joint. lbs. ozs.	Water Per cent.	Real Nutritive Matter (Fat and Lean) devoid of moisture. Per cent.
Home-grown Mutton :	10 7	43·78	40·53
Imported Mutton :	8 8	46·17	39·37

Practically speaking, the real nutritive matter present in each lot of meat, the home-grown and the imported, was identical, for the difference was but slightly more than 1 per cent. The quantity of meat in its moist condition was actually higher in the imported than in the home-grown legs of mutton, but the fat, which counts materially owing to its useful feeding properties, told in favour of the British meat, although a large proportion is invariably wasted.

These favourable verdicts, from persons so competent to pass them, make out a very excellent case for frozen meat, well worth the attention of the British public and the peoples of Continental countries. Indeed, the whole of this volume, when carefully epitomized, yields testimony, beyond question, not only of the dietetic value of the supplies of frozen and chilled meat which arrive at the ports of Great Britain, but also, speaking generally, of the enormous benefit to a nation engaged in industrial activities which these supplies confer, in being regular, skilfully handled, cheap, and, as to quality and soundness, guaranteed by Government veterinary certificate at the country of production, and by the examination of the inspectors of the Medical Officers of Health at the ports of importation.

Further tests made by English and foreign scientists of the dietetic value of frozen meat might be cited in this chapter, and additional evidence leading to a favourable conclusion of the frozen meat case would not be difficult to bring forward. But the authors think that enough has been written, bearing in mind that the general public of Great Britain have voted for frozen meat in an unmistakable manner, viz., by consuming 5,664,000 tons of frozen mutton and beef in the thirty-one years covered by the duration of the trade, 1880 to 1910.

The Testimony of Medical Officers of Health.

The reports issued by the Medical Officers of Health for the City of London and the Port of London (Dr. W. Collingridge and Dr. Herbert Williams) teem with statements testifying to the excellence of the imported frozen meats.

In looking through the Annual Reports of these Medical Officers one sees numerous references to the fine quality of frozen meat passing under the observation of their officials. The following is an extract from Dr. Williams's Annual Report for 1906 :—

"The arrangements for keeping the meat at a low temperature have been much improved, and it is only occasionally, when some defect in these arrangements occurs, that the meat on examination is found to be unsound. The bulk of the meat arrives in first-class condition, as is shown by the fact that only 4,279 carcasses of mutton and 844 quarters of beef (out of a total of 8,799,892 carcasses and 1,449,715 quarters) have been found unfit for food during the year as a result of unsoundness,* and *not a single carcass has been seized on account of disease.*

* *I.e.*, damage caused by accidents during transit.

The Medical Officers of Health have had occasion to find fault as well as to give praise. In their Reports for 1910 and 1911 they had to draw attention to a serious blemish in Australian frozen beef. The Commonwealth Government thereupon took measures, in the shape of a more organized inspection system, to meet the difficulty and remove the Medical Officers' cause of complaint.

The fact that Tommy Atkins abroad and at home is wholly or in part fed on frozen meat, and flourishes on the diet, is an excellent testimony to the sound quality of the fare. In this connection it should be mentioned that there is on record a statement made by the chief of the United States Commissariat at Manila that the American troops in the Philippines have found the Queensland frozen beef of first-class quality, suitable for keeping up the fighting power of men. And he ought to know, for he has bought a lot of it! As for the general utility of frozen meat as a war supply, a leading witness before the Royal Commission on the South African War said of cold storage that it "saved South Africa."

The Sound Quality of Imports.

Referring to the position of frozen meat on Smithfield Market, a well-known figure there has remarked: "The trade has now reached such a point that it is really difficult to conceive what our position would be were it not for frozen meat." The same suggestion occurs with more force with regard to the public. What would happen if there were no frozen and chilled meat to be had? The benefits of refrigeration have penetrated into all the conditions of life. Travellers recollect down to the eighties the wretched stock and miserable fowls carried on board steamers to supply fresh meat for passengers. Since frozen meat came along, the mail steamer is provisioned as easily and as well as the hotel.

If it may be stated that the vegetarian restaurant of our towns—the first of any size was that opened in Queen Street, Cheapside, London, in 1882—was, broadly speaking, a protest against meat scarcity and dearness, it may also be said that such food purveying institutions have probably been robbed of full development by the advent of cheap frozen meats. Though vegetarianism has its advocates, there is undoubtedly a consensus of scientific opinion that a generous meat diet is necessary for the hard worker.

CHAPTER XXI

WHAT THE TRADE HAS DONE FOR AUSTRALIA

(Specially Contributed by Mr. John Cooke, of Australia.)

It would be freely conceded by most that no man is in a better position to say what the frozen meat industry has meant to Australia than Mr. John Cooke, senior partner of the firm of John Cooke and Co., Australia. Mr. Cooke has contributed the following lines concerning the part which the frozen meat trade has played in developing the resources of Australia:

"Prior to the introduction of refrigeration in the preservation and transport of fresh meat, the only methods of disposing of the surplus sheep and cattle produced in Australia were :—

(*a*) Preserving and packing mutton and beef in cans ;

(*b*) Boiling down the carcasses for their tallow.

"Neither of these methods offered much encouragement to stock producers to increase their flocks and herds, as the returns from canned meat were frequently very trifling, while those obtained from tallow refining were small, after providing for the cost of manufacture, casks, etc.

"The values ruling for sheep almost entirely depended on their fleeces, the carcass on average representing very little indeed in the price. In the same way the value of any bullock or cow outside the category of a fat beast fit for local consumption depended entirely on the hide and the tallow that could be extracted through the digestor. There was accordingly little to stimulate station owners to improve their holdings or increase their flocks and herds, seeing that they could never reckon on getting a profitable market for their surplus stock when these were ready for market or when they wanted to dispose of them. Likely enough a dry spell would intervene when they were overstocked, and, having no profitable outlet whatever, they had simply to look on while their stock gradually disappeared. These conditions made stock-raising in Australia a very precarious proposition, and greatly retarded that development which the magnificent lands and healthy and genial climatic conditions ought to have brought about. There have been occasions when prices for sheep were so low that they were sold by the score instead of per head.

"The frozen meat export trade has changed all that, and has proved to be by far the most potent factor in the growth and prosperity of Australia during the past twenty years. The waste lands of the Crown have not only been rendered more valuable *per se*, but lessees have been encouraged to expend money in improvements, thus increasing the carrying capacity of their runs, and converting what might have become rabbit warrens or barren plains into wool- and meat-producing properties. All owners of rural land, whether of 100 or 100,000 acres, have had a permanent and increasing value placed upon their holdings owing to the fact that the stock they raised has had established for it a profitable minimum value. Landowners have accordingly been enabled to improve their holdings by fencing, water conservation, and cultivation, and furthermore, they have felt justified in spending money in the way of introducing improved strains of blood into their flocks and herds.

"It will be manifest that all classes of live stock, whether stores or fats, have benefited by the export of fat carcass meat; in fact, in many cases prices obtainable for store sheep and cattle have approximated, and, indeed, occasionally even equalled, those for prime slaughter animals.

"The long and terrible drought which culminated in 1902 and brought the total sheep stock of Australia from about ninety-eight down to about fifty-four millions, and cattle from about twelve down to seven millions, temporarily checked the export of frozen meat, and it is hardly necessary to point out that had that export industry not been maintained and extended, it would have been utterly impossible during the brief period of seven years to have restored the numbers of sheep to close upon 100,000,000 in 1910, the world's high water mark of pastoral activity. No more striking proof is needed of the enormous advantages accruing to Australia from this industry, and it is all the more extraordinary when it is realized that rabbits and other scourges have been prevalent during the last decade.

"Australian wool-growers have reason to be very thankful to meat exporters for the security it affords them to know that in meeting the ever-increasing demands for wool they possess two strings to their bow instead of one. Had it not been for the enterprise of the freezing companies, the sheep farmer would hardly have dared to enlarge his flocks so materially, and he would thus have missed the splendid market for wool which he has enjoyed on his increased output for several years past.

"The enormously increased employment directly arising from this trade, to station hands, butchers, freezing works labourers, fellmongers, tanners, wharf lumpers, and others, has been a factor of much importance in every State of Australia; while the indirect advantages to all local traders and manufacturers have been incalculable.

"The Government railways have reason to regard refrigeration as a source of great income to them, the transport of live stock by rail having enormously developed since its inception, while the carriage of frozen meat and the various by-products has also produced large revenues for the various States. No interest has realized, and benefited by, the growth of this frozen meat trade more than shipping, and one has only to turn to the expansion in the number and size of ocean steamships now trading with Australian ports to judge of the vast impetus that has accrued thereby to Australian production. And the cry is still for further refrigerated tonnage.

"The phenomenal increase in the export of dairy produce and fruit would not have occurred but for the facilities for effective transport initiated and mainly supported by the meat trade, and in this sense alone the amount of profitable labour that has been rendered possible can hardly be overstated. Further, while the exports of mutton and lamb have been the mainstay of the industry, the quantity of frozen beef shipped abroad has been steadily extending, and would have been much greater but for the long drought that ended about seven years ago. It can readily be understood that the restoration of herds of cattle takes a much longer period than the recovery of flocks of sheep.

"The commendable schemes of immigration now being so energetically promoted by all Australian States have manifestly been made much more practicable through the advertisement given to Australia by the ever-increasing distribution of Australian meat through the length and breadth of the United Kingdom.

"Other localities have taken their part in the trade, notably South Africa, the Philippine Islands, the Mediterranean ports, and it is not difficult to foresee that all the great European countries—France, Germany, Austria, etc.—must follow the examples of Italy and Switzerland, and, by admitting frozen meat free from impossible conditions and on a moderate scale of customs duties, enable their citizens to obtain wholesome meat at prices within their means."

THE RIGHT HON. SIR JOSEPH GEORGE WARD, P.C., LL.D., PRIME MINISTER OF NEW ZEALAND, 1906—1912.

MR. JOHN COOKE.

MR. HERBERT GIBSON.

THE CONTRIBUTORS OF CHAPTERS XXI., XXII., AND XXIII.

To face p. 302.

CHAPTER XXII

NEW ZEALAND'S GAIN FROM FROZEN MEAT

(*Specially Contributed by the Right Hon. Sir Joseph Ward, P.C., G.C.M.G., Prime Minister of New Zealand*, 1906—1912.)

THE experiments in the oversea carriage of meat in a frozen state were eagerly watched in New Zealand. In 1880 the pastoral interests of the Colony had reached a crisis. The flocks and herds had increased to numbers which provided meat in quantity far beyond the requirements of the Colony, while a greater revenue than that which was derived from wool, tallow, skins, and hides was required. The only forms in which the meat could be exported were preserved meat and tallow, and the market for the former was glutted, so that the boiling-down for the tallow was the only outlet for the surplus stock. The situation was intensified by the decline in the price of wool which set in about that time, and by several successive harvests being damaged by bad weather. The problem of profitably using not only the pastoral but also the agricultural lands of the Colony was one of vital importance, and means by which the surplus meat would be given a reliable market value were looked to as the only hope of restoring the agricultural and pastoral industries to a position of stability. When reports were received of the success of the initial frozen meat shipments from Australia to London, steps were at once taken to enter into the new trade thus opened. Before the close of the year 1880 preliminary action had been taken to form companies to embark in the refrigerating industry (the potentialities of which in regard to dairy produce as well as to meat were recognized later on), and early in the year 1882 the first shipment of frozen meat was despatched from New Zealand to London. The enterprise was crowned with success, and quickly every pastoral and agricultural district in the Colony entered into the trade, which has expanded to a volume exceeded in value only by wool, and which has not nearly reached its limits. This expansion has, no doubt, been greatly assisted by the popularity which New Zealand meat has gained owing to the efficient system of inspection by which the freedom from disease of all exported meat is assured.

From the beginning in 1882 with a small shipment, slaughtered on land and frozen in the hold of the carrying ship (a sailing vessel), the trade has grown by 1910 to the following dimensions :—

No. of meat freezing works in operation, 1911	31
Daily slaughtering and freezing capacity in sheep carcasses, 1911	82,000
Cold storage capacity, sheep carcasses, 1911	1,860,000

EXPORTS, 1910 :—

			Tons.
Mutton, carcasses...	1,997,633 ...	Weight	48,930
„ legs and pieces	— ...	„	2,587
Lamb, carcasses	3,515,001 ...	„	53,645
Beef, quarters	— ...	„	26,926
	Total meat... ...	„	132,088

Rabbits (number) 3,138,888. Total value (export), £3,883,065.

In addition, meat preserving is an adjunct to most of the meat freezing works, and adds considerably to the volume and value of the output.

The value of the land, buildings, machinery, and plant employed in the trade is approximately £1,625,000 ; the number of hands employed, 3,250 ; and the wages paid in the year 1910, £325,000.

When the question of entering upon the export of frozen meat was under consideration, it was held that if 1*d*. per lb. net could be realized by growers for their mutton, the prosperity of the pastoral industry would be assured ; the average prices actually realized have generally ranged from 2½*d*. to 3*d*. per lb. for mutton, 3½*d*. to 4*d*. for lamb, and 2*d*. for beef, besides the value of the skins, hides and fat.

The success of the frozen meat trade has been reflected in the rapid progress and great prosperity of the Dominion. This is to some extent shown in the following comparative official figures, though the progress is not entirely due to the frozen meat trade, but is largely contributed to by the kindred industry of dairying :—

Population	1880	484,864	...	1911	1,057,818
Capital value of land in New Zealand	1882	£101,000,000	...	1910	£277,630,083
Land in cultivation, area	1880	4,768,192	...	1908	15,679,943
No. of cultivated holdings	"	24,147	...	1908	75,152
Exports	"	£6,102,300	...	1910	£21,758,551
No. of sheep in the Dominion ...	1881	12,985,085	...	1911	23,996,126
No. of cattle	"	698.637	...	1911	2,020,171

From the inception of the trade in 1882 to December 31, 1910, there were exported from New Zealand 71,000,000 carcasses of mutton and lamb, besides millions of legs of mutton, notwithstanding which immense drain the number of sheep in the Dominion has in the same period nearly doubled.

The frozen meat trade has revolutionized agriculture in New Zealand. It has greatly increased the value of land and has caused the adoption of improved systems of farming and more thorough cultivation, though the capabilities of the land for production have not yet been more than very partially exploited. The imperfect results already obtained have demonstrated that New Zealand has a capability of production surpassing, area for area, that of any other country in the world, and requiring only population, skill, and energy to develop it. The trade, supplemented by dairying, bore the chief share in rescuing the country from depression, and enabling settlement to be promoted, with the result of a vast increase in production and also in the national wealth; and the improvement in the methods of farming has restored the fertility of a large quantity of land which had been exhausted by repeated cropping, and has increased the fertility of large areas which were previously considered to be not worth cultivating. Above all, the pastoral and agricultural industries of the Dominion have never before been in such a sound position as at the present time.

CHAPTER XXIII

ARGENTINA'S DEBT TO REFRIGERATION

(*Specially Contributed by Mr. Herbert Gibson.*)

It is not easy, even for those whose memory carries them back a generation or so, to step out from the throb and bustle of the great Argentine freezing factories, from the flocks of comely sheep of typical breeds, from the great paddocks in that mighty zone of twelve million acres of alfalfa where herd after herd of sleek Shorthorns graze, from the whole panorama of this modern pastoral industry, and to see things as they were when in 1877 the steamships *Frigorifique* and *Paraguay* received in their chilled holds the first consignment of River Plate fresh meat to be conveyed to Europe.

It is not, after all, so long ago. Men who still like to believe themselves young can talk of their stockbreeding experiences of those days. Wire fencing had scarcely yet come into general use and most of the live stock was herded and rounded up in open country. The water was drawn from the wells in a rude sleeve made from an ox hide. The thatch and wattle of the stockman's hut were bound with raw leather thongs, and often as not a horse-hide did service for a door. Cattle were unimproved. Here and there progressive breeders such as Juan Fernandez, Pereyra, Lezama, and others, had introduced Shorthorn blood to their herds, but the merits of English cattle encountered opposition. Their hides were too thin, and the hide on the ox's back constituted in those days half its sale value. Their flesh was too deep and fat to take the salt properly, and salted sun-dried *charqui* beef was the sole meat export. They wanted, too, more feeding than the hardy thick-skinned *creole* bullocks. So breeders would have none of these sleek roan and red English beasts, and the bulls bred from them were unsaleable except to a most limited circle of stockmen. The great majority preferred to go on producing six-year-old hide-bound bullocks sure of sale to the *saladero* where only a heavy skin and a good salting flesh were required.

The sheep, with few exceptions, were merinos of varying grades of quality. Some breeders, mostly Englishmen, had imported Lincolns to their flocks. Their carcasses gave a better return to the tallow boilers. Wool, tallow, and skin were all that sheep were there for.

Of course, those were the good old days. Every middle-aged man since Adam has said that. But the concern of the world is more with the welfare of the community than the *joie de vivre* of the individual. If that welfare is to be gauged by the pastoral wealth and revenue of the Argentine Republic then and now, it is as little difficult to find progress as to find the principal cause to which that progress is due. The following brief statement establishes the former :—

	1877.		1908.	
	Numbers.	Value.	Numbers.	Value.
Cattle	14,160,134	£18,340,686	29,116,625	£82,604,353
Sheep	59,226,300	17,898,740	67,211,754	25,287,598
Total	—	£36,239,426	—	£107,891,951
Exportation of animal produce	—	£8,586,289	—	£23,023,691
Official valuation per cow ...	—	25/6	—	56/6
" " " sheep...	—	6/-	—	7/6
Population of Argentine Republic	2,388,100	—	6,484,000	—

While cattle have doubled in numbers since 1877, and sheep have increased 13 per cent., their total local value has increased threefold ; and the revenue they produce, after supplying the local requirements of a population nigh three times as great as in 1877, has also increased threefold.

Now, this has not been brought about because ox hides and beef tallow have risen in value, or because wool and sheep-skins and mutton tallow have risen in value. The augmentation in Argentina's pastoral wealth and revenue is due, and is wholly and solely due, to the applied science of conveying overseas in a frozen or chilled condition fresh meat for European consumption. The improvement in the quality of the Argentine flocks and herds, the advancement achieved in pastoral methods and in laying down alfalfa and other permanent pastures, have been in response to the demand created by the refrigerated meat trade, and have only been made possible by the increased revenue that trade has brought. Progress has been attained, of course, in the general economy of the stock farm. In 1877 every 100,000 sheep produced 158 tons of wool, and in 1908 the same number produced 248 tons of wool. Wool still constitutes 40 per cent. of Argentina's total export of animal produce. But in 1877 the total value of meat exports from Argentina, being *charqui* and meat extracts, amounted to £480,000 ; and in 1908 the total value of meat exports, being frozen mutton and lamb, frozen and chilled beef, preserved meat *charqui* and meat extracts, amounted to £5,570,000. To this important sum *charqui* contributed £154,564 ! There is, therefore, sufficient evidence in the foregoing to justify the conclusion that the frozen meat trade has added four-and-a-half to five millions sterling per annum to the Argentine pastoral industry. Since the first shipment of frozen beef in 1877 the cattle stock of the country has doubled and the pastoral revenue has trebled, and the difference between the ratio of progress in revenue as compared to that in numbers must be credited to the business of refrigerating meat.

So far back as 1867 patents were granted by the Argentine Government for methods of preserving meat. Mysterious enough some of them were, as, for example, one whose sole ingredient was the oil of grape seeds. Another is somewhat vaguely described as " an application of heat or cold." In October of 1877 we find the first record of the refrigerating art. Mr. C. Tellier obtains a patent for the desiccation of meat at a temperature of zero by the application of methylic ether. Immediately following the name of this great pioneer in the refrigerating art comes another patent, granted in July, 1878, for freezing meat by means of sal ammoniac in a vessel or chamber with a temperature below zero. There is no need to follow the records further. The definition of the art of refrigerating meat had arrived.

In these experimental stages the cattle were slaughtered at the *saladero* of Don Eugenio Terrason, and the meat was laden in the chilled hold of the steamers at the mole-head of the port of San Nicolas, a town on the River Parana, between Buenos Aires and Rosario. Success did not attend these first ventures. The oft-told tale of the difficulties attending the crude freezing of beef need not here be repeated. Mr. Terrason constructed here his first freezing establishment in 1884. It was subsequently rented by the then three other River Plate freezing companies in order to shut it down, and its ruins by the side of the old landing stage of San Nicolas are a monument to the disaster that awaits the first pioneers in every industry.

In 1882 Messrs. Drabble founded at Campana the River Plate Fresh Meat Co., and a year later near Buenos Aires and on the banks of the Riachuelo Messrs. Sansinena and Sons erected what became at once, and has since continued to be, the most successful of all the freezing factories in the Republic. In 1892 Messrs. J. Nelson and Sons constructed at Zarate their Las Palmas Produce Co.; and ten years later, following upon the closure of the British ports to Argentine live stock, and the spurt given to the meat trade by affairs in South Africa, there arose in rapid succession the four more freezing companies that, together with the three existing previously, represent the total trade of the Republic in that industry. Their total capital amounts to £3,160,000, and they employ over 7,000 men in the works and yards.

The prominent place occupied by the Argentine in the British meat trade is one from which it is not likely to be displaced. Argentine flockmasters have experienced three years of drought, and the drought has been severest in the South where the sheepbreeding industry is still the predominant one. The effects of this drought will be most in evidence in the coming year, when supplies will fall short, but the shortage will not remain a permanent feature, nor is local consumption likely to overtake production in the visible future. Argentine breeders are giving more attention to the lamb trade, and new breeds of sheep are being introduced for that purpose.

The production of chilled beef will further expand. It is already three times as great as the production of frozen mutton, and it cannot fail to increase, for all three factors, the breeder, the freezer, and the consumer, are profiting by it. If the freezer has encouraged the estanciero, by buying his steers at prices unheard of in the live-stock export days, the estanciero has nobly responded to the call. There are already twelve million acres of alfalfa in the country and the area is rapidly increasing. England has been drawn upon for her best cattle (and has, incidentally, profited much thereby), and large sums have been expended to secure good blood. There are over 50,000 pure pedigree cattle now in the Argentine Herd Book. In the rural census of last year more than half of the total number of cattle were returned as improved, that is, of an approximately pure type of the English breeds, and principally by the Shorthorn.

There is abundant land both for breeding and feeding more cattle. It is sometimes said of the Argentine that agriculture is displacing cattle breeding, and sometimes the reverse is stated, that cattle breeding is ousting agriculture. Neither statement is true. They are component factors, sharing between them the process of occupying the West where formerly neither wheat nor alfalfa grew, but where a thin sprinkling of *creole* cattle eked out a starving existence on the unwilling native herbage. Not one-fifth of that country has yet been called into productivity either for agriculture or cattle raising. "Visible supplies" there may be, but the limits to increasing those supplies are not yet visible.

With the necessary material beside him, the land, the plough, the herd, and the willing market at his door, it would indeed be strange if the Argentine cattle breeder relaxed his energy and cried "Enough!" In 1877, when Don Eugenio Terrason shipped the first frozen beef at San Nicolas, a prime five-year-old bullock was worth to him in his *saladero* £4. This is what he could pay to turn the beast into *charqui* and tallow and salted ox hide. In this present year of grace the freezing factories have been paying the Argentine breeders £16 for prime three-year-old bullocks. If in doing so the freezers have made a profit, God yield them well of it! They have very materially enhanced the value of Argentine land and the grass that grows on it. They have put a bridge over seven thousand miles of ocean and brought consumer and producer within speaking distance of one another.

The United States with local demand outstripping their production of meat have already resorted to the River Plate to strengthen their supplies. In the fulness of time the Argentine, too, must cease to export meat. Before that stage is reached, before, indeed, the cattle land is fully occupied, the Argentine people will have learnt to economize in their own consumption of meat. The present consumption is enormous with relation to the population. It is probably over 500,000 tons of meat per annum, or, say, 200 lbs. per capita. Agriculture is still a new thing in the Republic, and though the country exports over two million tons of wheat, the Argentines have not yet become great bread eaters. Vegetables, poultry, and other items of food, are insufficiently produced. The Argentine *cuisine* would be described in Europe as an extravagant one. All this must change; is already changing. Since the freezing trade has made the storage and overseas conveyance of fresh meat possible, foreign and local prices have come closer to one another. The days are long since past when beef was sold by the piece and mutton by the carcass. A pound of prime steak costs as much in Buenos Aires as it does in London. It is not because prime beef is scarce, it is because the British consumer through the medium of the freezer buys his meat in the Argentine market.

But though the Argentine increases its production and economizes in its local consumption per capita, the increase of population will ultimately overtake the former. The greater part of the territory of the Republic is good agricultural soil, capable of sustaining a dense rural population. Progressing as it has done hitherto, the Argentine will have a population of nearly thirty millions by the middle of this century. By that time it will have ceased to produce meat in excess of its own requirements, and it will have commenced to draw on Paraguay and southern Brazil for its supplies. But there is no particular occasion to discount so remote a future. Suffice it that in the meantime there is yet abundant room southwards, as far as the Straits of Magellan, for the country's flocks to multiply; sufficient room in the farther west, in her northern territories and beyond their boundaries, in the rich pasture lands of the Paraguayan Chaco, and in the great cattle country of southern Brazil, to breed far more cattle than at present exist; and sufficient room in the core of the country to graze on the alfalfa lands and to grow on the wheat lands enough beef and enough corn to supply, even beyond the lives of the young generation, the demand of her foreign markets.

No doubt, too, there will still be then gentlemen who write articles. They will tell kindly of the good old days when, upon what have become the homesteads of the small farmer, there were paddocks of alfalfa each of which ran into thousands of acres; how in these paddocks there grazed herds of Shorthorn cattle to supply the export trade of chilled beef which at that time was one of the most important industries of the Republic; and how, even as in pre-historic times Don Eugenio Terrason loaded fresh beef in the chilled holds of steamers at the mole-head of San Nicolas, there were men who foresaw and worked for a new order of things in rural economy, and how sometimes the failure of the individual showed the way to the success of the collectivity. *Forsitan et nostrum nomen miscebitur istis.*

CHAPTER XXIV

HOW THE BRITISH PUBLIC HAS BENEFITED

When the mutual relations that exist between the great frozen meat industry and the many classes of traders engaged in its maintenance have been considered, there yet remains to be surveyed the effect of the trade upon the one great class on whose behalf it had its inception and upon whose custom its future rests. The public, and pre-eminently the British public, stands after all to gain most from the frozen meat trade, that is, if the 8,283,000 tons of frozen and chilled meat imported into the British Isles from 1874 to 1910 have advanced the physical welfare of the nation.

How has the community been affected? The opening chapter of this book dealt with the crisis of meat scarcity which culminated in the seventies, and the efforts of scientific men to meet it. A survey of ancient and modern history shows that nations eat meat according to their wealth, if this be not stating cause for effect. Conversely, John Bull without his roast beef would have made but a poor show during the last thirty years. He became unable to provide sufficient meat for himself when he became a manufacturer, and so it has come about that Britain in the Southern Seas and the Americas have had to turn wholesale meat suppliers for his benefit. The United States of America have furnished the expensive joints of beef for the tables of the well-to-do, and, not without refrigeration's aid, Australia, New Zealand, and Argentina did more than that. They have provided the meat food required by the great body of workers in city and town. The manufacturers upon whom the wealth and industrial fame of England and Scotland rest owe an enormous debt of gratitude to the pioneers of the frozen meat trade, by whose efforts their work-people have been able to get plenty of sound and cheap meat. When frozen meat was first introduced, there was great scarcity of fresh meat at reasonable prices in Lancashire and Yorkshire and in the cotton districts generally. It was frequently a matter of meat once a week ; now it is meat twice a day. The question remains, could the strain of modern industry be borne on the former allowance ? Taking the beginning of the industrial era in the United Kingdom as coincident with the middle of the nineteenth century, the following figures, supplied by Mulhall,* of the average price of "meat" —beef, mutton, and pork—for the quinquennia mentioned show clearly how prices rose as population increased, and as fresh meat became a more recognized article of diet :— 1851—1855, $6\frac{3}{8}$*d.* per lb. ; 1856—1860, $6\frac{5}{8}$*d.* ; 1861—1865, $6\frac{7}{8}$*d.* ; 1866—1870, $7\frac{1}{8}$*d.* ; 1871—1875, $8\frac{3}{8}$*d.* ; 1876—1881, $8\frac{3}{4}$*d.*

Now even in the rural districts the peripatetic meat man has his beat, districts which fifty years ago furnished nothing better to the agricultural population than a bread and cheese diet, with bacon as a luxury, and an occasional joint of butcher's meat to mark a red letter day. Then take the position of the professional and middle classes. When English and Scotch meat would not go round, what a struggle was that of people with small incomes and large families (a not uncommon association thirty and forty years ago) in battling with their butchers !

The position at the beginning of the eighties was indeed serious, with the English meat crop equal to the supply for no more than seven-and-a-half months of the year. The statistical writers of the period were keenly concerned in the problem. Mulhall, after the success of the *Dunedin* shipment was assured, pointed out in his pamphlet, referred to above, that the farms of Otago and Canterbury could send sheep to the London market more easily than could the Tweed farmers 100 years ago (1782), when meat was selling at 1*d.* per lb. in Scotland and 10*d.* in London. When statisticians forsake the past and present for the future, even a Mulhall may make mistakes ; here is an extract from the pamphlet : "We may expect in 1896 a population of 42 to 43 millions, and to feed our people

* England's New Sheep Farm ; pamphlet, published in 1882.

for five months in the year we shall then import over 1,000,000 tons of meat, the bulk of which must necessarily come from the Australasian Colonies"! In 1896 the population was 39½ millions; only 485,000 tons of mutton and beef (including live cattle) were imported, and out of the 87·3 lbs. of beef and mutton consumed per head of population, 59·9 lbs. were produced—equal to eight months' supply—by home farmers, and five-sixths of the imports came from the United States. Mr. Mulhall also wrote: "Twenty years hence New Zealand will have more sheep than Great Britain"; the flocks of the Dominion in 1902 were about 20½ millions against 25½ millions in Great Britain.

Popularizing Frozen Meat.

A notable event in the campaign to establish the frozen mutton trade from New Zealand occurred in 1891, when the late Earl of Onslow, then Governor of the Colony, did much good by despatching six prime Canterbury wether sheep to certain gentlemen in London for their report. Lord Onslow's desire was to find out if the difference in price of English and New Zealand frozen mutton could be accounted for by a relative difference in the quality of the two kinds, and the experiment was an excellent advertisement for New Zealand mutton. Lord Onslow, by request, himself described this interesting experiment. His words, written especially for these pages, are as follow:—

"So far as my records go, it was in the year 1891 that I sent some carcasses of sheep, which had been especially selected by Mr. John Grigg and Sir John Hall, as four- and five-year-old cross between merino and Down, to Messrs. Fitter, with injunctions to specially thaw and deliver them to certain gentlemen whom I knew I could rely upon to give me an unbiassed expression of opinion as to their fitness for the table. Those were the late Baron Henry de Worms, afterwards Lord Pirbright, Lord Rosebery, the late Sir Augustus Harris, then Sheriff of London and manager of Drury Lane Theatre, who used to entertain very largely at the theatre after the performances, the late Sir Henry de Bathe, who was chairman of the House Committee of the Beefsteak Club, the late Sir Morell Mackenzie, and M. Waddington, the French Ambassador in London. The other members of the Beefsteak Club who reported upon the mutton were Mr. Corney Grain, Sir Francis Burnand, Sir Squire Bancroft, and Mr. George Augustus Sala. All of them reported very favourably on the experiment, and I believe those interested in the trade in New Zealand made use of the reports for the purpose of recommending New Zealand meat."

All these connoisseurs pronounced in most enthusiastic terms on the mutton's merits. Baron de Worms reported: "The freezing did not hurt it in the least, in fact, the greatest epicure would fail to discover that it was not home-grown." General Sir H. de Bathe went further in saying that the mutton was "better than what I can buy at Chichester"! By the way, North Canterbury four- or five-year-old wether mutton is never seen at Smithfield nowadays. The difficulty in placing frozen meat on the footing in England which it deserves has not been with the "classes" nor with the "masses," but with the great and all-important section of the community, the middle classes, who are, unfortunately, to a great extent, despotically ruled by convention and prejudice.

The consumption of meat, including pork products, in Australia is put at 233 lbs. per head of the population, in the United States at 144 lbs., and in the United Kingdom at 122 lbs. Refrigerated meat accounts for 22 per cent. of this 122 lbs., say, 27 lbs. per head, and in a paper read by Mr. P. B. Proctor at the first Refrigeration Congress held in Paris in 1908, the remark was made that these 27 lbs. of refrigerated meat per head did not displace other meats previously consumed, but represented an additional supply. Could there be better evidence of the boon to the community as a whole through the introduction of frozen and chilled meats?

It has been possible for the British public to increase their consumption of meat in thirty years to the extent mentioned above owing to the low price at which frozen meat has been sold. The low price does not imply any inferiority whatever (the idea—if anyone holds it—that refrigeration has brought about any deterioration in the quality of meat consumed in this country must be dismissed); it simply indicates low cost of production in the country of origin. In cheapening the butcher's bill by 25 to 50 per cent., surely the farmer in Australasia and Argentina has conferred upon the English householder a boon of incalculable magnitude! The householder when he puts on his considering cap will perceive another point, viz., that had it not been for the advent of frozen and chilled meat, home-produced beef and mutton, not being able to keep pace with the population, must inevitably have been forced up to a prohibitive price.

The present generation obviously does not realize what frozen meat has done for them; the value of a great movement is never properly appreciated except in the perspective of after years. A new race has arisen since the days, over a century ago, when it was necessary to put a clause in the articles of the London apprentice binding the employer only to provide salmon so many days a week, and since the later days when it was the ideal of happiness of the Dorsetshire ploughboy to "swing on a gate all day long and eat fat bacon." Now the apprentice, were he here, could batten on beef steak from Queensland and Argentina, and the ploughboy, as it is, no doubt, revels in his succulent New Zealand chop.

In London, the biggest city in the world, and the centre of the largest Empire, refrigerated meat has its chief triumph, for the metropolis is mainly fed with meat from overseas. A certain proportion of English and Scotch meat certainly is consumed in London, but the West-end folk are very large customers for chilled beef of the highest quality and the best grades of New Zealand sheep and lambs. The huge population of London's suburbs absorbs great quantities of the frozen meat which arrives from all three sources, New Zealand, Argentina, and Australia.

Frozen Meat for Tommy Atkins.

One good thing frozen meat has done for the British public is to ensure a bountiful and economical diet system for Tommy Atkins. The meat issued to the troops at the home stations in the British Isles is beef six days a week and mutton one day; the daily ration for a soldier is $\frac{3}{4}$ lb. in barracks and 1 lb. in camp. The beef on four days must be home-fed or "town-killed," and on the other two days frozen beef may be used. The mutton issued on the one day may be frozen. So the British Army is fed on frozen meat for three-sevenths of the year: could there be a better evidence of its sustaining power? For many years the Army commissariat stuck out against the use of frozen beef, using chilled American instead, and the change made some years ago in favour of frozen is a welcome sign of the officials' conversion to reform and retrenchment. The War Office is a keen buyer; wether mutton from 50 to 70 lbs. only is accepted, beef must be 170 to 200 lbs. per quarter, and beef and mutton must have the original labels attached. Tommy Atkins wants beef all the time, but for his well-being the doctors modify this stimulating diet with one day of mutton. It was about 1902 that the War Office took the step of insisting that only home-bred and/or "Colonial" meat be supplied to the troops in the United Kingdom. This measure of preference to British produce, absolutely prohibiting the use of American and Argentine beef and mutton, came about through the influence of Mr. Joseph Chamberlain, then Colonial Secretary. The principle enunciated by this policy was so startling and revolutionary that many persons have wondered how it escaped the notice of Mr. Chamberlain's political opponents. Argentina suffered some inconvenience in consequence, and New Zealand ewe mutton jumped into favour with contractors. The supplying of ewe mutton from New Zealand spoiled the colonial contract trade. Argentina then shipped nothing but wethers and gave more satisfaction. Because Australasia was unable to supply the quantity required, in 1906 this contract clause became a dead letter. Since then sundry members of the House of Commons have from the Opposition benches inveighed against this letting in of the "foreigner," but owing to the irregular arrival of supplies from our own overseas Dominions the Government has remained under the necessity of maintaining the policy of the open door in Army frozen meat contracts.

The contractor is a patron of the frozen meat trade to some purpose. He supplies frozen meat in considerable bulk to unions, asylums, and other institutions in England, and some Smithfield traders make a speciality of catering for this trade, the profits in which, probably, are quite moderate. A saving of £1,000 a year is easily effected in the expenditure of one of these public institutions by substituting frozen for home-produced meat. It is amusing to note the protests of the inmates when such a step is taken; what is good enough for Peers and the Beef Steak Club will not do for the paupers, or can it be that the grumble of old-fashioned prejudice is aroused by "officious" guardians?

CHAPTER XXV

THE POSITION OF THE BRITISH FARMER

SEEING that the refrigerated meat industry in its forward movement naturally brings under discussion the question of home supplies, it is not out of place here to make a brief examination of the position of the English and Scotch grazier and sheep farmer in regard to the importation of frozen meats. In one sense the home producer has obviously been injured, because the imports of meat must tend to keep prices at a lower level than they would be if home supplies were alone available. But as the latter eventuality is out of the question, as far as England is concerned, one may at once turn to the real issue and consider whether the imports of meat have been of such magnitude and character as seriously to injure the British farmer or to make his operations altogether unremunerative.

The authors have consulted several gentlemen who are acknowledged authorities on this question. Mr. R. E. Turnbull, F.R.S.S., has been kind enough to work out a comparison between the years 1880 and 1910 as to the supplies of meat available for consumption.

1880 Compared with 1910.

1880.

Population of the United Kingdom, 34,772,000.

Supply of home-fed stock :—

		Dressed weight. Tons.
2,580,000	fat cattle and veal calves	468,700
10,860,000	sheep and lambs	300,800
4,800,000	pigs for pork and bacon	288,000
18,240,000	Home supply	1,057,500
	Imports	528,500
	Total supply of dressed meat	1,586,000

Home supply 66⅔ per cent., foreign and colonial 33⅓ per cent. Per head of population, 102 lbs., viz., home supply 68 lbs., imported 34 lbs. Average price of home-fed cattle (offal given in) about 7¾*d.* per lb., and of sheep about 9¼*d.*

The three years 1879, 1880, and 1881, were the most disastrous for owners of sheep in this country that any farmer now living ever experienced.

The total number of sheep and lambs in the United Kingdom in June, 1879, was 32,238,000 ; in June, 1882, the number was only 27,448,000 or 4,790,000 (14.85 per cent.) less than in 1879. Liver fluke was the malady that caused the loss. The harvest of 1879 was gathered in wretched condition, the rainfall occurring during harvest and being very excessive.

1910.

Population of the United Kingdom, 44,850,000.

Supply of home-fed stock :—

		Dressed weight. Tons.
3,226,000	fat cattle and veal calves	613,000
12,622,000	fat sheep and lambs	351,600
5,098,000	fat pigs	305,900
20,946,000	Home supply	1,270,500

Imported meat and meat from imported stock :—

	Tons.	
Beef and veal	446,500	
Mutton and lamb	277,800	
Pork, bacon and hams	264,500	
Unenumerated	45,800	
Imports		1,034,600
Total supply of dressed meat		2,305,100

Home supply, 55⅕ per cent., foreign and colonial, 44⅘ per cent. Per head of population, 114 lbs., viz., home supply, 63 lbs., imported, 51 lbs.

In 1910 the total supply of meat (beef, veal, mutton, lamb, pork, bacon, and hams) exceeded the supply in 1880 by 14,382,000 cwts., or by rather more than 45 per cent. The increase in population was 10,078,000 (assuming that in 1910 the population was 44,850,000) = 28·8 per cent. The increase in the supply of home-fed meat was 4,260,000 cwts., or 20·1 per cent. ; in foreign and colonial the increase in the supply was remarkable, being 10,122,000 cwts. = 95·76 per cent. In 1910 the supply of meat per head of population was 12 lbs. more than in 1880, being 114 lbs. as compared with about 102 in 1880. (Of the total supply about 20,000 tons were exported.)

In 1910 home-fed cattle averaged about $1\frac{1}{2}$*d.* per lb. less than in 1880, and sheep 2*d.* per lb. less than in 1880.

Mr. Turnbull concludes his statistical survey as follows :—"While the large and increasing supply of meat from abroad has pressed heavily on the farmers and landowners of the United Kingdom, the severe competition that breeders of cattle, sheep, and pigs, have been subjected to has undoubtedly resulted in a marked improvement in the quality of herds and flocks on nearly every estate in the kingdom. With regard to sheep, the number of sheep and lambs marketed in twelve months now equals about 40 per cent. of the number in the returns. During the ten years 1870—79 the number marketed only equalled on the average about 35 per cent. The quantity of mutton and lamb now marketed equals about 25 lbs. for each animal in the returns; in the earlier period the average was about 22 lbs. By securing earlier maturity and marked improvement in quality the home breeders have till quite recently held their own in face of increasing foreign competition.

"Some of the British breeds of sheep are now too heavy for the family requirements of the consumer. This is notably the case as regards the famous Lincoln and Leicester breeds. The joints of mutton cut from sheep of these breeds are too large for ordinary household use. (There is still a good demand for large joints for restaurants.) The demand is now almost entirely for joints cut from sheep of small or medium weights, 60 to 70 lbs. The Down breeds meet this demand admirably. The Lincoln and Leicester breeds thrive better than any other in the districts where they are bred, and having regard to their abundant production of wool they are not likely to be replaced by other breeds. With regard to cattle, the quantity of beef and veal marketed in twelve months now equals about $116\frac{1}{2}$ lbs. for each animal in the returns. In the twenty years that followed 1870 the average was only about 105 lbs. The improvement is much less marked than in the case of sheep. During the last forty years a greater general improvement has taken place in the quality of Irish and Scotch cattle than in English cattle.

"The number of fat cattle and veal calves now marketed in twelve months equals about 27½ per cent. of the number in the returns, as compared with 22½ per cent. in the earlier period. The total supply now includes a larger proportion of veal calves than formerly. The proportion of well-fattened calves is much less than was formerly the case; this is owing to the greatly increased demand for new milk for household consumption.

"The number of pigs marketed in twelve months equals over 140 per cent. of the number in the returns. The production of pork, bacon, and hams, equals about 190 lbs. for each animal in the returns.

"The live stock industry of the United Kingdom in the face of strong competition from abroad is still fairly remunerative, except in seasons of drought or when the general trade of the country is severely depressed, and this will probably continue to be the case even should the foreign and colonial supply of meat continue to be as large as it now is.

"To the wage-earning classes the abundant supply of imported meat has proved to be a great boon. I am of opinion that the consumer has largely benefited by the exportation of well-bred cattle and sheep to other countries, because this has resulted in an abundant supply of meat of fairly good quality, and also a considerable supply of excellent quality. The number of our farmers who have benefited has certainly not been large, only breeders of pedigree stock have benefited, the rest—a large majority—have suffered very severely."

A University Professor's Opinion.—Professor Robert Wallace, of the University of Edinburgh, on being asked for his views, sent the following lines:—

"With reference to the position of the British farmer in relation to the enormous increase in the imports of frozen meat from abroad, I am convinced that he would have been ruined but for the increased annual supply of gold which prevented prices falling as they would have done had the quantity theory of money not come to his assistance. On the other hand, the increased supply of meat from abroad has been a blessing in disguise to the farmer as well as to the consumer, for without the foreign competition, which the farmer has, I think, on the whole, very successfully withstood, prices of butchers' meat would have risen out of all proportion to its food value in relation to other commodities, and the ignorant people would have blamed the farmer and created disturbances and rioting—such as have recently occurred in France—which would have upset the social order of things and been bad for the farmer as well as for the country as a whole."

A Pessimistic Note from Yorkshire.—How the sheep farmers of Yorkshire and Scotland have been adversely affected is shown by an extract from the letter of one of them, Mr. John Mackenzie, North-Cote, near Skipton. "Speaking as a sheep owner myself, I should say that the importation of frozen mutton has reduced the profits of the rank and file of British flock owners to vanishing point. By 'rank and file' I, of course, mean sheep owners who are entirely dependent on the breeding and rearing of store sheep for their revenue. Indeed, over large tracts of the Scottish Highlands the business has had to be entirely abandoned; vast areas of excellent sheep land (mountain) having been cleared of sheep in recent years and put under deer, the former occupiers (most grazing runs are leaseholds) of those lands, in many instances, betaking themselves to Australia, New Zealand, the Argentine, and the United States of America, where, as a rule, they usually amass fortunes by following the occupation at which they could not make bread and butter at home. And I am quite within the mark in saying that the pressure on the home producer is not yet nearly so acute as it will be. Every man who has taken an intelligent interest in this question can have no two opinions on the matter."

The Splendid Pedigree Stock Export Trade.—The imports of frozen meat into Great Britain have steadied the prices of home-fed meat all round for the last thirty years; no more buying Scotch wethers (alive) at 1*s*. a lb.! But it does not follow from that that the home trade has been robbed of profits; in fact, taking an average of ten years, Mr. A. J. Hickman, of Court Lodge, Egerton, Kent, speaking a year or two ago, said that "sheep growing is the most profitable branch of British husbandry." Mr. Hickman, of course, did not include in his average the disastrous 1911 season for sheep farmers. One most favourable effect — as mentioned above — upon the home stock interests of the frozen and chilled meat trade has been the demand from South America and Australasia for pure-bred animals, and upon this demand the pedigree herds and flocks in England and Scotland have been largely built up, and brought to the present flourishing condition. Argentina particularly, as will be seen by figures given elsewhere in this book, has been a splendid customer, and no one can speak of the beef and mutton received from that Republic as a "foreign" article pure and simple, considering that it represents cattle and sheep descended from the Shorthorns and Lincoln sheep and other breeds imported from Great Britain. Some British breeders of live stock have in this way immensely benefited by the sale of their stud sheep and cattle for grading-up purposes to the pastoralists and graziers of the Southern Hemisphere.

The Evidence of Facts and Figures.—In order to prove the correctness of the view held by some men that the production of meat in the United Kingdom has been rendered unremunerative by importations it would be necessary to show that the numbers of cattle and sheep had been steadily reduced by the pressure of outside competition. But the returns show that with slight fluctuations the number of cattle and sheep in the United Kingdom has been maintained, and, indeed, increased.

	Cattle.	Sheep.
1880 . . .	9,871,644	30,240,722
1910 . . .	11,765,543	31,164,587

According to the Board of Agriculture returns, "Cheviot hill wedder lambs" averaged in 1876—1880, 14*s.* 2*d.* per head for top quality; in 1910 the price for the same animal and grade was 15*s*. As to store cattle, according to the same authority, a "Teviotdale stirk" averaged £9 8*s*. in the five years mentioned, and in 1910 the value is given as £10.

It only needs the working out of the argument of this chapter to show in how remarkably small a degree the large importations of chilled and frozen meat have affected the British stock business. It is evident that the farmer has perceived how best to meet his competitor, by striving to produce only the highest grades of meat. So it has come about that the market prices for the best qualities of home-produced cattle and sheep have well withstood the attack from overseas. Two distinct trades exist, English and Scotch meat, and chilled and frozen, respectively, and though there is a considerable overlapping, the higher retail rates paid for home-fed meat allow of a premium on the wholesale market and on the farm for stock fed in British pastures. The new industry of frozen meat has found out new consumers in England. Of course, at the start the British farmers were alarmed at the invasion of their tight little island by American beef and New Zealand mutton, and the discussions at the "market ordinary" must then have been portentous. The new era introduced for English and Scottish stock raisers about 1880 looked threatening enough. Rapid transit, cheap freight, and Governmental assistance and encouragement, helped the Australasian and Argentine farmers to launch their ventures in a style that promised badly for the British grazier, handicapped as he was with relatively dearly-rented and heavily-taxed land. That a difference in public estimation and price does exist between home-fed and imported dead meat we know, but it has not always been clear why it is quite so pronounced. Were the system of transit, handling, and cooking of frozen meat ideal, the margin of value between the two descriptions would probably be much less. American chilled beef—admittedly equal, or superior in some cases, to British in quality—often sells at the same rates as best English. Canterbury mutton and lamb have been mistaken for English Down meat by good judges when the properly thawed out joint has been placed before them hot from the oven ; when cold however home-fed has the advantage. But it is inevitable that home-fed meat should sell better than frozen Australasian or Argentine, and it is plain that the British farmer has been far less seriously hit by imported meat than by imported wheat.

The Farmer's Position under Free Imports.—It would be interesting to read an article from the pen of a competent person dealing with the question of the British farmer's position with lower rents and cheaper foods to-day as compared with the times of high prices for their produce a generation and more ago. In the preceding pages of this chapter the references concern only the stock breeder, but as the stock breeder's art is to a certain degree wrapped up in the agricultural industry as a whole, it may be worth while in passing to record the opinion of many high authorities in Great Britain concerning the alleged decadence of agriculture in England and Scotland. It is very difficult to contrast the past and present positions of agriculture, because everything has changed so much. In the early seventies the capital employed in farming in the United Kingdom was about £450,000,000; it is now about £400,000,000. The gross revenue, the average for the five years, June, 1872—1877, was about £255,000,000. It is calculated that the average for the last five years has been about £200,000,000. According to official figures, the gross income from agricultural lands in Great Britain sank from £59,500,000 in 1876—1877 to £42,156,000 in 1908—1909. Though these figures show a reduction in the capital invested in the agricultural industry to-day as compared with thirty to thirty-five years ago, such high authorities as Professor Wrightson and Mr. James S. Macdonald, who have been appealed to by the authors, hesitate to state that British farming is less prosperous than it was thirty to forty years ago. If this is so, then we must give the farmer credit for considerable smartness in changing his methods, bearing in mind the fact that in 1910 agricultural foodstuffs were imported into the United Kingdom to the extent of £206,650,566, of which huge sum chilled and frozen meats, £21,548,014, formed a part. A favourable change may come about when the Continental tariffs on meat are reduced or abolished, and the regulations, acting practically as a prohibition to the import of frozen meat, are relaxed or removed. Then some considerable part of the chilled and frozen meat imports, at present finding their market only in Great Britain, will be diverted to Continental States, and the British farmer should, in consequence, be able to secure an enhanced price for his stock.

CHAPTER XXVI

THE INCORPORATED SOCIETY OF MEAT IMPORTERS

SOONER or later in the history of a trade the efforts towards association among its members spring up, and it was early in 1894 that an attempt was made to establish a society in the frozen meat industry which should form a common ground for dealing with the differences of conflicting interests in the trade. Unrestricted competition amongst holders of stocks had led and was leading to serious losses, preventing the trade from being put on a proper basis at the selling point. Some of the leading importers of frozen meat (Australasian and Argentine) then formed a "Frozen Meat Importers' Association," and a set of rules was drawn up. But the *motif* of the whole movement, the regulation by mutual consent of prices on Smithfield market, proved too strong for the organization, which broke down when some members, who acted in dual or triple capacities as Smithfield salesmen, dealers, or direct consignees, found it impossible to reconcile the interests of their own businesses with those of the trade as a whole. So after a few months the Association was dissolved.

This preliminary failure, however, did not discourage those who had set their minds on the establishment of a trade organization, and the necessity for a trade association was so urgent that in 1895 the "Frozen Meat Trade Association" was formed, with ten members, on wider lines than its predecessor. The pitfall of adhering to the principle of price regulation was avoided, ample agenda being found in the shape of matters of moment affecting the whole body of the trade, such as insurance claims, market customs, and the relations of importers with carriers, etc. It was soon discovered that there were many questions which could be far better handled by a body representing the frozen meat trade than by individuals ; so the

Association gradually assumed an influential position and acquired important functions.

Disclosing Stocks.

One of the early schemes devised by the Frozen Meat Trade Association was that put forward on September 24, 1896, for the obtaining of reliable monthly returns of the stocks of frozen meat. This was outlined as follows :—

That the quantities held by each importer, whether in London or the provinces, be declared by the means of counters or coloured cards. The following numbers to be inscribed on the counters or cards:—Nil ; 500 ; 1,000 ; 5,000 ; 10,000. Different colours to represent the various classes:—Australian sheep by white counter ; New Zealand sheep by green counter ; Plates by red counter ; lambs by blue counter ; quarters of beef by yellow counter.

That the stocks be returned of the quantities held on the 1st day of the month. Stocks to include frozen meat on any ship or steamer, docked but not discharged.

That the counters be collected by a commissionaire calling at the various contributors' offices on the first Tuesday in each month, and that he be provided with a locked box or bag into which the counters can be deposited.

That each contributor be supplied with a stock of the necessary counters or cards, which may be put in the stock returns box, either when the commissionaire calls, or at the meeting of the Association.

That a meeting be held on the day of the collection, at such hour in the afternoon as may be arranged, when the box shall be opened by a member of the Committee of the Association, or, in the absence of a member of Committee, by a member of the Association in the presence of the secretary, and the stocks then declared.

Any contributor to the stock returns to have a general invitation to attend such meetings.

That contributors be requested not to impart to the Press, nor to any private person other than contributors, the state of the various stocks.

This scheme did not come into operation, but as the subject has received so much attention of recent years, the details may have present, and possibly future, interest.

The Weekly Prices Cable.

It was not until 1897 that the Association entered upon the most useful section of its operations, a phase of action which won for it prominence and appreciation in Australia and New Zealand. In that year Mr. R. E. N. Twopeny, the then editor of the *Pastoralists' Review*, arrived in London as the delegate of the Australian freezing companies, and the Association arranged with him for the weekly transmission by cable of a set of quotations. The prices were arrived at by averaging the London market quotations furnished by the members of the Association. From 1897 to the present day the weekly cable of Smithfield prices—amplified from time to time according to the needs of the trade—has been despatched. This official record has been found to be of incalculable benefit to all engaged in the trade in Australasia; it supplemented the erratic cabled advices on varying bases which had previously been despatched by London merchants, and the cable is now printed in every journal in Australasia which gives commercial news. Some misunderstanding has occasionally arisen; it has been supposed that the cabled figures represent prices as "fixed" by the Association. Nothing of the sort was ever contemplated, the cable simply giving a record, and as such, carefully arranged on the basis of figures supplied by the leading operators, no more informatory statement could be telegraphed. A reproduction of the first issue and the latest form appears in the Appendices.

The Charter and the New Name.

On October 26, 1909, a resolution was adopted winding up the Frozen Meat Trade Association, and passing over all its interests to "The Incorporated Society of Meat Importers." With this acquisition of its charter, the Association took up a variety of extended duties.

The following are some of the objects for which the Society was established, as named in its memorandum of association. Not only was it formed "to improve the conditions under which the trade of meat importers is carried on," but also "to watch over and improve the conditions under which imported meat is prepared, packed, shipped," etc., likewise to assist improvements in methods and appliances. An important function is "to officially record and register the current market prices of imported meat in the United Kingdom and to publish the same . . . provided that the Society shall not fix or attempt to fix prices, or impose terms of sale." Other purposes are the collection and circulation of information among members of the trade, the promotion of discussion, and the undertaking of work as a trade society in securing necessary reforms, etc., from the Legislature and other authorities. The Society also includes among its objects the conducting of arbitrations and the nomination of surveyors, valuers, arbitrators, and umpires.

Its scope and membership (including Australian, New

Zealand, North American, and South American firms, shipowners, bankers, underwriters, Australian Agents-General and Government Commissioners) now make it a very important body. All vexed questions affecting the interests of the frozen and chilled meat trade are taken up; where wrongs can be righted, and reforms effected, the institution takes action, and much of the improvement which has been introduced into the working of the trade during the last ten years or so has been due to its interposition.

One thing there is which the Frozen Meat Trade Association has been unable to do, viz., to introduce into the forward trade an approved c.i.f. contract form. For years this was hammered at, and at length a form, which seemed to meet all requirements, was welded into shape. But the great bugbear of the frozen meat trade—conflicting interests—has prevented the general acceptance of the form.

Past Presidents and Vice-Presidents.

The first president, in 1896, was Sir Montague Nelson, and successive presidents have been :—

1897.	Mr. George Goodsir.	
1898.	Mr. William Cook.	Vice-President, Mr. George Goodsir.
1899.	Mr. Richmond Keele.	Vice-President, Mr. William Cook.
1900.	Mr. George Goodsir.	Vice-President, Mr. Richmond Keele.
1901.	Mr. William Blankley.	Vice-President, Mr. George Goodsir.
1902.	Mr. J. N. Newman.	Vice-President, Mr. George Goodsir.
1903.	Mr. J. A. Potter.	Vice-President, Mr. J. N. Newman.
1904.	Mr. C. H. Inglis.	Vice-President, Mr. J. A. Potter.
1905.	Mr. A. W. Pottinger.	Vice-President, Mr. William Blankley.
1906.	Mr. W. Lane Mitchell.	Vice-President, Mr. A. W. Pottinger.
1907.	Mr. W. Lane Mitchell.	Vice-President, Mr. William Blankley.
1908.	Mr. Gordon H. Campbell.	
1909.	Mr. Gordon H. Campbell.	

INCORPORATED SOCIETY OF MEAT IMPORTERS.

1910. Mr. Gordon H. Campbell, first President.
Mr. William Blankley, first Vice-President.

1911. Mr. Gordon H. Campbell, President.
Mr. William Blankley, Vice-President.

1912. Mr. Gordon H. Campbell, President.
Mr. William Blankley, Vice-President.

Mr. Thomas Guy Nind acted as secretary till 1909, when the position was assumed by Mr. Louis H. Furniss. The registered office is at 15 and 16, West Smithfield, London, E.C. The Society, which is split up into sectional Committees and holds periodic general meetings, has now over a hundred members, and it is certain that on the lines of its charter it will in process of time become still more effective in operating for

SIR ALFRED SEALE HASLAM.

To face p. 332.

the general welfare of the trade as a whole. The Society will be able to take action where individual traders and firms cannot, and its functions will include watching the general course of the trade and guarding the interests of its members, who constitute the importers of foreign and chilled meats into London.

CHAPTER XXVII

MECHANICAL REFRIGERATION

ALTHOUGH a detailed and technical description of the processes by which the science of refrigeration is applied to the frozen meat industry would only appeal to a very few of those interested in the commercial side of the industry, a more important and wider function is fulfilled by giving a brief account of the elemental principles on which mechanically produced cold operates to preserve the great meat supplies of the world to-day, and of the development of the application of refrigeration in the service of man.

The Principle of Mechanical Refrigeration.—The present glance at the scientific side of refrigeration must be as brief and non-technical as possible. In the classification of the types of machine the divisions adopted by Sir J. Alfred Ewing may well be followed.

The first broad distinction that may be drawn is between those machines that use air as their working substance and those which use a liquid which is alternately vaporized and liquefied during the cycle of operations.

In machines using air, the air is compressed in a cylinder, the heat produced by the compression is removed by water circulation and the cooled air is allowed to expand in another cylinder against a resistance and becomes further cooled in virtue of the work it does, and is then discharged into the chamber to be refrigerated. These machines are usually spoken of as compressed air machines.

In machines using vaporized liquid the underlying principle is the using up of heat units when a liquid passes to a gaseous state, the heat used being said to become "latent."

In such machines simple mechanical compression may be adopted as the means of restoring the vapour to the liquid

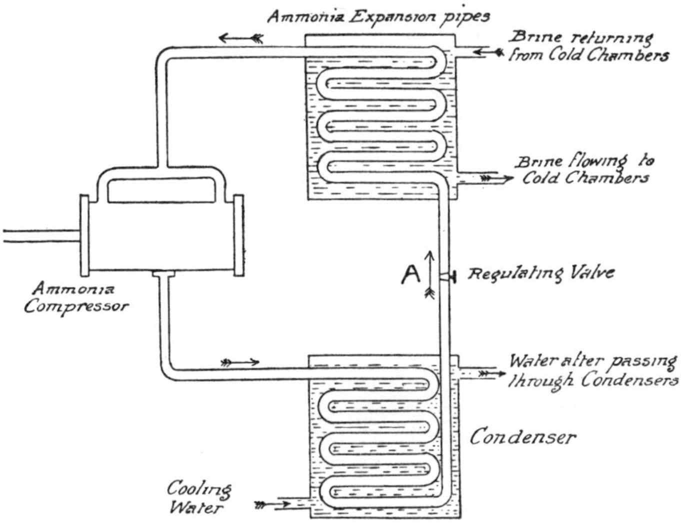

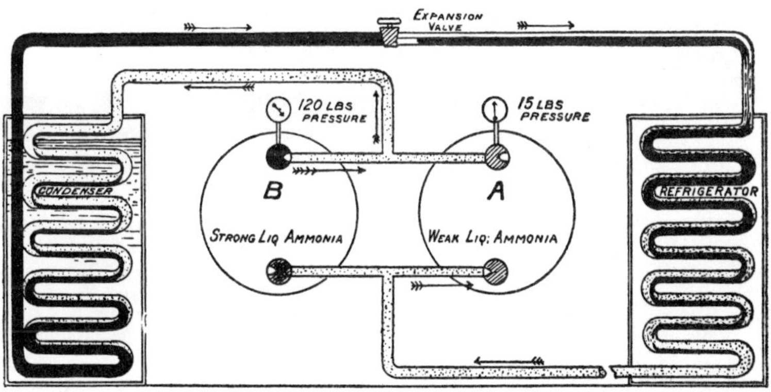

DIAGRAMS ILLUSTRATING (TOP) THE COMPRESSION SYSTEM AND (BOTTOM) THE ABSORPTION SYSTEM OF REFRIGERATION.

To face p. 334.

state, or the gas may be absorbed into water or some other liquid for which it has a chemical affinity, the machines adopting these alternative methods being grouped as "compression" and "absorption" respectively.

Broadly speaking, there are three methods by which the cold produced by either of the above methods is communicated to the chambers in which the meat or other produce to be refrigerated is stored. The direct expansion system is that in which, in the case of ammonia compression machines, the gas is allowed to expand direct into the pipes in the cold store. In the brine circulation system the cold is first produced in a vessel of brine by the immersion of the expansion pipes therein, the cooled brine being circulated through the pipes in the cold store. The third system of cooling is that known as cold air circulation, in which a current of cold air is passed into the store through ventilators called "air-ducts," such air having been passed for refrigerating purposes over a "battery" consisting of either direct expansion pipes, cold brine pipes, or an open flow of cold brine. The last method is said to have the advantage of not unduly drying the air, while it is also claimed to clean the atmosphere.

Reference should also be made to the vacuum process of refrigeration, which depends for the production of cold upon the evaporation of water. This is carried out by means of lowering the pressure of the atmosphere in the vessel containing water, and the absorption in strong sulphuric acid of the vapour given off by that water, but this system is of limited application.

Early Discoveries.—The earliest discoveries in the art of refrigeration date back much farther than its application to the preservation of meat, and the history of the developments which have produced the modern refrigerating machine, though it need not be told in the present volume, is a thrilling one. Mr. J. J. Coleman, himself a notable pioneer of mechanical refrigeration, as our previous pages show, stated in a paper read before the Institution of Civil Engineers in 1882 that "although Sir John Herschel and others had directed attention previously to the desirability of utilizing the expansion of compressed air for the production of cold, the credit of actually constructing such apparatus appears to belong in this country to Professor Piazzi Smyth. In 1839 he had commenced experiments. He seems to have worked for a long time with apparatus on the method of blowing air through loaded valves before the principles of the mechanical theory of heat as applied to gases were properly understood." A vacuum ice machine was invented by a Dr. Cullen over a century and a half ago, and Ferdinand Carré, a French chemist, invented the ammonia absorption machine sixty years ago. In 1834, Jacob Perkins invented a refrigerating machine on the compression system. The reputed father of the cold air refrigerating machine, the system of freezing so prominently associated with the early stages of the frozen meat export trade, was Dr. John Gorrie, an American, who lived and died at Appalachicola, in Florida. Gorrie introduced his machine in 1849, and now, sixty-three years later, United States citizens are proposing to erect a statue to his memory. There was also a long list of experimental work and commercial enterprises undertaken in Australia and New Zealand, and this pioneering effort is recorded at some length in another chapter.

In addition to the brief summary of early refrigerating patents recorded in Appendix XII., it may be well to give here a few particulars concerning some of those whose names are prominently identified to-day with refrigerating machinery.

Mr. J. J. Coleman, whose name will always be associated with the famous "Bell-Coleman" machine of refrigeration's early days, was, prior to his connection with that enterprise, one of the scientific experts in the employment of Young's Paraffin Light and Mineral Oil Co. In the paper which has been referred to above he gave some interesting particulars of the dry air refrigerating machine which he fitted on board the Anchor Liner *Circassia* in March, 1879. The plant consisted of two compressors, 16 inches diameter and 16 inches stroke, and was connected with a chamber of about 18,000 cubic feet capacity, including engine space and chamber walls. Mr. Coleman went to New York

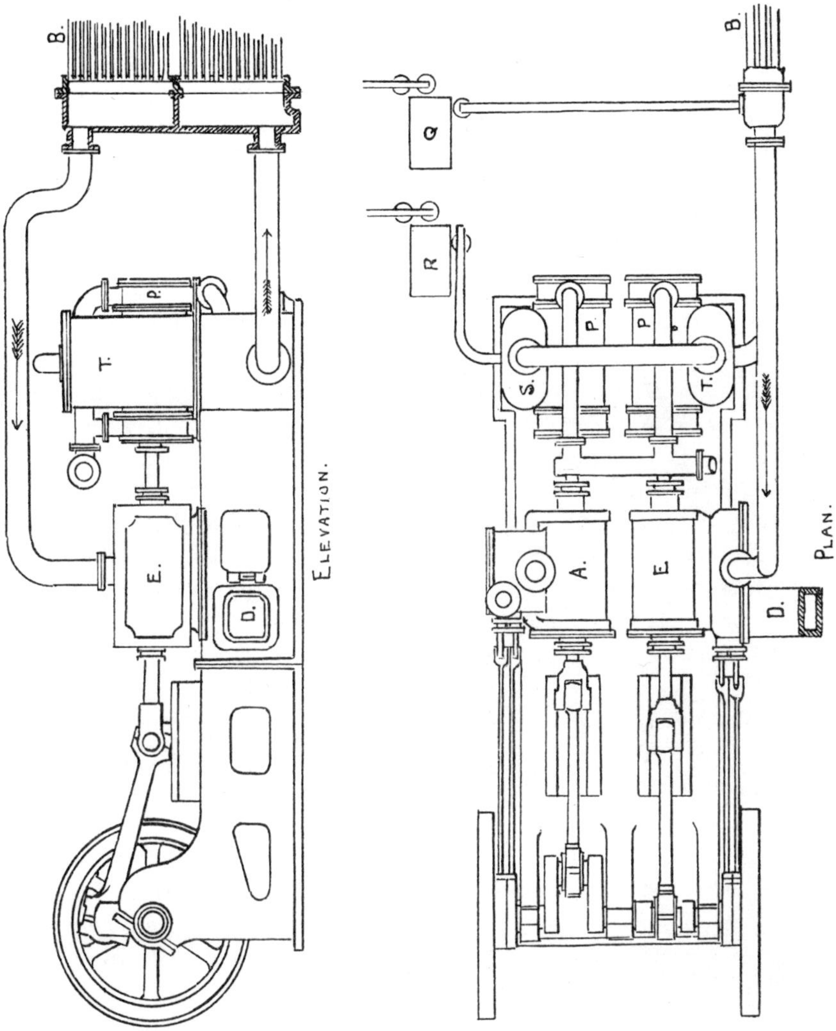

DIAGRAM ILLUSTRATING THE BELL-COLEMAN REFRIGERATING MACHINE.

To face p. 336.

twice with that machine, which was able to keep the chamber near freezing point at 60 revolutions per minute. Some 400 carcasses of beef and large quantities of mutton were carried on these voyages, the meat having been previously cooled to 35° F. in chill-rooms in New York, and landed in England at about the same temperature. The machine, in fact, ran a great number of voyages with similar cargoes, finally giving place to a larger one, and being transferred later to another steamer. The owners of the Anchor Line, said Mr. Coleman, deemed it desirable to fit up thirteen transatlantic ships with such machinery, it being resolved to abandon making the machines in duplicate.

A name connected with some of the earliest practical experiments in frozen meat export was that of Sir Alfred Seale Haslam, who was first identified with the science of mechanical refrigeration in 1876. Sir Alfred's first connection with refrigerating machinery was in developing the ammonia absorption machine in combination with his old partners, Messrs. Pontifex and Wood, whose business he afterwards amalgamated with his own. The name of Haslam is also identified with some of the earliest developments of the air compression refrigerating machine, and Sir Alfred became the purchaser of the Bell-Coleman patents. With regard to the early days of the cold air compression machine in marine work, the Haslam Foundry and Engineering Co., following the introduction of the Haslam "dry air" machine in 1880, carried the major portion of the meat imported into the United Kingdom for many years. The initial experiments made to perfect the Haslam dry air machine were carried out at Derby in connection with a large building fitted up by Sir Alfred Haslam so as to represent and give the working conditions of the freezing hold of a ship.

Dr. Carl von Linde was one of the earliest discoverers of practical mechanical refrigeration. Ammonia compression machine patents were first taken out by him in 1870, and the first Linde machine manufactured for freezing meat was patented in 1874.

Giffard was another early patentee of the cold air compression refrigerating machine, his first plant being produced in 1873, and improved in 1877. It was in the following year that the firm of Messrs. J. and E. Hall, Ltd., took up their first interest in refrigeration by bringing over the Giffard cold air machine which had been exhibited at the Paris International Exhibition of 1877; and in the early eighties this firm was at work fitting ships for the carriage of frozen meat, etc. In 1888, Messrs. J. and E. Hall, Ltd., brought out the carbonic acid compression refrigerating machine, a type of machine which has had a very extensive use on board ship. Messrs. J. and E. Hall, Ltd., have also carried out a great number of installations on land, and are in the front rank of manufacturers of refrigeration machinery. The *Oswestry Grange*, a sectional diagram of which appears on page 126, is one of the many refrigerated liners fitted with Messrs. Hall's machinery.

An early patentee of the cold air compression machine was Mr. T. B. Lightfoot, who became in 1890 the managing director of the Linde British Refrigeration Co., Ltd. In 1880 Mr. Lightfoot introduced an improved cold air machine in which the expansion was performed in two stages, and this machine did considerable work in meat freezing on land and at sea. Mr. Lightfoot was also the designer of what was probably the first commercially successful ship's refrigerating plant in which a chemical refrigerating agent was used, this being on Messrs. Turnbull, Martin and Co.'s s.s. *Perthshire* in 1893 (refrigerated capacity 240,000 cubic feet), all ships carrying frozen meat having up to that time been fitted with air compression machines. The first meat freezing plant designed by Mr. Lightfoot and supplied by the Linde British Refrigeration Co. was that sent out to the Wellington Meat Export Co., New Zealand, in 1891, and in the same year the two works of the Queensland Meat Export Co. at Brisbane and Townsville were fitted with refrigerating machines by the Linde British Refrigeration Co.

Among the earliest types of refrigerating machines applied commercially was that invented by Mr. J. C. De la Vergne, who introduced a patent ammonia compression refrigerating machine in the United States, and had several plants in opera-

PROFESSOR CARL VON LINDE.

To face p. 338.

tion by 1880, The De la Vergne British patents were taken up by Messrs. L. Sterne and Co., Ltd., in 1887, and it was in that year that Messrs. Sterne installed at the Leadenhall Cold Stores a machine of 40 tons' daily refrigerating capacity; this was the first De la Vergne machine erected in Great Britain, and it is still doing regular and efficient service. Machinery working on the same system is to-day very widely used in the frozen meat trade in all parts of the world.

A type of refrigerating machine popular in Australasia and elsewhere is the "Hercules" ammonia compression plant patented and manufactured by Mr. C. A. MacDonald, of Chicago, U.S.A., and Sydney, New South Wales. Mr. MacDonald's first connection with refrigeration was in the early eighties, when, for a number of years, he built machines for Mr. David Boyle, one of the pioneers of refrigeration in the United States. The first "Hercules" machine was built in 1885. In 1891 Mr. MacDonald fitted up the Cunard liners *Umbria* and *Etruria* for the carriage of chilled beef, and also five steamers for the Nelson Line, of Liverpool, for the same purpose. Mr. MacDonald went to Australia in 1894. Many of the freezing works in Australia and New Zealand have been fitted with "Hercules" machinery, which is installed in units as large as 240 tons' daily refrigerating capacity, a machine of that size being erected for the Municipal Markets, Melbourne.

Air Compression and Chemical Refrigerants.—As far as Australia and New Zealand were concerned, in the early days of the frozen meat industry—and it is only with the application of mechanical refrigeration to the meat trade that we are concerned in this chapter—what was known as the dry air process of refrigeration, viz., the compression and expansion of air, was adopted. The only important exporting company then using the ammonia process was the New South Wales Fresh Food and Ice Co., Ltd., of Sydney, which can justly claim the pioneership of ammonia freezing in Australasia, that company having had an ammonia system in use at its works in Darling Harbour long before its adoption by other companies. Honour is due to the late Mr. T. S. Mort, founder of the company, for his indefatigable energy in persevering with his experiments with ammonia for refrigerating purposes at a time when very little practical progress had been made with it in any part of the world.

During the early nineties, however, the old dry air process began to give way to the more economical system employing ammonia and carbonic acid gas, the former of these two chemical refrigerants receiving more favour so far as use in shore freezing works was concerned, and the latter being regarded with similar favour for marine refrigeration. Having become used to the system of circulating air through the rooms by the old dry air machines, which belched the cold expanded air into them to be pumped out again, re-compressed by the compressors, expanded, etc., in constant cycle, many meat experts then persisted that any system without such rapid circulation would prove injurious to the meat. The result was that, consequently, in the early days of the ammonia compression plants they were made to expend their energies upon cooling batteries connected with the freezing chambers by air ducts through which air was constantly circulated by means of fans, thus forming a complete cycle as before. After a time, however, this process began to wane, owing to some of the batteries frequently getting out of order and also on account of the immense amount of space taken up by the unnecessary plant and air ducts. Thus this system of refrigeration later gave way to the method of having the freezing rooms and stores fitted with direct expansion coils, and the Deniliquin works of the Riverina Frozen Meat Co., Ltd., can justly claim to be the first up-to-date establishment in Australasia to have its freezing chambers and stores specially designed for, and fitted with, a direct expansion pipe system operated by an ammonia compression plant, a combination now so widely adopted.

It is a good testimonial to the quality of refrigerating engineering work to state that at a considerable number of refrigerating stores to-day there may be found in operation machines erected when mechanical refrigeration was a new science, and these plants have been steadily discharging their

MR. T. B. LIGHTFOOT.

To face p. 340

duty for anything up to a quarter of a century. British makers lay claim to the fact that such survival is more frequently met with in their machines than with plant made elsewhere, and certainly British refrigerating machinery has a first-class reputation for finish, reliability, and efficiency. The long services of many of the old machines reminds one of the fact that the improvements which have been continually made in refrigerating machinery have not, generally speaking, involved the adoption of new principles as much as the perfection of detail.

Insulation.—But to revert to the remarks upon the elementals of mechanical refrigeration, it may be remarked that the low level of temperature is preserved in a cold store by filling the walls, floors, and ceilings of such chambers with heat-insulating materials. A variety of substances of low heat conductivity are in use, such as cork, in either granulated or slab form ; charcoal, in either granulated or flake form ; silicate cotton, or "slag wool," a fibrous material the thin threads of which are of blown, fused slag ; sawdust ; pumice ; felt ; rice husks ; kieselguhr, a mineral powder ; and other substances.

Modern Refrigerating Devices.—Experience has shown that the application of refrigeration to meat and other foods is successful proportionately to the care with which it is performed. The effect of an irregularly maintained temperature in a cold chamber is harmful to the goods stored, and not only have the mechanical arrangements of the refrigerating machines been improved to avoid irregularities of working, but the devising of air-locks and loading chambers to cold stores, improved means of air circulation, and arrangements of circulatory pipes and brine vessels, has been a feature of later practice. Tellier's early experiments in the carrying of meat at a temperature which would not congeal it—viz., 32° F.—have their later counterpart in the great chilled meat trade of to-day, and in those ships in which the small ranges of temperature have to be achieved for the carriage of the meat at a chilling level the mechanical arrangements to this end have to be very efficient and reliable. So exact have the regulating devices in the brine circulatory systems of ship's refrigerating installations now become, that the makers' claim to be able to keep the temperature of a meat hold to within a degree Fahrenheit, or less, is justified. Two systems of brine circulation are in use, viz., the open cycle and the closed cycle systems. In the former the return current from the various circuits in the system is open and visible. In the closed system this is not so, and it is claimed for the latter method that with it there is no possibility of air getting into the piping and by air-locks preventing the proper flow in some parts of the circuit. One of the latest systems of temperature regulation in meat chilling ships' holds is a patent method of brine attemperation introduced by the Liverpool Refrigeration Co., Ltd. This works on the closed system, and in this method there are two systems of mains, one conveying the zero brine, which is pumped directly from the brine cooler, and the other conveying attemperated brine at a much higher temperature, pumped from a container called the "attemperator." Specially designed valves and "headers" control the mixing of these brine currents and their distribution to the various parts of the installation.

Temperature Measurement and Recording.—The careful registration of the temperatures of cold stores is an important feature, and an up-to-date method is to have thermometric apparatus the indicator of which is at a convenient point outside the chamber or chambers, say, in a lobby or office. These "distance thermometers" are of great service in refrigerated vessels where varying conditions of temperature have frequently to be noted. Self-recording thermometers are in many cases used in the refrigerated holds of vessels, their tell-tale charts being taken as a log or supplementary log of the ranges of temperature experienced during a voyage. In the case of the Canadian export produce trade the Dominion Government has in force an admirable system in which it is part of the cargo inspector's duty at Montreal to place locked-up self-recording thermometers or thermographs in every cold storage chamber where produce is stored. These records are removed by the inspectors in Great Britain and at once mailed to Ottawa, while the instrument remains in the ship. Photographic copies are made of the thermograph records, one copy

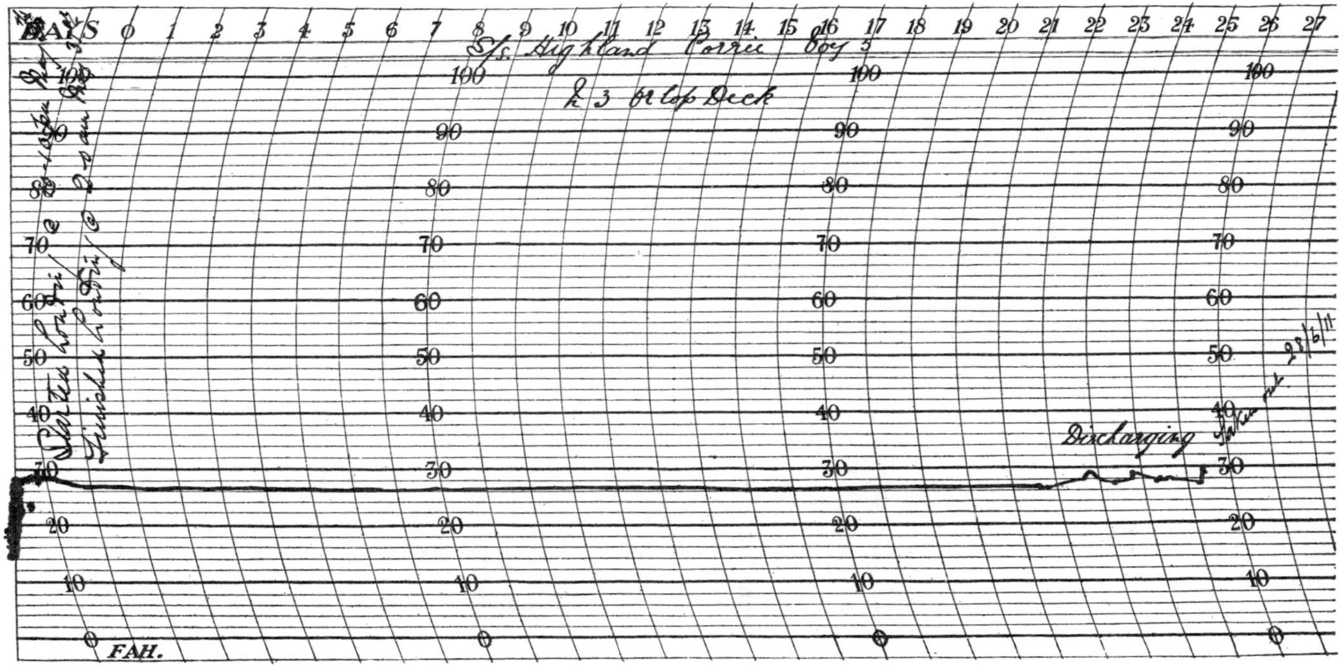

THERMOGRAPH TEMPERATURE LOG OF ONE OF MESSRS. H. AND W. NELSON'S CHILLED BEEF LINERS.

To face p. 342.

of which is filed in the exchange room at the Montreal Board of Trade (Chamber of Commerce), another copy is sent to the Montreal agents of the shipping companies, and copies may be had by any interested shipper. The accompanying illustration of a temperature chart, or "policeman record," of the thermometric conditions of the inside of a ship's hold during a voyage, concerns, as will be seen from the particulars written thereon, a recent South American chilled meat shipment in one of Messrs. H. and W. Nelson's Highland Line vessels. The record has been placed at the service of the authors by Mr. A. R. T. Woods, the marine superintendent of that well-known shipping line. It was this gentleman who devised the first arrangement for temperature regulation by brine attemperation installed in 1895 on the *Highland Chief*, a chilled meat vessel still performing efficient service. Messrs. J. and E. Hall, Ltd., fitted in this vessel a vertical CO_2 refrigerating machine, which was at the time the largest plant of its kind afloat.

Alarm devices to warn as to high or low limits of temperature being exceeded have been adopted in connection with cold stores on land and at sea, and it is worth while to record one of the earliest of these, a clever form of alarm thermometer, which was invented by Mr. Francis Arenas, of Christchurch, New Zealand. The instrument was in the form of a metallic horseshoe composed of five different metals, its action depending upon their varying tendencies to expansion and contraction. The horseshoe was put in a square case placed vertically against a bulkhead. The instrument was set in such a way that its movements produced no effect within normal limits of temperature. Directly, however, the temperature rose or fell beyond the maximum or minimum, the expansion or contraction of the metals caused electrical short-circuiting, a bell was set ringing, and could not be stopped unil the temperature had been adjusted. At the same time as the bell was rung, a disc was displayed at a little window showing the word "Heat" or "Cold," and clockwork was set in motion which automatically recorded the time when the temperature had been too high or too low. This alarum thermometer one of the authors inspected at 26, Jewin Crescent, London, in November,

1897, and shortly afterwards it was destroyed in the big fire which broke out in that quarter, despite its loud alarum!

The Cold Storage Chain.—One of the chief hindrances to the completely successful application of refrigeration to the preservation of perishable food products is the difficulty, nay, the impossibility, of having a complete cold storage chain from the time that the food is started on its way to the consumer to the time that it actually passes into consumption. In the case of meat there are many points at which this cold storage chain is broken. From the freezing chamber at the meat works it makes at least momentary acquaintance with the outer atmosphere on its being entrained, and even should the train journey to the ship be made under ideal conditions, there is the unloading and the putting either into the dock cold store or the ship's refrigerated hold. On its disembarkation, the outer air is again reached before the meat can be safely housed in the land cold store, and then again before it reaches the consumer the meat has several periods of exposure. Modern refrigerating practice has in several ways mitigated the deleterious effect which these imperfections in the cold storage chain are apt to have. Air-locks or ante-chambers are a common feature in cold store construction. When the produce to be stored is taken into these ante-chambers, the outside door is shut, and the outside air and heat are thus prevented from rushing into the refrigerating chamber on its being opened. A prominent example of the total avoidance of this contact of the produce with the outer atmosphere when the place of storage is changed is the use by the London Central Markets Cold Storage Co., Ltd., of patent "portable refrigerators," or large insulated boxes, into which meat or produce can be packed and conveyed either from ship to lighter, quay to wagon, or otherwise.

Railway Refrigerator Cars.—It is probable that if there is improvement needed in one stage of commercial refrigeration more than in another it is in the means provided on the railways for keeping frozen meat cool during transit. The modern railway refrigerator car is at the best nothing more than a glorified ice-box, and its interior is all too frequently an insani-

INSULATED MOTOR VAN.

THE SMITHFIELD AND ARGENTINE MEAT CO.'S TWIN-SCREW REFRIGERATED STEAM LIGHTER *El Zarate.*

To face p. 344.

tary chamber. Means for the mechanical refrigeration of such rolling stock have been devised before now, but nothing of the kind is in general use. The ordinary method of cooling railway refrigerator vans is by means of a tank containing ice and salt freezing mixture or ice only, fitted at one or both ends of the car. The wagons themselves are generally insulated, chief protection in this direction being overhead from the sun's rays. The efficiency of such vans quite depends upon their special design and fitting, and their importance is very great where meat or other produce has to be railed over long distances. Their value in the refrigerated trade, perhaps, cannot be better illustrated than by mentioning that in the United States of America there are altogether nearly 90,000 refrigerator cars in service on the railways, where journeys extend over several days.

Barges and Vans.—The barges which often form the intermediary between ship and cold store are now in a large proportion of cases thoroughly insulated, and the cold held in the frozen carcasses as they are packed tight in the barge converts the river craft into a veritable cold store. Where such barges make their trips only partly filled and unprovided with bulkhead compartments, it will be seen that their cargoes are liable to damage. In some instances where the barges or lighters have heavy duty to perform, refrigerating machinery is installed on them. A recent example of this is a barge, 207 feet long, built in England for the La Blanca Cold Storage Co., and towed to Buenos Aires in the summer of 1910. This craft has two steel decks, and is divided by bulkheads into two large insulated holds of a total capacity of 80,000 cubic feet. Steam driven duplex ammonia compression refrigerating machinery is installed, the barge's holds are lit throughout by electricity, and are also equipped with an elaborate system of meat runners and hooks which enable meat to be dealt with in the most convenient manner. The mammoth barge is named after the company which owns it. Another refrigerated craft of still later construction, built for barging work between the Smithfield and Argentine Meat Co.'s up-river works at Zarate and the ocean steamers bearing refrigerated meat from

Argentina to British ports, is the *El Zarate*, a fine twin-screw vessel, 210 feet in length, of which an illustration is given herewith. Her refrigerated capacity is 60,000 cubic feet, and Haslam's refrigerating machinery, cooling on the brine circulation system, is installed to maintain the vessel's cargo at either a chilling or freezing temperature.

It is universally accepted that the conditions of shipment or discharge of frozen meat cargoes are improved in ratio to the proximity of the cold store to the quayside and the amount of weather protection which the meat can obtain on the quayside itself. Where frozen meat out of a ship has to be sorted to marks, the best conditions for this work are under shelter. Sorting sheds are the ideal for this, and by and by refrigerated sorting shed facilities will probably be available at all the leading ports.

The importance of affording ample protection for refrigerated meat during its conveyance from ship to store, or store to market, etc., is now universally recognized, and efficient insulated vans, horse drawn or motor propelled, are in use in all quarters. Such wagons are, like other refrigerating apparatus, a subject of Lloyd's registration; a common pattern of horse-drawn vehicle being of about 2½ tons tare. Insulating material is used for packing the walls, roof, and floor of these vans, and air-tight doors are provided, some of the vehicles being equipped with meat runners and hooks. Experiments with motor traction for frozen meat were first made about 1900, and now the uses of this form of conveyance are many and widespread, both for town service and over comparatively long distances.

Defrosting and Thawing.—When frozen meat is being thawed out under ordinary conditions, the moisture from the air is apt to condense on the surface and thoroughly wet the meat, causing it to "weep" and become discoloured. Quarters of beef, with their great areas of exposed flesh, are particularly affected. The unsightly appearance of the meat is the chief harm done. Retailers have their own methods of dealing with this difficulty; and they generally wipe off the moisture as it collects. Many mechanical means have been tried for artificially thawing, or "defrosting," frozen beef, and in few sections of the industry have the engineer and inventor been more active than in this—the Patent Office records show a wealth of their ideas. Up to 1896 eight processes had been patented.

No. 1 subjected the meat to a continuous circulation of dry air, formed by mixing cold air at a temperature of 19° and dry air heated to 70°, the combined current at about 26°, increased to about 60°, being forced through the thawing chamber by a fan. Time required for thawing, two to five days.

No. 2 provided for the circulation of air, dried by an arrangement of pipes containing cooling medium and suitably heated by steam pipes, passing over the meat by natural means, and by gradually increasing temperature abstracting the frost without depositing moisture. Time required for defrosting beef four days, sheep two days. This process was in use in London for two-and-a-half years, and at Malta for meat supplied to troops.

No. 3.—Heated air was passed over fused or crystallized chloride of calcium and, mixed with the waste "cold" from the frozen meat, was drawn through the thawing chamber by an exhaust ventilator.

No. 4 provided for circulation, by means of a fan or otherwise, of dry heated antiseptic gas, preferably carbonic acid (CO_2) through a closed chamber in which the meat is hung, and when partly thawed, a quantity of carbonic monoxide (CO), for the purpose of giving the meat a good colour. Time required, with a temperature 80° to 120°, mutton six to eight hours, beef sixteen to twenty hours.

No. 5 consisted in a process for thawing meat by warm dry air, and the treatment used for warming and drying, the moisture being abstracted by contact with a deliquescent salt such as calcium chloride.

No. 6.—Thawing was effected by placing the meat in a closed receptacle served with heated compressed air.

No. 7 was a process for hermetically sealing and protecting the meat from the atmosphere by immersion in boiling fat, or highly refined, tasteless, colourless oil, so as to form an impermeable coating on which the moisture of the atmosphere was deposited.

No. 8 was a process similar to No. 1, except that dry air was saturated with a regulated amount of moisture before being propelled into the thawing room.

From 1897 to the present time the Nelson process has been in limited use. Noticing the accumulation of snow on the ammonia expansion pipes in the cold rooms at Nelson's Wharf, Lambeth, as the effect of freezing out the moisture from the air, it occurred to Sir Montague Nelson that it would be possible both to produce a dry atmosphere and one which could be varied in temperature while drying. This he accomplished by placing steam pipes under a grating in the floor and ammonia expansion pipes on the side walls of the roon, having screens in front of them open above and below for the circulation of air. Eighteen rooms of this construction were fitted up for hanging quarters of beef or sheep, which meat is put into them in a hard frozen state direct from the cold rooms. On steam being let into the steam pipes, warm air rises from around these to descend on each side of the room behind the screened ammonia pipes, on which the moisture of the air is frozen, and retained whilst the freezing pipes are at work. By regulating the steam and ammonia cocks, the temperature can be varied from about 38° F. to 56° F., gradually getting drier all the time. Beef requires four and sheep two days to defrost under this process. When frozen meat thaws under ordinary conditions it looks blue and wet, by reason of the moisture of the air condensing on the cold surface of the meat, but the effect of Sir Montague Nelson's "defrosting" process is to restore frozen meat to its original bright colour and dry surface as when killed, thus enhancing its value and at a comparatively small cost.

In 1894 Mr. Jacob Atherton, of Liverpool, patented a process for defrosting frozen meat by means of circulating an antiseptic gas round it in a dry and heated state. The British and Colonial Meat Defrosting Syndicate, capital £60,000, was formed to exploit Mr. Atherton's invention, and opened a defrosting chamber at 72, Cowcross Street, near Smithfield, doing a fair amount of business in thawing frozen meat for the market salesmen, at a charge of 3*s*. per quarter of beef, and 1*s*. per sheep.

In 1896 Mr. Peck, engineer of the Waitara Freezing Works, New Zealand, patented a plan for thawing frozen meat by blowing warm air upon it, and admitting atmospheric air in such a proportion as to approximate the dew point of the air in the chambers to the temperature of the meat.

Dr. O'Doherty, of Brisbane, was the inventor of a process of an ambitious nature for "preserving" fresh or frozen meat: he placed his meat in an air-tight chamber, and then pumped in a preserving gas, the temperature being kept between 30° F. and 40° F.

Mr. Postle devised a simple plan for the Sydney Fresh Food and Ice Co., his defrosting medium being hot dry air.

Messrs. E. Smethurst and R. W. Chapman, New Zealand, defrosted mutton by attaching the carcasses to the rim of a wheel 5 feet in diameter fixed to a vertical shaft. The wheel was revolved, and the carcasses swung outwards, and their rapid passage and centrifugal motion thawed the meat and prevented the forming of moisture. A machine was erected near Smithfield, and when put in motion, with fourteen carcasses extended horizontally from the circumference of the wheel whizzing round at great speed, the apparatus had a most weird appearance. It was dangerous withal, for had one of the flying bodies slipped its hook, the spectator might himself have become dead meat. The date of this invention was 1902. It was successful only when the atmosphere was dry.

Another thawing patent was that of Messrs. C. A. Lichtenberg and T. L. Washington, used in connection with the Mediterranean frozen meat trade—for the troops. Hot and cold air thoroughly dried and mixed in proper proportions was blown by fans on to the frozen beef, the temperature of which was gradually raised to 60° F.

Mr. A. H. Chapman, Otago, New Zealand, patented a close-fitting waterproof and air-tight bag; enclosed in this, until the atmospheric temperature was reached, frozen meat would, Mr. Chapman claimed, thaw out in good condition.

Mr. Fred A. Furlonge, Hawera, New Zealand, patented a non-porous bag for frozen mutton, but this was intended also to protect the meat as early as in the freezing stage.

The Macmeikan defrosting process is worked on the principle

of air sterilization. The supply of air is regulated by an electric motor, automatic inlet valves adjusting the air current according to the weight of the carcass hung in the defrosting chamber. The process, the proprietors of which are the Macmeikan Defrosting Process Co. Proprietary Ltd., is associated with a freezing system, and in respect of both defrosting and freezing very striking claims are made by the proprietors. A demonstration of the process took place in November, 1911, at Plymouth, and some meat defrosted by the apparatus was placed on view at the London Central Markets.

The chronicling of these various patents is interesting history, but the net result of all these attempts to solve the problem is that defrosting has failed commercially. The Nelson process has a limited scope in certain directions, but, speaking generally, mechanical thawing of frozen meat was not, and is not, successful, because buyers decline to pay a small extra price for defrosted meat ; frozen meat is an article that has to be sold at minimum rates, and a retailer will not pay the additional $\frac{1}{8}$*d.* or $\frac{1}{4}$*d.* per lb. that has to be charged to cover the cost of the process, and the New Zealand and other inventors in their patents did not take this fact into consideration. In addition, few of the processes were mechanically effective.

CHAPTER XXVIII

LEADING PERSONALITIES IN THE TRADE

Anderson, Gilbert, has a connection with the frozen meat trade which dates from 1891. Nine years had elapsed from the *Dunedin's* pioneer voyage in the interests of New Zealand's meat shipments, and the enthusiasm which followed that successful start having somewhat evaporated at the end of the first decade of the Colonial meat export industry, there was a felt want for the application of more organized methods in the business. In 1891 the Christchurch Meat Co. asked Mr. Anderson to take up the position of managing director. Mr. Anderson had had a varied mercantile career, and he applied his commercial experience to the management of the company with such success that Dr. Symes, speaking at a meeting of shareholders on March 2, 1906, said that Mr. Anderson had given the company an assured position and had converted it into an institution of colonial importance. Mr. Anderson made a point of working up the by-products so thoroughly that they became one of the chief sources of the shareholders' profits. He also developed the meat business on c.i.f. lines with standard grades. The capital of the Christchurch Meat Co. was £50,000 when Mr. Anderson took charge, and £300,000 when he resigned in 1906 ; the freezing capacity of 1,500 sheep per day had increased to 15,000 ; and the exports of mutton and lamb from 250,000 to 1,250,000 carcasses per annum. These figures include the capital and capacity, respectively, of the New Zealand Refrigerating Co., and also of a previous amalgamation, namely, that of the South Canterbury Freezing Co., and the Wairau Freezing Co., all of which had been amalgamated with the Christchurch Meat Co. In 1906 Mr. Anderson, having had a serious breakdown in health, resigned his position and went to London, where he now carries on business under the style of Gilbert Anderson and Co., colonial agents, frozen meat representing an important feature of the firm's operations.

Bell, John, and Sons, London, Liverpool and Glasgow, was a business started in 1827 by the late Mr. John Bell, and was subsequently carried on by his two sons, Mr. Henry Bell (now Sir Henry Bell, Bart.), and Mr. James Bell (now Sir James Bell, Bart.). Messrs. Bell turned their business in 1888 into a limited liability company, which a year later was taken over by Eastmans, Ltd., a concern formed to acquire this undertaking as well as the Eastman cattle and fresh meat business of New York. Messrs. John Bell and Sons began to open up meat shops in Great Britain about 1879, and the multiple shop company system may be credited fairly to their pioneering. When John Bell and Sons, Ltd., ceased to exist in 1889, the company had 330 shops in the British Isles. In another highly important matter is the name of Bell associated with the frozen meat industry. On p. 24 will be found information respecting the Bell-Coleman refrigerator, by means of which machine the first shipment of frozen meat was brought from Australia in the s.s. *Strathleven* to London by Messrs. McIlwraith, McEacharn and Co. in 1880.

Blankley, William.—Mr. William Blankley's connection with the frozen meat industry goes back to 1886. He left the business of his father, who was a Leadenhall Market salesman, in 1884, and after a year's sojourn in Australia returned to England and associated himself with the fortunes of Smithfield Market, moving to the Japanese Village (then quite a "Deserted Village") in 1890. Mr. Blankley was one of the earliest supporters of the Frozen Meat Trade Association, of which he was President in 1901, and Vice-President in 1906, 1908, 1910, and 1911, and has throughout been an active member of the Council. In the early days of the Association there was some little friction between buyers' and sellers' interests, and Mr. Blankley was instrumental in forming the Frozen Meat C.I.F. Buyers' Association in 1907. The

MR. GILBERT ANDERSON.

To face p. 352.

difficulties alluded to were, however, of a temporary nature, and the operations of the Buyers' Association were soon suspended, though as a body it still exists. Mr. Blankley has always been a practical supporter of the c.i.f. method of conducting the frozen meat trade, without which he does not believe that the industry could have reached its present proportions. He is satisfied with the development of the trade as a whole, but "deplores the obsolete and apathetic position in which the various insurance companies are content to remain." Mr. Blankley and other c.i.f. buyers are trying a new departure in connection with insurance, the results of which, if not equal to their sanguine ideas, must prove a great improvement, they maintain, upon existing conditions.

Borthwick, Sir Thomas, Bart., was born in Edinburgh, and started at Liverpool and Manchester in 1863 as live stock agent. When the *Strathleven's* cargo arrived in 1880, Mr. Borthwick was much impressed with the possibilities of the new trade, and he quickly discovered that the live stock business would be superseded by that in frozen meat. Foreseeing something of the enormous developments which awaited dead meat import from Australasia and Argentina, Mr. Borthwick decided to take a hand in the trade, and opened depôts at Liverpool, Manchester, Glasgow, and Birmingham, and in 1883 became selling agent for the New Zealand Loan and Mercantile Agency Co., Ltd. In 1892 he brought his head office to London, where he secured a stall at Smithfield. Assisted by his sons—whom he took into partnership—Mr. Borthwick transacted a large wholesale business as importer and distributor. Not content with importing and distributing in Great Britain, the firm turned their attention to frozen meat in Australasia, and opened freezing works at Waitara (1901) and Hastings (1905), New Zealand; and at Portland (1903), and Melbourne, Victoria, as well as at Brisbane, Queensland (1911). They also established branch offices at Christchurch, New Zealand (where the company now have offices in five towns), and in Sydney, Melbourne, and Brisbane, Australia, and greatly extended their Smithfield Market offices and stalls. In 1904, when the volume and scope of the firm's business had largely expanded, Thomas Borthwick and Sons, Ltd., capital £300,000, was registered. Mr. Thomas Borthwick, the eldest son, is managing director, and Sir Thomas's three other sons, Mr. Algernon, Mr. William, and Mr. James, have charge of the New Zealand, Australian, and Liverpool and Manchester businesses. This company's operations, it will be seen, control at every point the frozen meat it handles, from live stock market in Australasia to the dead meat market in Great Britain. Sir Thomas, who received his baronetcy in 1907, is a Past President of the Cold Storage and Ice Association. He has an estate at Whitburgh, Midlothian, where he farms about 1,000 acres of land ; he is on the Midlothian County Council Licensing Bench, and is interested in technical education. Sir Thomas has had time to take a prominent interest in national and local politics, and is considered in the North a tower of strength to the Liberal Party.

CAMPBELL, GORDON.—Mr. Gordon Campbell joined the firm of W. Weddel and Co. in 1895, having previously been associated with the Victorian Government offices in London, where he reported upon and generally supervised Victorian produce. Mr. Campbell first became connected with the produce business through the late Hon. Robert Reid, of Victoria, with whom he travelled to England in 1893, when that gentleman was on his mission to find fresh markets for the sale of Australian productions in Europe, and, more particularly, for those of Victoria, for which State he was, at that time, the Minister for Defence. Mr. Campbell acted under Mr. Reid in London, and later on, for a period of about eighteen months, under the Agent-General of Victoria, and it was through that source that he came into touch with Mr. William Weddel, who in 1895 offered him a position in his firm. The arrangement being mutually satisfactory, the agreement was from time to time renewed until 1906, when Mr. Campbell became a partner. Mr. Campbell was born in Sydney, but has spent most of his time in England, although for five years, from 1888 to 1892, he was on a sheep and cattle station in the north-west of New South Wales, so

SIR THOMAS BORTHWICK, BART.

To face p. 354.

that his practical experiences of stock matters have been of value to him in his work. Mr. Campbell has been from the first in close touch with all the sections of Smithfield Market, and at the present time is President of the Incorporated Society of Meat Importers, a position to which he has been four times re-elected.

CAMERON, HENRY CHARLES, who is Produce Commissioner in London for the New Zealand Government, comes of an old farming family, the Camerons of Balnakyle being well known in Scotch agricultural circles. Proceeding to New Zealand when a young man, Mr. Cameron became engaged in agricultural and pastoral pursuits, and has been closely identified with the development of the frozen meat industry ever since its inception. Returning to England after several years' experience in New Zealand, he, in 1894, opened a store in Manchester for the high-class retail sale of New Zealand meat, as mentioned at p. 203. Some years later, having worked up an extensive connection, he transferred his Manchester business to Messrs. W. and R. Fletcher, Ltd., by whom it is still carried on, and joined the New Zealand Government as Produce Commissioner for the Dominion in London. In this important appointment Mr. Cameron had to take in hand much responsible work in connection with the importation and sale of New Zealand meat. Some of the work was far from agreeable, although very necessary. The prosecution on behalf of the New Zealand Government of meat retailers charged with selling as "New Zealand" meat that had never left the shores of the Dominion, was successfully carried through by Mr. Cameron, although every difficulty that could be thrown in the way of prosecutions of this nature had to be encountered. Probably, on account of the fact that meat from Australia and South America has so much improved of late years, judicial procedure of this kind may not again have to be invoked. Mr. Cameron has always been a strong advocate of advertisement by demonstration and of meat marking as a means for the development of the New Zealand meat trade amongst the better class of consumers in Great Britain. Possibly, if the pigeon-holes of the Government office at Wellington were searched, many special reports by Mr. Cameron would be discovered dealing with practical issues and vexed questions relating to the frozen meat export trade of New Zealand. In shaping the destinies of the trade, as far as handling and marketing in Great Britain are concerned, and in the department of work which a representative of the Government may properly take, Mr. H. C. Cameron has not played an unimportant part.

Christison, Robert, now of Burwell Park, Lincolnshire, who was one of the earliest pioneers of meat freezing in Australia, went to Australia in 1852 and took up pastoral pursuits in North Queensland in 1865, acquiring Lammermoor Station in the following year. Mr. Christison established a pedigree herd of Hereford cattle, and it may be noted to his credit that he was kind to the aborigines on his property, and that he trained them successfully as station hands. He also was one of the first squatters to sink artesian wells. In 1881 Mr. Christison, observing the success of the *Strathleven* shipment, was instrumental in the erection of the Poole Island (Bowen, Queensland) refrigerating and slaughtering works for the export of beef to England (see p. 36).

Cook, the late William, was for many years general manager in Europe for the Compañia Sansinena de Carnes Congeladas. He was born in Argentina in 1851, and was associated with Messrs. Sansinena before they began freezing. Mr. Cook went to England in 1885—1886 mainly to fix up a freight contract, which after some trouble was arranged with Messrs. Houston. Messrs. Sansinena's business was in 1891 turned into a company under Argentine law. Mr. Cook managed the European interests with great ability and enterprise, and was a highly esteemed personage in frozen meat circles in London ; he had a clear perception of the problems of the trade, and had his own ways of settling them. He used to surprise the market at times by raising his prices when large imports of Argentine meat came along and lowering them when the stocks were getting low ! Mr. Cook was full of ideas. One of them shows that he thought out the theory of "pre-cooling." Insulated railway wagons had cold air pumped into them at Havre by means of a pipe from a refrigerating machine; so cold was meat thus kept on the journey to Marseilles that it arrived there with the frost on it. Mr. Cook in 1887 changed the English head office of Sansinena's from Liverpool to London, and at first had difficulty in selling Argentine meat against the primer New Zealand. He retired from Sansinena's in 1905, and died in Argentina in January, 1908.

COOKE, JOHN, senior partner in the firm of John Cooke and Co., Australia, whose activities at the beginning of the eighties were a large factor in the setting on foot of several of the leading meat freezing and exporting enterprises in New Zealand, and whose present connection with the Australasian exporting trade is also well known, as a youth served an apprenticeship with the linen manufacturing trade in Belfast, Ireland, his native town. In 1873 he left for New Zealand, a young man of twenty-one in search of more robust health. Within a few hours of arrival in Dunedin—after a voyage of 110 days in the *Warrior Queen*—he obtained employment in a wholesale warehouse, but left it after three months to join the staff of the Otago *Guardian*, which had then associated with it such men as R. J. Creighton, Vincent Pyke, and Thomas Bracken. While in that service, one of the newspaper company's directors, the late Henry Driver, induced him to join the New Zealand Loan and Mercantile Agency Co., Ltd., of which he was the Otago manager. Four years later, in 1878, Mr. Cooke was transferred to Christchurch, and immediately became manager of the company's Canterbury business. At that time New Zealand was in a very bad way, having felt the influence of bad seasons and very low markets for stock and produce, besides which the City of Glasgow Bank troubles very prejudicially affected many New Zealand financial institutions. The efforts being made in Australia to transport frozen meat attracted Mr. Cooke's attention, and when the New Zealand and Australian Land Co. began its experimental shipments by sailing vessels from New Zealand he determined to take a vigorous hand, recognizing that it was the sole hope for permanently enhancing values of rural lands and live stock. From 1880 until 1889, when he was transferred to the managership of the New Zealand Loan and Mercantile Co.'s Melbourne branch, he worked day and night in promoting and extending the New Zealand frozen meat trade, especially in Canterbury. The banquet given in his honour in July, 1889, by the Canterbury Agricultural and Pastoral Association and the Canterbury Chamber of Commerce was a unique compliment, attended as it was by most of the leading pastoralists and farmers in Canterbury, and also by the leading merchants of Christchurch. Mr. Cooke became manager in Australia in 1891 of the Australian Mortgage Land and Finance Co. He was a leading spirit in resurrecting the Australian frozen meat export trade, which had slackened off and threatened to disappear. Mr. Cooke helped Sir Thomas McIlwraith in the formation of the Queensland Meat Export and Agency Co. In 1895, Mr. Cooke retired from company management, and commenced business on his own account, devoting himself chiefly to frozen meat export; he acquired the Newport (Melbourne) freezing works from Mr. Hotson. The tide of demand for meat and other frozen food products that beat on Australian shores from South Africa during the war was taken at the flood by Mr. Cooke, and doubtless led on to fortune; he supplied about 90 per cent. of the Australian meat issued to the British troops. As beef was called for in that connection, Mr. Cooke erected the Redbank Freezing Works, on the Brisbane river, and acquired a large shareholding interest in the Burdekin works in North Queensland; he also acquired the output of other works in Queensland to build up his beef export trade to various parts of the world; Mr. Cooke, in 1902, took over the Sandown Works and other establishments owned by the Austral Freezing Works. It was in the early eighties that Mr. Cooke became intimately acquainted with Mr. William Weddel, then associated with the New Zealand Grain Agency Co., Ltd., and Mr. Cooke professes pride in the fact that he exercised some influence on Mr. Weddel in his subsequent determination to devote his future energies to the frozen meat trade. Some record of Mr. Cooke's work in connection with the industry is to be found in the chapters on Australian and New Zealand freezing works.

Cox, E. Owen, general manager in Australia of Birt and Co., Ltd., originally entered the Bank of Australasia, New Zealand, and was eleven years in that institution. When he left the Bank to enter commercial life he was travelling manager. Very many men who now occupy important positions in Australia in the mercantile world learned their A.B.C. of business in the Australian Banks. Mr. Cox was one of the first men in Christchurch to set the c.i.f. meat sales going. He came to England in 1895, and joined Birt, Potter and Hughes, Ltd., and took charge of their frozen meat interests. In 1897, Mr. Cox went to Sydney in the interests of the firm and settled there as managing director of Birt and Co., Ltd. During the South African War, Mr. Cox had, on account of his firm, a large share of the conveyance of troops and horses, and has been prominently connected right along with all departments of the frozen meat trade. Mr. Cox has a genius for details, which make him master of every department of the meat and meat shipping business, and what he does not know about it is not worth knowing.

Coxon, Frank, set sail from England for Dunedin, New Zealand, on November 11, 1881, a time when "refrigerating engineers" were not so numerous as they are now. He went out to supervise the erection of the first meat freezing works in the Colony, those of the New Zealand Refrigerating Co., and this work, which was successfully completed at Burnside in 1882, made Mr. Coxon one of the pioneers of the frozen meat trade. Mr. Coxon prepared plans for the buildings of the Belfast works, Christchurch, opened in 1883, and had a hand in the planning of most of the early meat freezing works. Mr. Coxon has been associated all along with the development of the meat freezing industry, and many of the Australasian works have been designed by him. It is interesting to recall some of his pioneering experiences. One of his early pieces of work was the dismantling and equipping of a wooden sailing ship the *Jubilee*, for use as a freezing hulk for the Gear Meat Freezing Co., of Wellington. The vessel had been in the slave trade, and large numbers of old handcuffs were found in the fore peak when clearing it out. The craft did many years' good work in the freezing capacity, and is still in service as a coal hulk in Wellington Harbour. A considerable number of sailing ships including the *Opawa*, *Wellington*, *Marlborough*, *Lady Jocelyn*, *Northumberland*, *Turakina*, etc., were installed with freezing machinery after the success of the pioneer voyages. The s.s. *Fenstanton*, which was a new tramp steamer, was chartered by the New Zealand Shipping Co. in 1883. The vessel was not fitted to carry meat, but she was sent to Lyttelton, and, under Mr. Coxon's supervision, a Haslam freezing plant was installed. Gangs of labourers working the whole twenty-four hours were employed, and the job was finished in twenty-eight days; that is to say, the *Fenstanton* was fitted with machinery, insulated, docked, cleaned, and loaded with 11,000 carcasses of mutton and general cargo, and was off to London within a calendar month. She was the first vessel to enter Port Chalmers dry dock. Mr. Coxon mentions that after this time the meat carrying steamers arriving at New Zealand ports exceeded in capacity the output of the works on shore, and that it was no unusual occurrence to have mail steamers lying off the wharf freezing a portion of their own cargoes on board. Mr. Coxon was employed by the New Zealand Shipping Co. to supervise the freezing arrangement of their vessels, and was constantly consulted by the Shaw, Savill and Albion Co. also. The first steamer of the last-named company intended for the trade was the *Triumph*, she was wrecked under the lighthouse on Tiri Tiri Island, Auckland Harbour.

Dalgety and Co., Ltd., who have neglected no branch of Australasian produce in the scope of their all-embracing business, were, of course, one of the pioneer houses to help forward the frozen meat trade in the days of experiment and doubt. A tragic incident associates Dalgety's and the frozen meat industry. Mr. Richard Blackwood, for some time acting as manager for the firm of Dalgety, Blackwood and Co., Melbourne, when on a visit to England fell dead with heart disease on the deck of the *Protos*; that vessel had arrived in London with the second cargo of frozen meat on January 22, 1881, and Mr. Blackwood had gone to the docks to inspect the meat.

DAWES, THE LATE SIR EDWYN SANDYS, K.C.M.G., was a pioneer of the frozen meat trade, perceiving its possibilities at an early stage, and applying his organizing genius to, and embarking his capital freely in the new business in connection with his shipping enterprises in Queensland and elsewhere. Born in 1838, the son of the Rev. Charles Dawes, rector of Dilhorne, in Staffordshire, he entered at the age of sixteen the service of the Peninsular and Oriental Steam Navigation Co., and in this way was brought into contact with the late Sir William Mackinnon, founder of the British India Steam Navigation Co., with which concern Mr. Dawes became associated. In 1881 the British India Associated Steamers was formed to establish the Queensland Royal Mail Line by Sir William Mackinnon, who had been approached by the late Sir Thomas McIlwraith, then Premier of Queensland. Mr. Dawes took a prominent part in this enterprise, and a regular monthly service between London and Queensland was established via Torres Straits, which did enormous service in stimulating the trade of Queensland ports and the agricultural and pastoral resources of the Colony. The Queensland Steam Shipping Co. was formed soon after as an offshoot of the British India Associated Steamers. This development brought the new coastal line into conflict with the old-established Australian Steam Navigation Co., and, as the result of prolonged competition, the Queensland Steam Shipping Co. absorbed in one transaction the fleet of upwards of twenty steamers of the older coastal service, and the combined companies constituted the present Australasian United Steam Navigation Co., which rapidly gained ground, under the chairmanship of Mr. Dawes. In continuing to refer to his work in connection with Anglo-Australasian shipping, it may be stated that Sir Edwyn Dawes in 1890 took over from Sir William Pearce his controlling interest in the New Zealand Shipping Co., which then owned fourteen sailing ships and five steamers. Sir Edwyn's foresight suggested that the days of sailing ships were numbered, and he accomplished what might justly be stated as the greatest triumph of his commercial career, the complete reorganization of the New Zealand Shipping Co., and the solidification of its position in the Colony. He first disposed of all the sailing ships and subsequently of the five costly steamers, replacing them by powerful and well-equipped passenger and cargo steamers, and this raised the company in a short space of time to a foremost position in the New Zealand trade, with a magnificent fleet. Sir Edwyn Dawes was a director of the Suez Canal, and one of the leading lights generally of the shipping community in London. His death in December, 1903, caused great grief to a wide circle of friends in England and the Colonies. In reviewing his life, it may be said that Sir Edwyn Dawes was a pioneer, whose passion was for developing the undeveloped, bringing order out of chaos, and breathing new life into old bones ; he was ready to hand over to others the carrying out of schemes which he had initiated, whilst himself turning to some fresh task. Concerning especially Sir Edwyn Dawes's association with the frozen meat trade, both as to freezing works and transit, the following information may be given :— Poole Island, Queensland.— This is referred to on p. 36. Austral Freezing works.—This syndicate was formed in 1900 to take over the Sandown Freezing works, New South Wales, and the Newport Freezing Works, near Melbourne, from Messrs. Turnbull, Hotson and Co. Sir Edwyn was chairman, and Mr. William Weddel, Sir George Mackenzie, and Mr. James Caird, directors in London, Mr. John Cooke undertaking the local control. The company's operations unfortunately synchronized with the great drought, and after incurring very heavy losses, the company was liquidated. Central Queensland Meat Co.—Sir Edwyn Dawes was indirectly connected with this concern, formed in London in 1901, along with Sir

THE LATE SIR EDWYN SANDYS DAWES, K.C.M.G.

To face p. 362.

Montague Nelson, Sir George Mackenzie, and Mr. James Caird. It took over, and still runs, the Lakes Creek Freezing Works, Rockhampton, Queensland. Queensland frozen meat shipping trade.—From the early eighties up to the present time the British India Co. has been engaged in the shipping of meat from Queensland, sometimes via Torres Straits, sometimes via the Cape of Good Hope ; associated with them were the Ducal line of steamers, owned and managed by Messrs. J. B. Westray and Co., and the Shire Line, owned and managed by Messrs. Turnbull, Martin and Co. Sir Edwyn, as director of the British India Co., and chairman of the British India Associated Steamers, identified himself particularly with this branch of the company's operations. South African Meat Trade.—In 1902, Sir Edwyn undertook the chairmanship of the South African and Australasian Meat and Cold Storage Co., since liquidated. The concern was an offshoot of the old Dutch company of Combrinks, since merged into the Imperial Cold Storage Co., which is in existence to-day.

Drabble, the late George Wilkinson, chairman of the River Plate Fresh Meat Co., Ltd., from 1882 until the time of his death in 1899, was born in 1823. He went out to the Plate in 1847, but proceeded to Monte Video first, and on to Buenos Aires in 1848. In both places he established the firm of Drabble Brothers and Co. He came home from South America for good in 1868, only paying two short visits to the Plate after that date. What Mort discerned in Australia Drabble saw in Argentina, and, moreover, developed into a successful industry. Perceiving Argentina's possibilities as a cattle and sheep producer, Mr. Drabble in 1882 conceived the idea of starting the meat freezing and exporting industry in the Republic. He was then chairman of the London and River Plate Bank and a very large estancia owner, and he engaged a young Scottish engineer, John Angus, to go out to the Plate and fit up a freezing plant. Mr. Drabble had formed in London the River Plate Fresh Meat Co., Ltd., the original board of which company was composed as follows : Mr. G. W. Drabble (chairman), Mr. James Anning, Mr. George R. Davies, Mr.

Charles Gunther, and Mr. (afterwards Sir) Joseph Pulley. Of this board only Mr. Davies survives. The beginning of the great Argentine frozen meat export trade was the slaughter of 350 sheep on October 15, 1883, and 7,500 carcasses of frozen mutton were shipped to London in November from the completed works of the company in the s.s. *Meath*, which was fitted with Haslam cold air refrigerating plant. The Campana works were on the smallest scale, their area being 200 by 100 feet, and Haslam machinery being used. Shipping operations were harassing in the early days; the company chartered the *Zenobia* (1884), *Zephyrus*, *Zeta* (1886), and *Zarate* (1887), and these vessels proceeded direct to Campana for their meat loading. Mr. G. W. Drabble was a member of the municipality of the city of Buenos Aires; was one of the originators of the Buenos Aires Great Southern Railway, and the City of Buenos Aires Tramways, and the first chairman of the Buenos Aires Western Railway, which post he held until the time of his death. Although not one of the original directors of the London and River Plate Bank, he was invited to join the Board when he came to Great Britain in 1868, and remained its chairman up to the time of his death.

Fitter, Henry Shipley, became interested in the frozen meat trade on its initiation, as a member of the firm of Messrs. Henry S. Fitter and Sons, of which he is now the head. The business was founded in 1856, and was carried on in the old Leadenhall Market. Originally, Messrs. H. S. Fitter and Sons occupied Nos. 192 and 193 in a side avenue at the Central Meat Market, but soon moved to No. 142 in the main Avenue, and later added No. 105. When the old fish market was shut up, Messrs. Fitter applied for space there, but the Corporation at first thought they would reserve it for provisions and pork. Afterwards, however, the authority gave way, and Messrs. Fitter started at Nos. 358 and 364 in "The Japanese Village," in which section of the London Central Markets such a large proportion of the frozen meat business is transacted. The other members of the firm are Mr. Fitter's two brothers, Messrs. Percy and Lewis Fitter, and Mr. Fitter's two sons, Messrs.

MR. HENRY SHIPLEY FITTER.

To face p. 364.

Sidney and Eric, will probably shortly join the firm. Messrs. Fitter in their connection with the frozen meat trade have strictly adhered to the business of commission salesmen. In the early days of the trade the business was almost entirely transacted on commission; but as the industry grew rapidly, the agents outside the Markets found that it suited their purpose better to sell on c.i.f. terms. Many of the market salesmen also found it more convenient to buy on these terms, rather than trust to the uncertain supplies which they were getting on commission, but Messrs. H. S. Fitter and Sons depend almost entirely on the commission business for their trade, both in Australasian, home-grown, and Continental meat, the steady increase in the volume of their commission business having proved conclusively that their efforts have been appreciated by their numerous senders. Mr. Fitter's early recollections of the meat trade date back, of course, to pre-refrigeration days, when dealings were chiefly in home-killed meat, with Dutch mutton and pork in the season. A popular trade at the time was in what were known as Dutch "gigots," *i.e*, the haunches of sheep (not eaten by some of the strict Jews), cut up in Holland and shipped over to this country. This meat was not preserved in any way, but made a fair appearance on the market, being sponged down with salt and water when necessary before marketing. Mr. Fitter has recollections of the very first shipment of American refrigerated beef which came from New York. He also has keen recollections of the early misgivings of the public and of the meat trade itself as to the possibilities of ever firmly establishing the frozen meat trade. Messrs. Fitter handled some of the pioneer consignments of Colonial mutton.

FLETCHER, MESSRS. W. AND R., LTD., is a company originally established as a private concern in the year 1888, which has grown continuously until it is now recognised as one of the most progressive and highly successful businesses in the frozen and chilled meat trade of the United Kingdom. It has about 400 retail depôts situated in the populous centres, besides wholesale establishments in London, Liverpool, Manchester,

Bristol, Newcastle, Hull, Sheffield, Southampton, and Leeds. They are also owners of freezing works at Geelong, Victoria. The managing directors are Mr. Samuel H. Fletcher (son of Mr. Robert Fletcher, founder of the company), Mr. W. J. Kempson, and Mr. William Blagburn—the manager of the retail shops. The two last named gentlemen have been actively associated with the business since its inception. The head office of the company is situated at 19 and 20, King Street, West Smithfield, London, E.C. In February, 1912, a public company entitled the Proprietors of Fletcher's (Meat Importers), Ltd,, was formed to acquire the share capital interest of Messrs. Fletcher's business. "It may be interesting to note," write Messrs. Fletcher, "that in the early days of the frozen meat trade, imports consisted solely of carcasses of mutton, which, as everyone knows, have been augmented in later years by heavy imports of frozen lamb and frozen beef." Notwithstanding the increased shipments of chilled beef, they are of opinion that frozen beef will still remain a permanent feature of the frozen meat trade.

GEDDES, THE LATE J. H., was prominently connected with the trade for the last twenty years of his life. About 1890, Sir Patrick Jennings, the Hon. G. H. Cox, Mr. Henry White, and others, were associated with Mr. Geddes in the establishment of the Pastoral Finance Association at Sydney, particulars of which are given elsewhere. In 1894 Mr. Geddes suggested to Mr. Henry White that he should try the experiment of shipping a few of his fine cattle alive to London as an object lesson for the Australian beef trade. Accordingly, 18 cattle, and 1,000 quarters of beef were dispatched from Sydney on August 20, 1894, in the s.s. *Port Pirie*. Unfortunately, the engineers in charge of the main meat cargo and of the experimental parcel of "chilled" beef quarrelled (there were separate refrigerators), bilge water was pumped into the pipes, and the "chilled" beef had to be frozen. Shortly after this Mr. Geddes experimented in Sydney with chilled beef, holding beef for seventy-two days at 32° F. A dinner to the members of the Queensland Club, Brisbane, was given, at which some of

THE LATE MR. J. H. GEDDES.

To face p. 366.

this meat was eaten. In October, 1894 " J. H. Geddes, Birt and Co., Ltd.," was registered in New South Wales (in 1899 the name was changed to Birt and Co., Ltd.). This represented the introduction of the important shipowning firms of J. Potter and Co., J. Gavin, Birt and Co., and Allport and Hughes. In 1897 Mr. Geddes took up his residence permanently in England. At this period he made further efforts to open up the Continent for frozen meat. Retiring from his association with the Sydney company, Mr. Geddes, supported by Mr. Joseph Moore and other Smithfield gentlemen, negotiated with the Corporation for a lease of the extensive series of cold storage chambers under the Central Markets. The Smithfield Markets Cold Storage Co., Ltd., was formed, with Mr. Geddes as its managing director, and in 1899 this became the London Central Markets Cold Storage Co., with which he was similarly associated until the time of his death. For years he was consulted by gentlemen in various parts of the world engaged in establishing systems of cold storage, markets, or meat distribution. Mr. Geddes was mainly instrumental in forming the Imperial Food Supplies, Ltd., and the General Produce Co., Ltd., and one of his sons, Mr. J. Robertson Geddes, is now managing director of the latter concern. One of the resolutions passed at the first International Congress of Refrigeration at Paris in 1908 was brought forward by Mr. Geddes : " That the refrigerating industry having attained world-wide importance, it is highly desirable that exact scientific data be obtained for determining the condition as to time and temperature under which perishable produce can be satisfactorily kept." For many years before his death Mr. Geddes had in view the application of scientific tests to determine how long bacteria could be kept at bay in their action in meat, and one of his last letters to a friend in the trade contained the following sentence: "My own impression is that chilled beef can be brought from Australia. I shall be very glad to place space at your disposal in the London Central Market Stores if you care about trying an experiment." Chilled beef was brought in 1909, and it was in that year, on March 22, that Mr. Geddes died.

Goodsir, George, director of Messrs. W. Weddel and Co., Ltd., is one of that comparatively small band which can claim not only an active interest in the frozen meat industry, but a connection with it which dates back to the very start of the trade, for his introduction to the frozen meat business took place when he wrote a report upon the outturn of that famous pioneer cargo of meat that arrived in the *Strathleven* in January, 1880. He did this when attached to the produce department of the New Zealand Loan and Mercantile Agency Co., Ltd., whose service he entered in 1878, when that company was under the management of the late Mr. H. M. Paul. That early trade report of Mr. Goodsir's concerning frozen meat was the parent of many, for the Market Circulars, and "Annual Reviews of the Frozen Meat Trade," which have undoubtedly been an appreciable factor in the success of his present firm, as well as informatory to the trade as a whole, have long been under his editorship. Issued regularly for the past twenty-three years, these "Reviews" have been published in recent years also in German and French, and they have attained a world-wide reputation for fulness and accuracy of information. Written from an impartial point of view, these "Reviews," it is understood, are dependent for much of their value upon the goodwill of some of the firm's trade rivals, who are good enough to furnish their quota of information in order that the statistical records of the trade as a whole may be complete. It was in 1884 that he left the service of the New Zealand Loan and Mercantile Agency Co. to take charge of the London produce business of the National Mortgage and Agency Co. of New Zealand, Ltd., but returned to the former company in the following year in order to take up a more responsible post. In 1888 he became a partner with Mr. William Weddel, and his brother, the late Mr. P. G. Weddel, with the object of developing their frozen meat and general produce business. From the outset the combination proved successful, and the firm of W. Weddel and Co. forged ahead rapidly, extending its colonial and foreign connections until it handled a large percentage of the meat imports of the United Kingdom. Mr. Goodsir recently visited almost all the

MR. GEORGE GOODSIR, F.R.S.S.

To face p. 368.

freezing establishments in Australia and New Zealand, and in this way has completed his acquaintance with practically the whole gamut of experiences in this many-sided and world-wide industry. A firm believer in the fact that the frozen meat trade is still in its infancy, Mr. Goodsir has taken an active part in the pioneering efforts of his firm to secure new markets on the Continent of Europe and elsewhere, and also in the recent successful experiments in the carriage of chilled beef from Australia. He has worked all along for the establishment of the frozen meat trade on well organized and systematic business lines, and was to a great degree the pioneer of the system of "grading" frozen meat, upon which system the now extensive c.i.f. or "forward" trade in frozen meats was gradually built up. As one of the promoters and a past President of the Frozen Meat Trade Association (now the Incorporated Society of Meat Importers), Mr. Goodsir has taken his full share in the general development of this trade, apart altogether from any personal interest. He is a Fellow of the Royal Colonial Institute and of the Royal Statistical Society.

GORDON, WOODROFFE AND CO., whose interests had previously been confined to India and the East generally, opened a Colonial department in 1895, and devoted themselves specially to the frozen meat trade, in which as agents they hold a prominent position. In the great extension of the "forward" trade, and the improved organization and methods by which it is worked, the firm have taken a leading part. The partners in the Colonial department of the business are Messrs. G. W. P. Woodroffe, W. A. Wigram (brother of the Hon. H. F. Wigram of New Zealand), E. H. Robinson, and Robert Galloway. Mr. Robinson visited New Zealand in 1894, and has since identified himself very thoroughly with all the rapidly developing features of the trade in the interests of his firm. Mr. R. Galloway, formerly the London representative of the Christchurch Meat Co., joined the firm in 1900, and was made a partner in 1909.

GREENSTREET, CAPTAIN HERBERT EDWARD, master of the New Zealand Shipping Co.'s *Remuera*, has extracted a few particulars from the log of the *Mataura*, the first vessel of the line to bring frozen meat from New Zealand. Captain Greenstreet is a most popular skipper on the Anglo-New Zealand line; he is indeed a veteran, having made eighty round trips, and sailed or steamed over 2,000,000 miles. "November, 1881. Appointed master of the barque *Mataura*, about 900 tons register. The barque was being insulated at the fore end and fitted with Haslam's dry air machinery. Mr. (now Sir) Alfred Haslam was accidentally shut up in a small refrigerated chamber, and would have been frozen to death had he not been discovered in time. Left London December 15. Crossed Equator January 15. Caught albicore, 120 lbs. weight, 5 ft. 1 in. in girth. This fish, and birds shot in Southern Ocean, were put in the chamber, and on arrival at Lyttelton were presented to Sir Julius Von Haast for the Christchurch Museum. April 27, made fast to Port Chalmers wharf, 150 carcasses per day were sent on board and frozen in 'tween decks for twenty-four hours, then bagged and stowed in lower hold. The meat cargo consisted of 3,844 carcasses of mutton, 24 quarters beef, and 77 pigs: total weight, 322,092 lbs., freight, £3,340. The voyage home lasted 103 days, and great worry was experienced as the boiler feed pumps would not act on one tack when the ship heeled over. The voyage was a success, and the meat was delivered in excellent condition." Captain Greenstreet's second voyage in the *Mataura* with meat was not so successful. Whilst he was taking delivery at Auckland of sheep carcasses, a vessel was discharging bone dust near, causing a fearful stench, and, doubtless, distributing microbes, germs, and bacteria; about 4,000 of the carcasses turned out bad in consequence. On the homeward passage a game of snowballing was indulged in at the Equator. Ten years later Captain Greenstreet commanded the s.s. *Ruahine*, with a carrying capacity of 120,000 carcasses.

Grigg, the late John, of Longbeach, New Zealand, can claim a pioneer's part in the frozen meat trade. A portion of the cargo of the sailing ship *Dunedin*, which took New Zealand's first frozen consignment to England, consisted of some half-

CAPTAIN HERBERT E. GREENSTREET.

To face p. 370.

bred Shropshire wethers and lambs sent by Mr. Grigg from Longbeach, which has been spoken of as the "best farm in the world." From the first, Mr. Grigg expressed great confidence and hope in the future of the frozen meat trade, and with others he built the first freezing works in Canterbury, at Belfast, afterwards forming a small farmers' company. His early efforts among the sheepowners to get them to send sheep for shipment was an honourable part of the uphill work done for a great industry, and his pluck and determination were ill-rewarded by the loss he sustained in the failure of the consignment on the *Mataura*, which he chartered to carry his own fat stock from Longbeach, as well as a small quantity of meat from other farmers. Unfortunately, the temperature of the freezing chambers was allowed to rise too high, and the meat was all condemned when it arrived in London, which failure cost him a large sum of money; as no insurance company would take the risk of insuring frozen meat in those days of trial (notwithstanding the fact that the meat by the *Dunedin* was covered for £5 5*s*.), the whole cargo was uninsured. The freight for this boat was $2\frac{1}{2}d$. per lb., besides other charges, and now $1\frac{1}{5}d$. will pay all charges to London. In many cases it would have been better from a point of view of profit if some of the sheep had jumped into the sea instead of going on board a ship, as in many cases the prices fetched in London did not pay the expenses of freezing and freight. At first there was no use found for lambs' pelts, and they were buried to get them out of the way, and several of the by-products (paying well now) were not paying then. So unprofitable was the trade of buying and shipping meat at first that either in 1886 or 1887 Mr. Grigg bought thousands of prime fat lambs at from 5*s*. 3*d*. to 7*s*. 6*d*. Many farmers in the Ashburton County followed the policy of selling their lambs forward for the whole season at 7*s*. 6*d*. per head. The brand under which Mr. Grigg shipped was known as the "Plumes" on the London market, being three feathers over No. 16. This brand made a good name, and was well known in the early days of freezing for export, representing as it did extra quality and strict, even grading. During the early years of freezing Mr. Grigg used to buy up to 80,000 fat sheep and lambs in the Ashburton County and ship them to London every season. In growing rape, turnips, and clover, with Italian rye-grass, to top off lambs after weaning, he was probably one of the leaders. In one season from December to the end of May 18,000 lambs were fattened on Longbeach and averaged well over 40 lbs. each. One line of 4,400 lambs, which were all sold on the same day, averaged over 42 lbs. Mr. Grigg was a strong advocate of the principle that the freezing companies should not deal in meat. He advocated the open door for the farmer and his stock, and the advantage of the consignment principle. The Canterbury Frozen Meat Co. right throughout its career has felt the good effect of the energy with which Mr. Grigg devoted himself to the starting of the Canterbury trade.

Haslam, Sir Alfred Seale, has played a part as pioneer in the introduction of refrigerating machinery best indicated, perhaps, by reference to the fact, set forth in another page in this book, that his "dry air" machine was introduced in 1880, and that following this the Haslam Foundry and Engineering Co., formed by Sir Alfred as early as 1868, had what amounted practically to a monopoly of British marine meat refrigeration for fourteen years during the life of the patents concerned. In 1880, when the trade between Australasia and Great Britain was opened for the carriage of frozen meat, Sir Alfred fitted ships to carry approximately 150 tons of meat per cargo, and this was considered a great achievement. The fact that at the present time he is fitting up a number of vessels to carry from 3,000 to 4,000 tons of meat or dairy produce gives an idea of the enormous development of the trade. Sir Alfred's early training as an engineer was at the Midland Railway Works at Derby, and he was at work later under Lord Armstrong's company. His first association with the manufacture of refrigerating machinery was in 1876, and that association, like his connection with Derby, has since seen no interruption. He received his knighthood when, as Mayor of Derby, he received Queen Victoria on her visit to Derby in 1891. Like many other prominent engineers, Sir

Alfred found time to enter Parliament, representing Newcastle-under-Lyme as a Liberal Unionist from 1900 to 1906. Sir Alfred in 1896 presented London with the bronze statue of Queen Victoria and the granite base standing near Blackfriars Bridge. This statue, unveiled by the late Duke of Cambridge, is the only one of her late Majesty in the City of London.

HENDERSON, WILLIAM, late London manager of the Christchurch Meat Co., was born in Aberdeen. He emigrated to New Zealand in 1867, and went on to the Mataura and Edendale Estates (Southland) of the New Zealand and Australian Land Co., and later resided in Invercargill. In 1872, he joined the Bank of Otago (afterwards taken over by the National Bank of New Zealand) at Dunedin, and later entered the firm of Guthrie and Lamach, in 1879 going to London as manager for this firm. About 1888 Mr. Henderson was invited to assume the management of the National Mortgage and Agency Co. of New Zealand, Ltd., and in that position he had the supervision of the Longburn Freezing Works, which the National Mortgage Co. had taken over. Mr. Henderson resigned from the National Mortgage Co. in 1904, and went to London as manager for the Christchurch Meat Co. in 1905. In September, 1910, a farewell banquet and presentation at the Café Royal, London, were given by the trade to Mr. Henderson, who, on resigning his connection with the Christchurch Co. and retiring from business life, left for New Zealand in October of that year.

HOTSON, JOHN.—Mr. Hotson was born at Langholm, Dumfriesshire, in 1851, and received his early business training in the National Bank of Scotland, and the London and Westminster Bank. Later he joined the firm of Dennistoun, Cross and Co., London. Mr. Hotson sailed for Australia in 1879, and arrived at Melbourne at the time when pastoralists and merchants were keenly interested in making the export of frozen meat a commercial success. The Australian Frozen Meat Export Co. was formed soon after Mr. Hotson arrived in Melbourne, and he became secretary, and later manager and secretary. In 1887 the works of the company—at Newport—

were acquired by the Victorian Government; Mr. Hotson subsequently leased the establishment, and during his period of occupancy about 1,200,000 frozen sheep and lambs were exported to London. In 1896 Mr. Hotson passed over his business interests at Newport to Mr. John Cooke. During the last few years Mr. Hotson has resided in England, at Bournemouth.

Johnson, the late E. L., became associated with the frozen meat trade as a surveyor in 1885. Mr. Johnson was the pioneer of the surveying of frozen meat on defined lines, and his force of character and ability, combined with tact, made him quite a figure in frozen meat circles. As he had to deal with a new business, he had, naturally, to establish his own precedents, and only a strong man could do that. Many battles over meat and mouldy rabbits did he have with claimants. Captain T. R. Mowat, who began surveying frozen meat in 1893, joined in 1901 the firm of E. L. Johnson's Sons and Mowat, the successors of the connection of the subject of this brief memoir. Mr. Johnson was wont to use some quaint and expressive phrases at Smithfield in the course of his work. "As sound as a bell of brass," he would say when frozen meat was submitted to him for damage allowance, which he disputed. "Stone fed" was another of his expressions in alluding to carcasses the poorness of which, amounting to emaciation, admitted of no dispute. Mr. Johnson died in December, 1900.

Keele, Richmond, who has been called the "Grand Old Man of the Meat Trade," was for many years the frozen meat manager for Nelson Brothers and the Colonial Consignment and Distributing Co., Ltd. He was a well-known and highly respected figure at Smithfield, where, prior to his retirement, he was to be seen every morning. With good general knowledge of live stock, country born and country bred, Mr. Keele finished his education by two years at an agricultural college, which had numbered the late John Tyndall and Edward Frankland among its professors and Henry Fawcett among

THE LATE MR. E. L. JOHNSON.

To face p. 374.

its students. In 1849, as a lad of eighteen, Mr. Keele arrived in Adelaide, South Australia; in 1851 he joined his brother, a Canterbury Pilgrim, in New Zealand, and later visited the East, residing for several years in India, Japan, and China. In Shanghai the food supply of the foreign settlement was under his supervision, and he ran a British dairy, supplying Shanghai with milk and butter, and Hong Kong with butter. Mr. Keele returned to the Old Country in 1874, and took up dairy farming, with a large shed of milch cows and several shops in London. About this time Mr. Keele became associated with Sir E. Montague Nelson in the farming business. In 1881, when on a voyage to the East, Mr. Keele met at Niagara Mr. Robert Campbell, of New Zealand, and travelled with his party through Canada. At Quebec Mr. Campbell then told him he had just concluded arrangements with Mr. Davidson for providing from his stations 10,000 sheep for the *Dunedin* experimental shipment, and New Zealand's pioneer cargo. In 1886 Mr. Keele joined Nelson Brothers, Ltd. He retired from the C. C. and D. Co. (which had taken over the distributing business of Nelson Brothers) in 1903, and on October 20 in that year he was entertained at the Café Royal by seventy gentlemen representing the import and wholesale departments of the trade, who made him a presentation of a massive silver bowl.

MARTIN, EDWARD, was a distinct force in the pioneering days of the frozen meat trade, on the shipowners' side of the industry. As managing owner of the Shire Line (Turnbull, Martin and Co.), he perceived at an early date the importance of the developing trade to the shipowner, and had a clear idea in his mind of the type of vessel that the trade required, "the refrigerated provision tank," as he termed it. Mr. Martin became Chairman of the New Zealand Loan and Mercantile Agency Co., Ltd., in May, 1894, when that company was reconstructed. The affairs of the company were naturally then in a depressed state, and the new chairman's vigorous personality helped in no small degree to restore the business to a firm footing. Mr. Martin remained chairman until his death, which took place suddenly on February 6, 1900: he was but fifty-one years of age when he passed away. Mr. Martin married a daughter of the late Sir Edwyn Dawes in 1891. Had he lived, there can be no doubt that he would have taken a keen interest and played an important part in the settlement of the many interesting problems accompanying the growth of the frozen and chilled meat trade, occupying, as he did during the last six years of his life, the dual position of shipowner and chairman of a company interested very extensively in handling refrigerated produce at producing and marketing points. Mr. Martin had a wonderfully energetic temperament, and was particularly good at attacking and overcoming difficulties.

McIlwraith, McEacharn and Co., Ltd., is a firm of which the two founders, Mr. Andrew McIlwraith and the late Sir Malcolm McEacharn started business together in the year 1874, trading as McIlwraith, McEacharn and Co., the well-known mercantile and Anglo-Australian shipping house. The firm built a fleet of sailing ships, chiefly for the purposes of taking emigrants to the northern parts of Queensland. These vessels, after some years, became obsolete, and the company then acquired a fine fleet of steamers, which are now principally occupied in the Australian intercoastal trade. In 1879 the firm created a sensation by chartering the s.s. *Strathleven*, and fitting her with refrigerating machinery for the purpose of sending her to Australia to bring home to England a cargo of frozen meat, which enterprise, as recorded elsewhere in this book, was eminently successful. McIlwraith, McEacharn and Co., Ltd., was formed in 1891, the company being registered at Melbourne, Victoria, and there now exist, in addition to offices in London and Melbourne, branches at Adelaide, Sydney and Newcastle (New S. W.), Fremantle, Perth, and Albany (West Aust.). The head office is at Melbourne. Messrs. McIlwraith, McEacharn and Co., Ltd., act as agents in Australia for several of the important lines of steamers carrying refrigerated produce to Great Britain.

Martindale, Colonel C. B., as the general manager of the London and St. Katharine Docks Co., had a great deal to do

SIR EDWARD MONTAGUE NELSON, K.C.M.G., D.L.

To face p. 376.

with the early days of frozen meat. He was with the company named before the union took place which brought about the London and India Docks Joint Committee, and had to erect and instal the store at the Victoria Dock, and initiate and organize a working system for receiving and storing and delivering to consignees who used his cold storing facilities. A very considerable part of the work fell to the share of Mr. H. W. Williams, the assistant general manager, who was subsequently the manager of the Joint Committee above-mentioned, and he threw himself with vigour into this new industry of handling perishable goods with the many tough problems that it brought with it for solution. A huge "sorting shed" for frozen carcasses at the Victoria Docks store was a scheme proposed at one time, and Mr. Williams took much interest in the proposal.

Nelson Brothers.—The personal connection of Messrs. Nelson with New Zealand dates back to 1862, when Frederick and William Nelson arrived in Auckland. A sheep station was purchased in the Hawkes Bay district, where their relatives, Bishop Williams and his family, were the earliest settlers, and later a noted flock of Lincoln sheep was established. Mr. William Nelson returned home in 1875, and as the result of experiments carried on at the well-known gelatine works of Nelson Dale and Co., at Warwick, machinery was perfected by him for a process of meat preservation and the preparation of tallow; hitherto the only method for dealing with the carcass in New Zealand had been boiling down. On Mr. Nelson's returning to the Colony in 1880 (in conjunction with Mr. J. N. Williams as partner), machinery was installed, and the establishment at Tomoana, Hawkes Bay, was in full working order the following year. Conducted on scientific principles, this was far in advance of the then existing boiling-down works in Australia, which were crude both in their appliances and methods. In 1882 the success of refrigeration in the carriage of perishable products had been established, and it was decided to form a company that should take over the existing business, and add refrigerating machinery to the works at Tomoana. In May,

1883, the prospectus of Nelson Brothers, Ltd., was issued, and mainly owing to the personality and influence of Mr. Edward Montague Nelson (who was in charge of the business in London) the subscribed capital of £160,000 was immediately raised. The refrigerating works started under the management of Mr. William Nelson, and the three shipments of frozen meat made by the company in 1884 marked the first stage in a business which developed to very large proportions. From time to time additional amounts of capital were raised and spent in developments, and this expenditure was a great factor in establishing the freezing trade on a strong and lasting basis in New Zealand. Both in the Colony, and, as regards distribution at home, Nelson Brothers deservedly occupy the position of pioneers of the frozen meat trade. Selling the meat in Great Britain was just as important as freezing and shipping, and the system had to be initiated. For the first two years the cargoes were almost entirely sold at Smithfield, but in 1885 the stores in Thames Street, under Cannon Street Station, were opened by Nelson Brothers, and the opportunity was afforded of sending the meat farther afield ; the country business was started, and quickly attained big proportions. England was mapped out into districts, an army of travellers being employed to push the trade, not only in the cities but in the country towns and villages. In addition to butchers, other provision retailers were induced to give the new commodity a trial, and were supplied in small quantities, even to a single carcass, the great end in view being to get the meat known throughout the country. To assist in this object Nelson Brothers purchased the barque *Prince of Wales*, fitted her with refrigerating machinery, and sent her to Plymouth as a distributing station for the West of England. With their up-river store supplied by barges from the vessels at the docks, and served as to deliveries by railway vans, no difficulty was experienced in getting away the Colonial produce. So much progress was made that before long there was hardly a town in England without its frozen meat shop. The system was to send out price lists on Saturday to the various customers offering the different classes of meat, and the orders came

MR. WILLIAM NELSON.

To face p. 378.

along in the following week. The chairman at a general meeting, questioned by a shareholder as to the advisability of opening shops, answered that having some 5,000 butchers as customers purchasing meat, the directors thought it a preferable system to continue supplying them and not come into competition with them. In 1893 the fine and extensive premises, known as Nelson's Wharf, embracing warehouse, cold stores, and distributing station, were opened at Lambeth, with frontage on the river Thames. These stores are fitted with Haslam and De la Vergne machines, and there are ten miles of refrigerating pipes in the building. The works are said to have cost over £150,000. Nelson Brothers' meat when first introduced into London was sold at Smithfield by Messrs. Black and Stimpson; Mr. William Stimpson—who died in 1900—looked after the frozen meat department of the business. To show the enormous risks of the trade, it may be mentioned that in the early nineties Nelson Brothers, Ltd., who had established a system of annual contracts with the farmers in New Zealand in order to secure regularity in supply and freight arrangements, lost no less than £102,000 on their shipments in one year. The sheep farmers of Hawkes Bay met the situation by a reduction on contract price. The system was found inapplicable to a trade suffering from such great fluctuations in market values, and it was not continued. In 1895 Nelson Brothers, Ltd., sold their English distributing business to the Colonial Consignment and Distributing Co., Ltd., of which Sir Montague Nelson is also chairman and managing director, and, with the fresh capital introduced, every effort has since been made to develope possible outlets for the produce of the Colonies. Nelson Brothers consign all the frozen and general produce which they control to the Colonial Consignment and Distributing Co., which company acts as general agents for Australian and New Zealand clients. The Central Queensland Meat Export Co., Ltd., (which company acquired in 1901 the undertaking of the Central Queensland Meat Export Co., Ltd., of Rockhampton), and the Australian Chilling and Freezing Co., Ltd. (works at Aberdeen, New South Wales), are domiciled at the Colonial Consignment Co.'s

London office. Mr. W. A. Porter is secretary of Nelson Brothers, Ltd., and of the Central Queensland Meat Export Co., Ltd., and Mr. P. B. Proctor is secretary of the Colonial Consignment and Distributing Co., Ltd., and of the Australian Chilling and Freezing Co., Ltd.

Nelson, Edward, started in 1879 in the live cattle business, and when the frozen meat trade developed he associated himself with it, and had considerable experience with the trade up to the time of the building of the Las Palmas works in Argentina, connected with Messrs. James Nelson and Sons, Ltd. As regards that company, Mr. Edward Nelson was joint managing director with his brother William Nelson in its early days, and continued the joint managing directorship with him after the death of Mr. Hugh Nelson in 1893. Mr. William Nelson retired from that position in 1904, and some time afterwards Mr. Thomas Nelson, another brother, at that time resident in Glasgow, and engaged in the live stock trade, joined Mr. Edward as joint managing director. Both have since continued to manage the affairs of the company.

Newman, J. N., the London manager of the National Mortgage and Agency Co. of New Zealand, Ltd., joined his company in 1885, and at once got into close touch with the frozen meat trade. The National Mortgage Co. was then receiving consignments, and, in fact, was more or less connected with the trade from its beginning, owing to its close relation with the New Zealand and Australian Land Co. For eighteen years, up to the end of 1909, Mr. Newman represented his company daily on Smithfield Market, the whole of the consignments received including the output of the company's works at Longburn, New Zealand, passing through his hands, the bulk being actually sold by him to dealers and salesmen on the market. Mr. Newman was a member of the Committee of the Frozen Meat Trade Association, and filled the part of president during the years 1901 and 1902. In these years some progress was made in settling the terms and conditions of c.i.f. contracts, and in the better handling of meat at the selling end.

Mr. Newman is now a member of the council of the Incorporated Society of Meat Importers, and continues to take an active interest in everything connected with the trade.

Pasman, Miles A., managing director in Buenos Aires of the Compañia Sansinena de Carnes Congeladas, was first connected with the firm as sindico, and was appointed managing director in 1899. Mr. Pasman occupies a very influential position, and is one of the few men in the Republic to whose enterprise and genius the tremendous growth of the Argentine frozen meat trade may be attributed. He is president of the Estancia y Colonias Curamalan, a huge pastoral property now being broken up for settlement, is also president of the Bristol Hotel at Mar del Plata, and is chairman of the local Board of the Argentine Estates of Bovril.

Paul, the late Henry Moncreiff, was born at Glasgow in 1834. He went to Melbourne when a young man, and later had a sheep station near Wagga Wagga, New South Wales. He joined the New Zealand Loan and Mercantile Agency Co.'s service in 1871, and retired therefrom in 1899; he died in 1907. The first consignment of frozen mutton received by the company was in 1881, and Mr. Paul specially interested himself in introducing it to the public, both through the trade and by personally bringing it under the notice of his friends and business acquaintances. In the matter of the erection of refrigerated stores in London Mr. Paul was largely concerned, as well as in the compilation and dissemination in Australia and New Zealand of information regarding the selection and preparation of meat for despatch to Great Britain. A shipment of frozen meat was made to France under Mr. Paul's supervision in 1886, and though everything was done to establish the trade there upon a satisfactory footing, owing to the heavy duties imposed by the French Government business with that country was found to be impracticable. In all questions relating to frozen meat and other Australasian interests Mr. Paul took a keen interest, and under the pseudonym of "Pomingalarna" he wrote numerous letters to the Press, as well as frequently taking part in discussions at the meetings of the Royal Colonial Institute and elsewhere. The New Zealand Loan Co. has played a very important part in the development of the natural resources of New Zealand and Australia, especially the frozen meat business. Mr. Paul was a prominent and popular figure in Colonial circles.

Pearse, Albert William, J.P., editor and managing director of *The Pastoralists' Review*, has taken a prominent part in suggesting and introducing reforms in connection with the systems under which the frozen meat trade in Australia, Sydney in particular, has been worked, and has written many pamphlets on this and other branches of the business, *e.g.*, "The Right and the Wrong Way," Letting in Daylight," etc. Mr. Pearse advocates an improvement in many directions in methods of handling meat for export in Australia and New Zealand, and urges the removal of the Commonwealth 15 per cent. duty on meat wrappers, so that Australian shippers may be on equal terms with their competitors in New Zealand and South America in this respect. Sydney has an unenviable reputation regarding the conveyance of stock by the railways, the long distance between sale yards and abattoirs, and general handling of freezing stock, as well as regards the carting of the meat to the ship's side, and Mr. Pearse has been hammering away at these things for many years. He was the first man to discuss in public the Argentine systems of running the frozen meat trade, and in "Our Great Rival," published in 1910, he described these methods, and instituted a comparison —by no means flattering to Australian methods—between the Argentine and the Australasian ways of working.

Rose, John, and Co. dates back in its provision and tea business to 1822, and Mr. John Alexander, a pioneer frozen meat retailer, and the chief partner in this firm, saw early in the nineties the possibilities of handling the retailing of frozen meat in a new and special way. The daily Press had in the early days of the frozen meat industry expressed anxiety as to the food supply of the United Kingdom, and although for ten years frozen meat and lamb had been imported from New Zealand and Australia, the public appeared to know little about the new supplies, and seldom had the chance of making a purchase. One of the first efforts to retail New Zealand mutton and lamb placarded as such was that made by John Rose and Co., who on Friday, November 25, 1892, made an initial trial with six carcasses at one of their branch establishments in London; on the following day, Saturday, eighteen carcasses were ordered and sold. On the Friday and Saturday in the three following weeks, 36, 57, and 75 carcasses of New Zealand mutton were dealt with. This was the beginning of Messrs. John Rose's business in the retailing of high-class mutton and lamb, a trade to which they have adhered in the great business which has been established. The firm have relied to a great extent upon household orders. Mr. Alexander states that the most difficult part of the business is the "extraordinary fluctuations in the prices of imported meat." He maintains that "New Zealand's best" is still the best of all imported lamb and mutton, and it is important, according to him, that New Zealand farmers should see to it that they continue to hold this position.

SINCLAIR, RUSSELL, is one whose connection with the Australian meat freezing industry, in the capacity of an advisory constructional and refrigerating engineer, has been of the widest. The firm of J. Wildridge and Sinclair, Ltd., first became associated with the meat freezing development in Australia in connection with Mr. Robert Hudson, who was instrumental in having the Tenterfield and Narrandera Chilling Works erected in New South Wales in the early nineties. In 1891 Messrs. Wildridge and Sinclair fitted out the Melbourne Markets with two 50-ton Linde refrigerating machines. They also carried out the erection of the large Kirribilli Freezing Works (Sydney) in 1891—1892, and the extension of the Queensland Meat Export Co.'s Works in 1894. An idea of the difficulties connected with the establishment of freezing works in the tropics is given in the following notes from Mr. Sinclair:—"In 1894 and 1895 the Bowen Meat

Works were formed, and we carried out the design and erection of the whole of this plant, fitting machinery to handle 100 head per day. These works are situated about four or five miles from the Bowen Jetty, over on the other side of the Burdekin River, and during the erection of the works very heavy rains came down, causing a flood, which washed away the railway bridge, and I remember well the difficulties we had to carry out the erection of the plant, slinging a wire rope across from one tree to another on each side, fitting up a temporary cradle. All our men and ourselves had to get hauled across, at times the cradle dipping perilously near the river. A portion of the machinery weighing several tons had to be taken across the river in this way." Some years later Mr. Sinclair's firm altered the Darling Harbour Meat Markets cold storage chambers (Sydney), so as to fit them for export purposes. In 1895—1896 they considerably enlarged and modernized the Lakes Creek Freezing Works, Rockhampton, and in 1896 fitted up the Charleville (Queensland) Meat Works for inland chilling. This establishment has never been in commission.

Spearing and Waldron, of London, chartered the sailer *Hengist* and shipped frozen mutton from San Carlos and Port Howard, ports in the Falkland Islands, to London from 1890 to 1895, shipping 100,000 carcasses in six voyages. The *Hengist* was wrecked in the Straits of Magellan. Messrs. Spearing and Waldron were largely interested in frozen mutton shipping from Patagonia to London from 1896 to 1899 (see p. 86).

Stephenson, the late R. Macdonald, became connected with the frozen meat trade in 1881, when he joined the Australian Co., Ltd. as secretary; of this concern he was general manager in 1884. The Australian Co. owned the meat export works at Poole Island, Bowen, Queensland. It was wound up in 1888, but prior to that Mr. Stephenson became managing director of the New Zealand and Colonial Consignment Co., Ltd., which took over the consignment business of the former concern in 1885. In 1886 Messrs. Nelson Brothers, Ltd., purchased the goodwill and business of the New Zealand and Colonial Consignment Co., Ltd., and Mr. Stephenson went over and was appointed country manager. He visited Australia twice, and South Africa, and in 1897 took the position of secretary to the Colonial Consignment and Distributing Co., Ltd., on the retirement of Mr. V. S. Hervey. Mr. Stephenson vigorously tackled all the problems which presented themselves in the early days of the trade, when precedents were few. He took a very active part in the promotion of the Australian Chilling and Freezing Co., Ltd., (works at Aberdeen, New South Wales) in 1890, and his ambition was to raise Australian meat to the standard of that from New Zealand. His sudden breakdown of health put a premature end to a career of unflagging energy and enterprise. He died in 1901.

SWIFT AND Co. is the great meat business which, with a gigantic capital of 75 million dollars, is now so closely identified with the frozen meat trade in all parts of the world, and had its foundations laid some fifty years ago by the late Mr. G. F. Swift, who started with the purchase of a heifer with 20 dollars given him by his father, a farmer. In the interval that separates that day from this the great industry known by the man in the street as Swift's has become a huge concern, possessing in about 400 cities in four continents distributing houses employing over 30,000 hands. As many as 11,875 cattle, 16,553 sheep, and 34,562 hogs have been transformed into dressed meat in a single day in the seven great packing houses of Swift's. These figures are sufficient to show the proportions of the great business, but the direction in which it qualifies for mention in these pages has chiefly been in the leading part it has played in the great North American transatlantic chilled beef export trade, which, although now dwindling because of the growing internal meat consumption of the United States, has in the past supplied Great Britain with the class of refrigerated meat that by its unfailing quality and good condition has always commanded in English markets top refrigerated prices. This has been achieved partly on account of the shorter trip from the United States as compared with the voyage from South America and the much longer journey from Australia and New Zealand. This has permitted chilled transport, in which the surface of the meat is not hardened by frost; but much of the trade's success has been due to its excellent organization and conduct. Mr. G. F. Swift's first meat sale, that of the heifer already referred to, was transacted at Barnstaple, Massachusetts, and brought him a profit of 50 per cent. The founder was a born judge of beasts, and an early partnership with James A. Hathaway, who was in the dressed beef line at Boston, improved the fortunes of both of these men. At Buffalo, where the two went, there were other "starters," including Armour and Hammond, and soon, with the centre of gravity of the American live stock business shifting westward, Swift ended his partnership with Hathaway and went to Chicago in 1875. This was the beginning of the great dressed beef business of Swift's, and the foundation of the establishment of the seven big packing-house centres of the same company at Chicago, Kansas City, Omaha, St. Louis, St. Joseph, St. Paul, and Fort Worth, Texas. Mr. Swift was in 1875 solely a cattle buyer, the preparation and shipping of dressed beef from Chicago then being little more than an embryonic idea. In 1868 a refrigerator car was invented, and in 1869 the first consignment of dressed beef was shipped from Chicago to Boston, but it was not a success. Personal supervision on the part of Mr. G. F. Swift himself brought about the arrangements which made the transport successful, and then later came the great struggle with the railways and the eventual operation by Swift's of its own private car lines, which have been an important feature of the ramifications of the business. Following the refrigerator cars came the refrigerated steamships that carried the meat across the Atlantic. Ten years after Mr. G. F. Swift arrived in Chicago the corporation of Swift and Co. was registered with a capital of 300,000 dollars in the State of Illinois. In the early days of Swift's business only from 56 to 58 per cent. of the beef animal was available for food, the remainder being sheer waste. The refrigerator car was

THE LATE MR. G. F. SWIFT, FOUNDER OF THE FIRM OF SWIFT AND CO.

To face p. 386.

the first means of avoiding the necessity of transporting the waste remainder, and the by-products industries which came later have employed to the fullest extent that part of the animals not transported with the meat. In the case of the pig, only the squeal, we are told, remains to be utilized (even that was once utilized by Mr. Gilbert Anderson on a gramophone, just to go one better than the Yankees). The story of the gradual climb of Swift's and other American capitalist concerns into the English markets has been the subject of much debate. Regarding the first efforts of Mr. G. F. Swift in this direction, it is said that he probably crossed the Atlantic twenty times before he got his Chicago dressed beef firmly established on British soil. When in London he personally supervised the selling of his beef. These are the methods which have characterized the onward march of the Americans right through, and relentless energy and untiring organization have been responsible not only for the strong position which they hold at West Smithfield, but also for their success in capturing an important share of the Argentine export trade. It may be as well to mention here that the Swift Beef Co., Ltd., with offices at West Smithfield, is an English company, with a capital and board of directors separate from those of the American company known as Swift and Co.

Tellier, Charles, of Auteuil, Paris, who may claim to be the pioneer of refrigeration as applied to the preservation and transport of meat, is a civil engineer, who has been engaged for the last fifty years on the problems connected with the chemical production of cold. He was born in 1828. Particulars appear in Chapter II. concerning M. Tellier's early efforts to import fresh meat from South America under refrigerating conditions, and a few particulars may be added here, based on a communication received by the authors from M. Tellier. In 1860 M. Tellier considered the matter of the public food supply; he perceived that it was for the engineer to bring about a balance between the over-production of meat in the new countries and the under-consumption in the Old World. His first idea of transport of fresh meat was by means of air-tight chambers, but on proceeding to experiment in this direction he discovered that the elements of decay were present in the meat itself from the earliest stages. Pasteur was teaching at that time the doctrine of the "pre-existent presence of organic germs" (*l'existence préexistant des germs organiques*). So M. Tellier adopted the refrigerator and was faced with the problem of fitting up plant on a steamer at a cost of 120,000 francs: following on the necessary efforts came the *Frigorifique* enterprise. In considering future discoveries and developments in refrigeration, M. Tellier speaks of the possibility of producing cold without combustion. Even in the year 1911 M. Tellier favours his old "desiccation" method of handling meat for long transport, "grocers'" meat, as he calls it, "*la viande d'épiciers.*" Experience, he states, has taught that when meat has been deprived of 15 to 20 per cent. of its weight of water the following results occur: In appearance, no change; "preparation" can be effected in twelve hours; absence of hygrometrical conditions; taken out of the machine after twelve hours' treatment, meat can be kept in the open air indefinitely. It must be noted that no cold is required for this process. To accomplish these results, fresh meat has to be dried in the open air by means sufficiently rapid to allow the operation to prevent either chemical action or fermentation setting in. "*Voilà le qui réserve l'avenir,*" concludes M. Tellier. In 1910 was published a work entitled "Histoire d'une Invention Moderne: le 'Frigorifique,'" written by M. Tellier. It is given to few men at the age of 82 to possess the intellectual capacity to supplement their life's labours by penning a technical work of 450 pages.

THOMSON, JAMES JOHN, chairman and managing director of Eastmans, Ltd., which company has about 1,400 shops in the British Isles retailing frozen and chilled meats, began his long connection with the frozen meat trade when he joined the firm of Messrs. John Bell and Sons in 1878, when they opened at Liverpool, and was associated with them till they were formed into a limited liability company. Mr. Thomson has presided over the destinies of Eastmans, Ltd., since the formation of

MR. J. J. THOMSON.

To face p. 388.

the company, and is one of the leading figures in the imported meat trade of Great Britain. Mr. Thomson is well and favourably known in Smithfield Market and other business circles, and is always to be found at gatherings and meetings connected with the trade. He is equally at home presiding at an ordinary general meeting or sharing in the convivialities of a festive gathering.

TWOPENY, RICHARD ERNEST NOWELL, was editor of *The Pastoralists' Review* from its establishment in 1891 to 1910. He went to England in 1896 as the delegate of the Australian Meat Export Trade Association, and there was formed by him a committee of the London representatives of the Australasian meat shipping interests, and considerable work was done by this body in endeavouring to establish reforms in the working of the trade. Fuller reference to this matter will be found on p. 268. A full report of his mission, containing much useful information as to London methods in the frozen meat trade, was written by Mr. Twopeny on his return to Australia. He was instrumental in adding one of the most valuable functions to the operations of the Frozen Meat Trade Association, when, in conference with the committee of that society, he suggested the despatch of a weekly cabled statement of the Smithfield current prices to the Australasian Press (see p. 330). Mr. Twopeny is an *officier d'Académie*.

WAGSTAFF, THE LATE CAPTAIN F., who went as an apprentice in a sailing ship to Australia in 1853, will be pleasantly remembered in Smithfield circles as surveyor to the Australian and New Zealand Underwriters' Association. In the nineties he was on most days to be met with in the Central Markets. He issued a useful annual list of vessels engaged in the refrigerated produce trade, and in 1897 devised a method for the discharge of frozen meat direct into barges or trucks through small portholes 18 inches square in the 'tween decks of vessels, for the purpose of saving damage in the hoisting of carcasses up the hatchways.

Ward, John J., who was from 1904 to 1910 general manager for Europe of the Compañia Sansinena de Carnes Congeladas, first sold S. G. Sansinena and Co.'s sheep in Birmingham in 1886, and in 1887 joined the company as its Liverpool manager. Developing the business, he held the position until 1904, when the Buenos Aires board invited him to assume the position of general manager for Europe, with residence in London, on the resignation of Mr. William Cook. Mr. Ward may well be called one of the trade's pioneers, for it was he who introduced New Zealand frozen sheep into Yorkshire in 1884 and sold them on Leeds Bridge. Prior to that date he sold frozen meat on the Birmingham wholesale market, and introduced it into Staffordshire at Hanley. In 1891 Mr. Ward visited Chicago, and in 1895 he went to the Argentine and dressed the cattle at La Negra works to show the Argentine what was wanted in England. In the year 1896 he met with a very serious accident in the course of his duties. During discharge of the s.s. *Elstree Grange*, he was struck by the steamer's coaling bucket and knocked down the hold, sustaining such grave injuries that his life was despaired of. He, however, recovered completely and resumed his duties after a long illness. After a life of strenuous work, Mr. Ward retired at the end of 1910, being honoured with a complimentary banquet and presentation by the meat trade of the United Kingdom, with its banking, shipowning, and marine insurance connections, at the Café Royal, on November 29 of that year.

Weddel, William, director of W. Weddel and Co., Ltd., has been one of the most prominent figures in the frozen meat trade, and there are few persons connected with this industry who have grappled more vigorously and successfully with all the difficulties which have marked its growth from 1880 onwards. Educated at the Edinburgh Institution, Mr. William Weddel came to London in October, 1872, and in 1874 entered the service of Messrs. Russell, Le Cren and Co., of London, agents for Messrs. Russell, Ritchie and Co., of New Zealand. With his brother the late Mr. P. G. Weddel, Mr. William Weddel

MR. WILLIAM WEDDEL.

To face p. 390.

established in 1887 the firm which bears his name, being joined by Mr. George Goodsir in 1888, the partnership extending to this day. Later, in 1904, Mr. D. J. Goodsir and Mr. Gordon H. Campbell became partners, and in December, 1910, the firm was registered as W. Weddel and Co., Ltd., under the Companies Acts. Concerning the operations of Messrs. W. Weddel and Co., it is, perhaps, only necessary to say here that the firm is one of the largest importers from Australia and New Zealand. It is about twenty years since Mr. Weddel's house initiated c.i.f. business, on which basis, of course, a very large proportion of the frozen meat trade from Australasia is now carried on, to the advantage of both buyers and sellers. The firm in 1895 opened up a trade in frozen meat with Austria and Germany. These operations, though quickly brought to a close at that time, clearly marked the first stage of that great fight for new markets on the Continent in which the frozen meat industry is now engaged. Fair quantities of meat were sold during the campaign, but the inspection and other regulations enforced by the authorities, as well as the high import duties imposed, put a stop to the business. In co-operation with the board in South Africa, and Mr. John Cooke, of Melbourne (a large shareholder), Messrs. W. Weddel and Co. organized and worked for some years the La Plata Cold Storage Works in Argentina, an undertaking still boasting the most up-to-date equipment in that part of the world, but now owned by Swift's Originally the principal shareholders in the La Plata Works were South African meat importers, but when the importation of frozen meat into that colony fell off to small proportions these parties had not much interest in retaining the works, and so decided to sell their property, which they did at a fair profit. Mr. Weddel's firm imports considerable quantities of meat from Uruguay and Patagonia. Mr. Weddel visited South Africa several times during the war, and he did business so satisfactorily for the War Office in Australasian produce that a very considerable portion of the orders for these commodities was placed with his firm, so far as imports from Australasia were concerned. His firm also secured nearly all the orders for shipments from Australia to China at the time of the Boxer trouble, and they also sent to Taku some frozen meat for the United States Government. Mr. Weddel and his partners have been prominent figures in connection with the settling of the problems arising as the frozen meat trade grew, and many of the commercial methods applied to the industry as it now exists are on the lines of procedure established in the early days by him and his partners. Mr. Weddel is a member of the Port of London Authority, and has always been a staunch supporter of the Royal Colonial Institute and other Australian and New Zealand social enterprises in Great Britain.

Wood, John A., joined the service of the River Plate Fresh Meat Co., Ltd., in 1882, on the formation of the company, as secretary, being the first official appointed in the service. He was made manager and secretary in 1888, general manager and secretary in 1895, managing director in 1903, and the chairman in 1904. Mr. Wood had to initiate, supervise, and control the various developments of the company's business, and the successful carrying out of the work has, of course, only been possible owing to the assistance of a most capable staff both in the Plate and in Great Britain. Mr. Wood has paid many visits to the Argentine in connection with the business of the company. It is fitting that Mr. Wood's name should be coupled with the system on which the River Plate Fresh Meat Co. has been managed. This system has been as far as possible for the company to control and work with their own men the whole business from the buying of the live stock in the Plate until delivery to the retailers or consumers on this side, the aim all through having been to meet as far as was possible the consumers' requirements, and the control of the company being in the hands of those who were acquainted with what was required on this side enabled that end to be fairly well attained.

CHAPTER XXIX

THE FUTURE OF THE INDUSTRY

DRAWING an outline of the future is always a speculative proceeding, but there are few industries that have shown the same potentiality of development as the frozen meat trade. The rule hitherto, with few exceptions, has been that nations as they rise in status become greater meat eaters. Such countries, as they in the course of time progress in the industrial arts, stand continually less chance of furnishing their meat supply from within their own territory.

The frozen meat man might truthfully say " The world's mine oyster." That the oyster is sometimes hard to open is seen in the struggle to gain Europe for frozen meat, but the possibilities are none the less for that. Great Britain herself has been an ever-expanding market, as the following round totals of imports of mutton and lamb carcasses bear witness : 1885 (United Kingdom), 1,000,000 ; 1890, 3,000,000 ; 1895, 5,000,000 ; 1900, 7,000,000 ; 1910, 13,000,000. The market in Great Britain has shown a wonderful power of expansion, and the country's need to import further meat will become still more keen as the industrial scope waxes.

The democratic movement in Germany will of a certainty find expression by-and-by in the relaxation of the official attitude of refusal to import the meat which the workers there so sorely need. That the manufacturing peoples of Europe cannot produce within their own borders the flesh food which they require is an economic truism, and sooner or later the refrigerated meats from the two great continents in the South must find eager customers in these states. The United States of America, which have played such a preponderating part in the past in supplying Europe with food of all kinds, are finding the needs of their own population so insistent that it is plain that Argentina and/or Australia, instead of finding a competitor in that quarter, will ere long find a customer.

Refrigerated Meat the Import Trade of the Future.

We may, perhaps, reasonably dismiss, as a negligible factor, one important avenue of competition in the past with frozen meat, the live cattle trade. The transit of live stock for slaughter has become obsolete with the erection and virtual perfection of the vast system of plant belonging to the frozen and chilled meat trade. The distance between the producing and consuming points of frozen and chilled meats, with the cheapening of freights, is no stumbling block to the industry, and no disadvantage to the consumer. The struggle for existence as the population of the world increases will of necessity be so keen that a stimulating meat diet will be essential; a world-wide demand is certain. Whence are supplies to come? Not from the consumers' countries; there will hardly be elbow-room by-and-by for men and women in these hives of industry, let alone pastures spacious enough to feed cattle and sheep. The 251 vessels that now ply across the ocean from the South freighted with meat supplies will, when that period comes, have doubled—trebled—who can say? Vessels with chilled and frozen South American beef now arrive at English ports as regular as clockwork, and imports of New Zealand lambs and Australian mutton and beef will be so arranged and systematized that the periods of "famine and feast," and various shortcomings and irregularities in transport, which have weakened the frozen meat trade during its thirty years, will be swept away.

Underwriting will play a less prominent part, and insurance may become a question of cover against total loss, as greater skill and care are exercised in all departments. In the *Pastoralists' Review*, the question was recently asked—"Why should there be *any* damage in the transit and handling of frozen meat?" This sounds academic, but it is a healthy ideal to work up to. At any rate, vast improvement is certain to take place in this respect, as the progress made in the past ten years indicates, and we may confidently look forward to the time when the condition of all Australasian and South American meats on being marketed will be as sound as their intrinsic quality.

British capital, already heavily invested in the industry, will be likely to be embarked still further in Australia, New Zealand, and the meat shipping countries of South America. Not only will it be sunk in plant, but there will be floating capital or credit on a large scale made available by British merchants at the point of meat production, for the purchase on the spot of produce for refrigerated shipment to Europe. As the system of retail trade combination spreads in the great centres of population in England, in the form of multiple provision-shop companies, the necessity for buyers to cover their requirements ahead will become more pressing. All along the line we may expect the wholesale purchasing and importing of frozen meat to be conducted on more systematized lines, as the demand grows larger and the competition keener. Dependence in a casual way upon farmers' consignments will not do for this. The proprietors of these companies, knowing precisely their requirements, will, by agent or cable, arrange their contracts in a precise manner for quality, grade, and time of shipment.

New Zealand Lamb.

New Zealand may be expected to show a gradual increase in her meat export trade, unless her flourishing dairying industry should present obstacles to the further development of meat freezing. It is not expected that the increase will be in beef. One may fairly anticipate that New Zealand will in the future maintain and improve the supreme position she holds in the export of lambs. The capacity for expansion of the frozen meat business in the aggregate in the Dominion is limited, but it is difficult to place a limit upon the production of the fine meaty lamb exported. From 1887, when the shipment of New Zealand lambs exceeded a million, the yearly despatches have steadily expanded to 1911, when about 3,500,000 were landed at English ports. The excellent quality of the New Zealand frozen lamb, so widely acknowledged that it freely enters houses where other descriptions of frozen meat are tabooed, has been well maintained, although the numbers have increased so rapidly. There is nothing in the figures of mutton exports from New Zealand of late years to encourage the idea that any considerable expansion will take place in that direction; and there is much in the fact that the New Zealand standard of quality in mutton has fallen and is falling, and that Argentine is taking its place, to confirm the suggestion that the New Zealand freezing works will in the future cater for an increased output of lamb rather than mutton.

Australia's Possibilities.

Australia will, probably, in periods of abundant rains, hold a lead in mutton; it may be that in the future it will be a long lead. But it would be difficult to speak with much confidence when forecasting Australia's future in the frozen meat industry, on account of climatic fluctuations and the influence of wool values. Taking Australia as a whole, when one considers that the yearly income from wool was estimated in 1910 at £30,000,000, it is easy to see that frozen mutton is a secondary product to the sheep farmer. It is quite possible that the meat shipping States of the Commonwealth may play a very leading part in years to come. With her small population, and huge unopened grazing territories only waiting for railways to make them operative, the possibilities of Australia, given good seasons, in the way of supplying meats to the Old World, and, helped by her geographical position, to the East, can only be guessed at. The Gulf country of Queensland, north-western Western Australia, and parts of the Northern Territory coastal country, when exploited by capital and furnished with means of freezing and shipping beef, should add in no small way to the Commonwealth's power of supplementing the roast beef of Old England. And what limit can be placed on Australia's possibilities in the way of exports of refrigerated produce to Eastern countries, when we hear of Tasmanian apples going to China! The imports of Australian mutton and lamb into Great Britain in 1910—over 4,000,000 carcasses—which showed an increase over the figures of the previous year of 58 per cent., give an idea of the possibilities of extension in mutton and lamb freezing in the Commonwealth States, especially in view of the expansion of the farming element, now being brought about by the division of the big pastoral properties, under which system the Australian flocks will be increased, if the seasons are propitious. The prospect for sheep breeders in Australia is full of promise. The demand for Australian mutton and lamb is certain to increase as time goes on, not only because of the growth in the population of the United Kingdom, but also because the tendency is for flocks to decrease on the Continent of Europe and in the United States.

South America's Prospects.

Since the invasion of Argentina by the United States meat packing firms and the taking over of two of the finest factories by the Americans, the department of most interest has been that of chilled beef, which inclines one to expect much of that article in regard to Argentina's future part in the meat export industry. Swift's descent upon La Plata and the subsequent capture of La Blanca was a fine coup, and it is seen that these developments have brought about a change in venue of the chilled beef business, for which the Swifts, the Morrises, and the others made the meat centres of the United States of America so deservedly famous. But, of course, chilled beef was a profitable trade in the Argentine before the two works named went over to the Americans. As exporters of chilled beef the United States of America are now a negligible factor, and one associates this industry specially with Argentina because in that Republic it is so rapidly waxing. Chilled beef is such a high-class trade that the future of the meat export of any country which can grapple successfully with the various problems concerned therein is certain to be brilliant.

Though, on the average, the financial results to the Argentine shippers of chilled beef in 1911 were probably unsatisfactory—owing to English markets being swamped with excessive quantities—there can be no doubt that the part which Argentina will play in the future in the meat export trade will involve increased attention to the shipment of this article. Notwithstanding the steady improvement in the quality of Argentine frozen mutton, one's attention in this chapter is drawn to the great preponderance in beef shipments, which in 1910, referring to imports into Great Britain from South America, totalled over 250,000 tons, frozen and chilled, 40 per cent. of the total import of frozen and chilled beef and mutton imported in that year. The decline of sheep stocks in Argentina of late years cannot escape attention. In 1895 there were 74,380,000 in the Republic, and this number had sunk to 67,212,000 in 1908. At the present day, says the *Review of the River Plate*, it is doubtful if there are 60,000,000 sheep in Argentina, as a large number perished in the drought of the last two years.

In treating such a subject as the future of the industry, one cannot neglect allusion to the possibilities of further expansion in the Americas. The two existing works in Chilian Patagonia are a useful means of handling the surplus of fat mutton and lamb in the bleak regions round Punta Arenas. Not much development of this trade is expected. Sheep are not available in sufficient numbers in Argentine Patagonia from Bahia Blanca southwards to tempt the exploitation of this part of South America for the freezing industry—and the general circumstances and lack of settlement in the country are discouraging. The freezing works at Rio Gallegos are on the point of commencing operations. Much is being said about additional freezing establishments to be erected in Uruguay, but so far the Frigorifica Uruguaya alone represents the industry in La Banda Oriental. The proprietors of the freezing works in Venezuela, in tackling the industry in the tropics, probably had in view the supplying of a special demand in England for a certain grade of beef. The cattle available for their purposes are small. It is not the idea of the most experienced men in the frozen meat trade that much is to be done in planting freezing works for export in such countries as Mexico, Paraguay, Brazil, etc. When the cattle of these countries have been improved, we may see Great Britain and other parts of Europe drawing supplies of beef thence. Still, there is little doubt that one of the developments of the future, possibly, when Argentina's fat cattle run short, will be the tapping of the more northern parts of South America, and Mexico in North America, for beef supplies for Europe. Again, South Africa has sent some little lots of frozen meat to English markets, and although the results were discouraging, there need be no doubt that frozen meat supplies will some day be available from the Union of South Africa. Pedigree stock are being imported there, and the ranching business is opening up well in Rhodesia.

Looking Forward.

As bearing upon this subject, there may be given the last paragraph in the paper, "The Meat Supply of the United Kingdom," read on May 18, 1909, before the Royal Statistical Society, by Mr. R. H. Hooker, M.A. "In the future our dependence on North America will steadily diminish, and I look to the Southern Hemisphere for our extraneous meat supply ; mainly to South America for our beef and to Australasia for our mutton, the latter probably exhibiting great fluctuations." Mr. John Clay, of Chicago, an undoubted authority, recently stated that he did not believe that the United States would be able to send much more beef to the United Kingdom. "But when the South American supply has ceased to grow, there seems every probability that population will increase faster than live stock, and the next generation will have to pay dear for its meat unless some radical change is made either in the method of producing meat or in the diet of the United Kingdom."

A Comparison.

In an Appendix to this book will be found some tabular statistics of the freezing works of Australia, New Zealand, and South America. The Australasian system, with its numerous works scattered all over the place, compares unfavourably with the centralization which marks the Argentine freezing industry, as in the former there must be great waste in the various processes, both in the erection and management of works, and in shipment, etc. Looking at the total capacity of the Australasian freezing works and of those in Argentina, in relation to the total shipments for a given period from the two sources, we see how, relatively, much more economical is the Argentine than the Australasian plan ; and as the meat from both has to be sold in the same market, it is easy to perceive that the producer in Australasia must necessarily get a lower net return than must be secured by his rival in Argentina. But it is difficult to see how things could have been different in Australia, and New Zealand—where the farmers have shown an inclination to take a hand in the work of distribution. The frozen meat trade was established, and has since been carried on, commercially in the Argentine Republic. A handful of business people, English, Scotch, Argentines, experienced in trade and industry, worked out the whole freezing cycle without being hampered by a multitude of counsellors swayed by conflicting interests. Buenos Aires was a ready-made centre for the business in all its branches, helped by a fine railway and ocean shipping system, and no other choice was possible. And the South American freezing industry, as to the modern part of it, has had the shoals upon which it might have grounded well buoyed by the mistakes made by some of the pioneers in Australasia, both in preparing the meat at one end and disposing of it at the other.

The producer in Argentina, that is, the grower, has no direct concern in the frozen meat trade. He sells his fat stock to the frigorificos, and there is an end of it. But in Australia and New Zealand the grower, whether he consigns or sells on the spot, takes a keen and personal interest in all the stages through which the meat passes. He follows the frozen carcass from meat works to market and has tried to master the technique of the trade. The close season for the operations of the Australian and New Zealand freezing works has no parallel in Argentina, where the frigorificos do not stop except for overhauling and repairs. Argentina has had a taste of drought of late years, but, speaking broadly, the climate there and the up-to-date methods of farming enable stock to be offered fat at the markets practically all the year round. The freezing industry has been helped by the co-operation of the wealthy estancieros in breeding up suitable stock, men who are content to sell and who do not embarrass the companies by any interference. The fact that the whole of the operations, from slaughter of stock to selling meat in Great Britain, have been mainly in the hands of the Argentine companies has been greatly in their favour, and this control has made and still makes it a comparatively easy task for them to contend successfully with their Australasian competitors.

But there was no possibility of the New Zealand freezing works and their collateral processes developing as did the Argentine, because the business was taken up in various districts of the Colony, and farmers, not commercial men, were at the head of affairs. From the Bluff to Auckland is a far cry, and to apply the principle of concentration in a drastic fashion to so delicate an industry as frozen meat in a country where the sources of raw material supply range over about 800 miles would have been impossible. *Quot homines tot sententiæ.* And in Australia, handicapped with droughts in the pastoral districts, freezing is at any time liable to be interfered with for years, as it was during the period 1899—1905, following the record drought.

Whilst recognizing the fact that the conditions under which the frozen meat industry in Australia and New Zealand, and in Argentina, had its origin were and are different, it may not be out of place to call attention to the consideration that, if the meat shippers of Australia and New Zealand desire to compete in a really effective manner with their South American rivals in the struggle for the markets of Great Britain, and later on for those of the Continent of Europe, attention should be devoted to the consolidation of efforts in the leading centres of Australia and New Zealand. The meat export trade in Australasia appears to require broader commercial lines of management than it possesses at present, so that the whole business may be worked in the most economical and scientific way.

And, whilst combination and concentration are desirable and indeed necessary, continuity of supply is equally vital. Cannot the farmers of Australia and New Zealand modify their present system of breeding and feeding so that a continuous instead of the present seasonal supply of fat stock for freezing can be secured? This has been brought about in the Argentine Republic by an intelligent system of breeding and feeding, and it is clear that, before the Australasian shippers of meat under refrigerated conditions can put their business in a thoroughly fit shape to compete with South America for the markets of Great Britain, the farmers of New Zealand and Australia must concentrate their energy upon this problem.

APPENDIX I

CATTLE AND SHEEP IMPORTS INTO GREAT BRITAIN FOR FOOD

RISE AND FALL OF THE TRADE

An important collateral subject is that of the supplies of live cattle and sheep brought into Great Britain to supplement home-grown beef and mutton. One has to go back to pre-statistical days for the beginnings of this business. Cattle plagues occurred in 809—10, 1348—49, and 1480, but there is lack of evidence to show that these attacks were rinderpest. In 1715 there was a limited outbreak, and again in 1745 a more extensive one, of the veritable plague, which continued to 1757. This was brought into the country either by two white calves from Holland, or in a parcel of distempered hides. Even so early as 1348, when the attack of cattle disease alluded to occurred, the stamping-out system was understood in England. The diseased cattle were slaughtered, and infected herds and the herdsmen attending them were kept from coming into contact with sound animals. In 1770 an ordinance was issued prohibiting imports of cattle and sheep, and also of hides, horns, etc. Great Britain long continued to be free from all kinds of contagious cattle disease; rinderpest, pleuro-pneumonia, and foot-and-mouth disease were unknown, though every country in Europe had been constantly ravaged by these scourges. In the Napoleonic wars rinderpest followed the track of the various armies, especially those coming from Russia, or those having been in contact with Russian troops; France, Italy, Spain, and Portugal all suffered, while Germany, Austria, and all South-Eastern Europe were never free. The only immune countries were the British Isles. And this happy state of things continued until the change took place in the national system of Great Britain in the early forties, and import regulations were relaxed. From that time onwards cattle diseases prevailed for pretty well fifty years, during which period the United Kingdom was never free from disease ; pleuro, foot-and-mouth disease, etc., and in 1865 rinderpest, ravaged the herds and caused irreparable losses. This awful attack of rinderpest was introduced by 331 cattle from Revel, landed at Hull. The total number of cattle attacked in Great Britain was 73,549. The disease spread till an Order in Council made it compulsory to slaughter and bury all diseased animals. A slight attack in 1872 and a still more local one in 1877 or 1878 were quickly suppressed, and with that Great Britain said good-bye to rinderpest. The beneficent Order of 1892 prohibited the importation of cattle, sheep, and goats from nearly all the Continental countries, Malta, and Morocco, but cattle were permitted to be landed for slaughter from the United States and Canada. Sheep were allowed to be landed from these two countries without being subject to slaughter. This Order was made under the Contagious Diseases (Animals) Acts of 1878 to 1892. In 1896 another Act was passed consolidating these Acts and making it compulsory that all cattle and sheep imported be slaughtered within ten days of being landed. This Cattle Diseases Act of 1896 has enabled Great Britain to keep free from the serious forms of stock disease ; a few attacks of foot and mouth disease have been promptly suppressed, and England and Scotland have become the nursery of pure-bred cattle, sheep, and pigs ; and breeders have been enabled to supply the world, under the protection of the Act. These foot-and-mouth disease outbreaks caused much loss to farmers. Some of the visitations were very serious, and ruined many herds of dairy cattle by injuring their udders: £20 fat cattle ran down to £15 in value. The mortality was small, but the loss of income was large.

Prior to *Strathleven* days (1880), and for twenty-five years after, Great Britain was greatly dependent upon live animals imported from Continental countries and North America for immediate slaughter at port of landing to supplement the insufficient home supply of fresh meat. Holland, Denmark, and Germany, before dead meat came along, and, in fact, right up to 1892 (when the Order above referred to entirely prohibited imports of cattle and sheep from Continental States), were the principal countries drawn upon. From that period Great Britain has imported her supplies of fat cattle and sheep from the United States of America, Canada, and Argentina (for the circumstances connected with the stoppage of the Argentine live stock trade see p. 75); and the present generation of Englishmen have forgotten to what degree their country was indebted to Continental supplies of cattle and sheep up to twenty and thirty years ago. The dead meat —mutton and beef—now received from the Continent is but a dim shadow of the trade formerly existing.

The landing in the United Kingdom of live animals from abroad for food supplies is governed at the present time by the Foreign Animals Order of 1910. The landing of live animals from certain countries—including swine from the United States of America—is entirely prohibited, and animals brought from other countries can be landed only at specified foreign animals wharves for the purpose of slaughter at the wharf within ten days. There are at present six of such wharves, one at each of the following ports :—Liverpool (Birkenhead), London (Foreign Cattle Market, Deptford), Glasgow (Merklands), Bristol (Avonmouth Dock), Cardiff (Bute Docks), and Manchester (Old Trafford). In practice, live animals are imported into this country at the present time only from the United States of America and Canada.

Under the Order of 1910 the following is the schedule of "prohibited countries" :—

Argentine Republic, Austria-Hungary, Belgium, Bolivia, Brazil, Cape Colony, Chili, Columbia, Denmark, Ecuador, France, Germany, Gibraltar, Greece, British Guiana, Dutch Guiana, French Guiana, Italy, Malta, and Mexico.

The Channel Islands and Iceland export a limited number of cattle, and sheep, respectively, to Great Britain. Australia and New Zealand are not scheduled, *i.e.*, prohibited, but there is no likelihood of stock from those countries being imported into Great Britain.

When frozen meat first came to London, in 1880, this was

the position regarding imports of live cattle and sheep in that year :—

Source of Supply.	Cattle.	Sheep.
Belgium	—	11,140
United States of America . .	154,814	66,722
Denmark	64,788	94,882
Canada	48,103	78,074
Holland	38,795	307,119
South Holstein	25,889	48,416
Germany	—	327,760
Spain	23,450	65
Portugal	14,736	94
Sweden	10,616	4,741
Channel Islands	2,632	—
France	1,572	511
Norway	1,031	2,754
Other Countries	6	1
Totals	386,432	942,279

The "rise and fall" will be observed in the following figures of importations into Great Britain. As the Continental trade was stopped in the early nineties, only the imports from North and South America are given in detail :—

—	1880.		1885.	
	Cattle	Sheep.	Cattle.	Sheep.
United States of America	154,814	66,722	137,826	11,829
Canada	48,103	78,074	68,524	39,684
Argentina . . .	—	—	—	100
Other Countries . .	183,515	797,483	166,728	699,273
Totals . . .	386,432	942,279	373,078	750,886

—	1890.		1895.	
	Cattle.	Sheep.	Cattle.	Sheep.
United States of America	384,139	3,906	276.307	453,250
Canada	121,326	42,664	95,747	214,310
Argentina . . .	653	22,082	39,436	308,094
Other Countries . .	136,629	290,705	1,847	89,816
Totals . . .	642,747	359,357	413,337	1,065,470

—	1900.		1905.	
	Cattle.	Sheep.	Cattle.	Sheep.
United States of America	350,209	142,905	414,906	150,095
Canada	104,328	35,663	148,718	28,240
Argentina . . .	38,562	178,969	—	—
Other Countries . .	2,035	25,285	1,515	4,749
Totals . . .	495,134	382,822	565,139	183,084

—	1910. Cattle.
United States of America . .	138,387

The high-water marks of imports were as follow :—

—	Cattle.		Sheep.	
Total imports of live stock for food into Great Britain . .	(1890)	642,747	(1895)	1,065,470
Imports from United States of America	(1897)	416.299	„	453,250
Imports from Canada . . .	(1903)	190,815	„	214,310
„ „ Argentina . .	(1898)	89,368	(1898)	430,075
„ „ Continent . .	(1883)	260,651	(1876)	1,038,103

Australasia's Abortive Experiment.

In 1894 shipments of live cattle and sheep to London were made from Australia and New Zealand, the promoters of which enterprise had great hopes that they would be able to secure a higher range of values than those prevailing for frozen meat. Messrs. Bergl and Brabbin entered largely into this trade and invested a considerable amount of capital in it. As far back as 1872 two live bullocks were imported from Adelaide, but the commercial movement began with the despatch of 20 head of cattle, of which five were in store condition, by the *Maori King* from Sydney: this experiment was made on account of the owners, and the vessel sailed round the Horn. After being sixty-seven days on board, the cattle were landed at Deptford on September 12, 1894, in fair condition, one having died on the voyage. Though this pioneering venture showed that live stock could be taken from Australia to England, the financial result was a heavy loss; the bullocks cost, landed, about £19, including insurance, and the sales averaged out at £13 14*s.* 6*d.*, gross. The shippers wished to finish the store animals on the English pastures, but the Agricultural Department refused to allow this privilege—possessed by the American store stock shippers—to the Australian cattle. The next shipment from Australia was that made on the *Port Pirie*, 18 cattle and 48 sheep: the vessel's course was through the Red Sea, and during the voyage one ox and six sheep died. The cattle were Shorthorns and Devons, and arrived in excellent order, making £21 10*s.* per head, and the sheep made £1 4*s.* 9*d.* each. The shipments of stock continued to November, 1895. Amongst the most successful was that of 250 sheep per *Banffshire* from Dunedin; only one animal died on the way. The sheep were landed fresh and healthy, ready for slaughter, and they were sold at from £1 18*s.* to £2 3*s.*; the average per head was £1 19*s.* 5*d.*, and the sheep cost £1 17*s.* 1*d.* with all charges added, the freight being 10*s.* The freight charged on cattle shipped was £6 per head.

As the shipments continued to be made in 1895, there were

terrible losses, so that it became clear that the underwriters would soon stop the trade; sentiments of humanity had weight in the same direction. The shipment of 1,208 sheep per the *Buteshire*, made by the Shipping Conference, resulted successfully; the sheep made 41*s.* 6*d.* on average. The horses imported sold from 20 to 50 guineas for hacks, and the draughts made from 27½ to 45 guineas; 42 horses ex *Gulf of Lions* fetched 1,100 guineas. Towards the end of the campaign the cattle deteriorated. The large-framed old animals shipped were quite unsuited to compete with the nuggety American beasts handled in English markets. Also no improvement took place in the arrangements made for housing and feeding the cattle on board. The *Angers* shipment of 381 cattle from Gladstone, Queensland, clapped an extinguisher on the trade. The vessel left the port on November 22, 1895, and arrived at Deptford on February 9, 1896, with 32 animals, some of which were in a maimed condition. The Royal Society for the Prevention of Cruelty to Animals took the matter up, and the Board of Agriculture inspected the ship and issued an order prohibiting the *Angers* from carrying live stock from or to any port in Great Britain for a twelvemonth.

It is evident that the whole enterprise was badly planned; to have been successful, cattle should have been conveyed in specially built and fitted boats, with skilled men in charge, as in the Atlantic system. Lawsuits marked the closing scenes of this "episode," which for a short time threatened the frozen meat trade. The imports covered a period from September, 1894, to February, 1896, and the following table is useful as a record:—

Carrying Vessel.	Cattle.	Sheep.	Horses.
Maori King . .	20, lost 1	—	
Port Pirie . .	18, ,, 1	48, lost 6	
Port Stephens . .	—	20, ,, 4	
Echuca. . . .	40, ,, 6	—	
Maori King . .	30, ,, 2	—	
Culgoa . . .	5, ,, 0	—	
Gulf of Bothnia .	120, ,, 45	—	

Carrying Vessel.	Cattle.	Sheep.	Horses.
Buteshire . .	26, lost 5	120, lost 3	
Celtic King . .	18, „ 0	—	
Port Chalmers . .	80, „ 6	—	
Nairnshire . .	101, „ 9	—	
Banffshire . .	—	250, „ 1	
Gulf of Lions . .	77, „ 4	—	
Port Victor . .	72, „ 17	—	
Morayshire . .	—	60, „ 0	
Wooloomooloo . .	—	268, „ 58	
Hawke's Bay . .	—	100, „ 3	
Gulf of Genoa . .	60, „ 3	—	
Perthshire . .	160, „ 22	—	
Gulf of Siam . .	71, „ 9	20, „ 0	18, lost 11
Warrigal . . .	20, „ 4	231, „ 6	
Echuca . . .	20, „ 12	100, „ 6	
Tekoa . . .	—	102, „ 1	
Southern Cross .	550, „ 52	488, „ 82	29, „ 1
Urmston Grange .	230, „ 20	—	
Buteshire . . .	—	1,208, „ 20	
Gulf of Lions . .	—	—	102, „ 45
Maori . . .	183, „ 20	—	
Maori King . .	117, „ 6	—	
Nairnshire . .	90, „ 1	—	
Hubbuck . . .	—	201, „ 145	
Gulf of Bothnia .	80, „ 4	—	
Port Chalmers . .	85, „ 9	119, „ 12	3, „ 0
Yarrawonga . .	—	300, „ 4	
Angers . . .	381, „ 349	—	
Warrigal . . .	—	247, „ 26	

Cattle shipped 2,654, losses 607; sheep shipped 3,882, losses 377; and horses shipped 152, losses 57.

APPENDIX II

BY-PRODUCTS OF THE SLAUGHTER-HOUSE

EARLY in the history of the frozen meat trade it occurred to practical men that there might be a market in Great Britain for clean offal. In a letter written on March 14, 1884, by the late Mr. Henry Moncreiff Paul, manager in London for the New Zealand Loan and Mercantile Agency Co., to the New Zealand offices of the company, information was given as to the methods of preparing "lamb kidneys," "sheep tongues," "throat and heart sweetbreads," and "running gut," for export in a frozen state to England. This was the first suggestion of working up the slaughterhouse by-products in the frozen trade. Little material as to this department of the industry is available, no statistics being published; it is not considered a high-class business, and there has been trouble at times with the inspectors when these "oddments" have not arrived in good condition. Some of the exporters in Australia, New Zealand, and South America make a feature of preparing and shipping slaughter-house by-products, and prior to the regulations put in force in 1909 in Great Britain the imports of these articles rose to a considerable volume.

But the industry in the countries named is trifling compared with the by-products trade in the United States of America, especially pork products. Imports thence of meat sundries into England, chilled and frozen, were formerly enormous. The condition of the offal from the United States of America on arrival in Great Britain at times left much to be desired, and occasionally kidneys, etc., from Australasia and Argentina went bad, so the sanitary authorities were not in love with the trade. Some of the New Zealand freezing works ship to England joints of mutton in bags, a special trade of limited dimensions, useful, possibly, where it is not convenient for the whole carcass to be exported; also useful sometimes for the English butcher, who, at certain seasons of the year, likes to supplement carcasses with extra joints. Meat offals most commonly shipped from Australasia and South America to Great Britain are ox and sheep tongues, kidneys, hearts, and livers; ox tails, skirts, ox cheeks, tripes, sweetbreads, and trotters.

Sheep kidneys have been exported from New Zealand and Australia for many years past, but owing to the faulty condition of much of the goods on reaching the English market this department of the trade has not been so satisfactory to exporters as it should have been; kidneys have always been troublesome. An estimate has it that kidneys from stock frozen in Australasia exported to Great Britain would, if properly prepared, be worth from £30,000 to £40,000 annually.

It was in the packing house of the "beef trust" firms, Chicago, Kansas, and now in Argentina, that meat by-products were first handled on a thorough and scientific scale. When the export of chilled beef to the United Kingdom grew to very large proportions, the offal was too much for local consumption, so that also was shipped: the trade followed the improvement in the refrigerating processes, as the freezing and shipment of offal could not be carried on with freezing apparatus in its earlier stages.

In Argentina the trade in meat oddments was started in a small way when meat was first shipped, and the business is regarded as a necessary adjunct to the main trade. In recent years it has developed very largely, but not much dependence is placed upon this department. The chief offals shipped from South America are, (cattle) tongues, tails, hearts, kidneys, and livers; and (sheep) tongues, hearts, and kidneys.

Owing to the decrease in the slaughter of pigs in England, and the gradual reduction of all sorts of meat products from the United States, the prospects for exporters of offal from Australia and New Zealand would have been bright but for the regulations under the Public Health Act, 1907, which inflicted a severe blow upon the trade in these articles, and upon frozen pork also, boxed or carcasses (see p. 122). These regulations may be said to have severely affected the Australasian trade, though South American meat "oddments" may survive on account of the shorter voyage. In fact, it has been stated that the regulations are an advantage to Argentine meat "oddments" shippers now they have found out what is really wanted by the medical officers. It is a pity that some statistical record of this trade has not been kept; the quantities shipped from Australasia and Argentina to England must have amounted to quite a respectable proportion to the carcass and quarter meat.

American Meat By-Products.

The following list of meat "sundries" and offals, which exhaustively represents the range of slaughter-house by-products of the packing houses of the United States, may be useful as a record, and shows what a huge and finely organized industry this export business was in the hey-day of its prosperity:—

Beef kidneys.
„ livers.
„ tails.
„ tripe in pickle (fresh).
„ frozen tripe (fresh).
„ tripe in pickle (cooked).
„ tongues L/C, S/C, and C/T.
„ cheekmeat.
„ cheeks.
„ hearts.
„ kidney knobs.
„ caul butts.
„ brains.
Cow udders.
Beef rounds.
„ rounds and rumps.
„ sirloin butts.
Boneless beef.
„ „ chucks.
„ „ shoulder clods.
„ „ butts.
„ „ rolls.
Spencer rolls.
Manhattan chucks.
Beef chucks.
„ rolls.
Square cut chucks.
Hanging tenderloins.
Beef skirts.
Corrugated rolls.
Smooth rolls.
Skirts and kidneys.
Beef strips.
„ trimmings.
Muscles.
Solid chuck butts.
Loose chuck butts.
Beef sausages.
Shin beef in cases.
Ox suet.
„ tenderloins.
„ reeds.
„ weasands.
Chuck butt rolls.
Frozen tripe, cooked.
Pork loins.
„ „ roasts.
„ kidneys.
„ plucks.
„ tongues.
„ ears.
„ hearts.
„ snouts.
„ feet.
Boneless ham butts and Boston butts.
Cala butts.
„ hams.
Pork cheek meat.
„ trimmings.
Fat backs D/salted, also fresh and frozen.
Pork livers, salted.
„ „ frozen.
„ hams, hock on.

Pork hams, hook off.
,, tails.
,, legs.
,, cauls.
,, sausages.
,, sausage meat.
Skinned hams.
Green skinned hams.
Pork shoulders.
,, bellies.
,, hams, boneless.
Minced hams.
Green hams.
Hog chitterlings.
,, brains.
,, melts.
,, faces.
Neck bones.
Lamb plucks.
,, livers.
,, tongues.

Sheep plucks.
,, hearts.
,, tongues.
,, brains.
,, livers.
,, kidneys.
Mutton necks.
,, legs.
,, shoulders.
,, loins.
Calf tongues.
,, sweetbreads.
,, livers.
,, heads and feet.
,, heads.
,, brains.
Boneless veal.
Frozen tripe in boxes.
Suet.
Trotters.
Frozen poultry and game.

APPENDIX III

THE WORLD'S SUPPLIES OF FROZEN MEAT IN 1910

These figures are based upon the ARRIVALS into the United Kingdom for the twelve months ending 31st December, 1910, plus SHIPMENTS to the CAPE and other markets during the same period.

	Mutton (carcasses).	Lamb (carcasses).	Mutton and Lamb (tons).			Beef (quarters).	Beef (tons).		
AUSTRALIA									
into United Kingdom . .	2,723,000	1,496,600	=(B. of T.)	76,300		537,500	=(B. of T.)	44,000	
to Cape . . .	75,000	6,100	=(Est. wt.)	2,000		22,200	=(Est. wt.)	1,800	
to Mediterranean . .	28,000	6,700	=(„ „)	800		17,600	=(„ „ (	1,400	
to other Ports . .	83,000	14,600	=(„ „)	2,300		71,400	=(„ „)	5,750	
					81,400				52,950
NEW ZEALAND									
into United Kingdom .	1,991,000	3,416,000	=(B. of T.)	105,200		344,000	=(B. of T.)	26,700	
to Mediterranean . .	4,500	—	=(Est. wt.)	120		3,000	=(Est. wt.)	250	
to Vancouver . .	438	—	=(„ „)	10		—	=(„ „)	—	
					105,330				26,950
SOUTH AMERICA									
into United Kingdom .	2,838,700	515,050	=(B. of T.)	81,560		3,078,000	=(B. of T.)	252,900	
to Cape . . .	—	—		—		—		—	
to other Ports . .	466	—	=(Est. wt.)	10		20,600	=(Est. wt.)	1,650	
					81,570				254,550
					268,300				334,450

Grand total, mutton, lamb, and beef: 602,750 tons.

APPENDIX IV

INCORPORATED SOCIETY OF MEAT IMPORTERS. (THE FROZEN MEAT TRADE ASSOCIATION.)

The first of the "Weekly Quotations," which is reproduced on this page, was made on July 2nd, 1897.

To THE MANAGER, AUSTRALIAN PRESS ASSOCIATION.

THE FROZEN MEAT TRADE ASSOCIATION.

WEEKLY QUOTATIONS.

57, *Charterhouse Street*, *London*, *E.C.*, 2nd July, 1897.

The Association's Smithfield Quotations on this date for the undermentioned classes of Frozen Meat, based on actual sales of not less than 100 Carcasses of Mutton or Lamb, or 25 Quarters Beef, are as follows:—

						Per lb.
1.	NEW ZEALAND SHEEP	Canterbury	Crossbred Wethers and Maiden Ewes		55/65 lbs. Fair average quality	$3\frac{1}{2}d.$
2.	Do.	Dunedin and Southland	Do.			$3\frac{3}{16}d.$
3.	Do.	North Island	Do.			$3\frac{1}{16}d.$
4.	AUSTRALIAN SHEEP	Heavy, viz., over 50 lbs.	Crossbreds, and/or Merino Wethers		Fair average quality	$2\frac{3}{4}d.$
5.	Do.	Light, viz., under 50 lbs.	Do.			$2\frac{13}{16}d.$
6.	RIVER PLATE SHEEP	Heavy, viz., over 50 lbs.	Do.		Fair average quality	$2\frac{13}{16}d.$
7.	Do.	Light, viz., under 50 lbs.	Do.			$2\frac{15}{16}d.$
8.	NEW ZEALAND LAMBS	Prime Canterbury	32 to 40 lbs.		Prime quality	$4\frac{5}{8}d.$
9.	Do.	Fair average, viz., including Dunedin, Southland, Wellington, Secondary Canterbury, &c.	32 to 40 lbs.	These quotations do not include small Lambs or heavies, or inferior quality.	Fair average quality	$4\frac{3}{8}d.$
10.	AUSTRALIAN LAMBS	Prime	32 to 40 lbs.		Prime quality	$4\frac{1}{4}d.$
11.	Do.	Fair average	32 to 40 lbs.	Do.	Fair average quality	$4\frac{1}{16}d.$
12.	AUSTRALIAN FROZEN BEEF,	Prime Ox Fores				$2\frac{5}{16}d.$
13.	Do.	Prime Ox Hinds				$3\frac{7}{8}d.$
14.	Do.	Secondary Ox Fores	Excluding Beef under 140 lbs. per quarter			$2\frac{1}{8}d.$
15.	Do.	Secondary Ox Hinds				$3\frac{5}{8}d.$

(Sgd.) G. NIND,
Secretary.

For THE FROZEN MEAT TRADE ASSOCIATION, (Sgd.) WILLIAM MILNE, (Sgd.) PHILIP HOLMES, *Members of Committee.*

Fourteen years later, July 7th, 1911, the quotations had grown to much greater proportions, as will be seen on the next page.

THE INCORPORATED SOCIETY OF MEAT IMPORTERS

Tel. No.: Holborn 1250.

WEEKLY QUOTATIONS.

Smithfield Market Quotations for the undermentioned classes of Frozen Meats, based on actual sales of not less then 100 Carcases of Mutton or Lamb, or 25 Quarters of Beef of fair average quality.

These quotations are not for selected lines, but for parcels fairly representative of the bulk of the shipments now on the market.

No.	Class	District	Description	Weight	@ per lb.	No.
1.	New Zealand Sheep	Canterbury	Crossbred Wethers and Maiden Ewes	48/56 lbs.	4*d.*	1.
2.	Do.	Do.	Do.	56/64 "	$3\frac{7}{8}$*d.*	2.
3.	Do.	Do.	Do.	64/72 "	$3\frac{11}{16}$*d.*	3.
*4.	Do.	Southland	Do.	56/64 "		4.
*5.	Do.	North Island	Do.	55/65 "	$3\frac{5}{8}$*d.*	5.
*6.	Do.	Do. Best Brands	Do.	55/65 "	$3\frac{3}{4}$*d.*	6.
7.	Do.	Ewes		48/64 "	$3\frac{1}{4}$*d.*	7.
8.	Australian Sheep	Light, Crossbreds and/or Merino Wethers		40/50 "	$3\frac{5}{8}$*d.*	8.
9.	Do.	Heavy, Do.		50/70 "	$3\frac{3}{8}$*d.*	9.
10.	Do.	Ewes		30/50 "	$3\frac{9}{16}$*d.*	10.
11.	River Plate Sheep	Light, Crossbreds and/or Merino Wethers		40/50 "	$3\frac{3}{4}$*d.*	11.
12.	Do.	Heavy, Do.		50/70 "	$3\frac{1}{2}$*d.*	12.
13.	Do.	Ewes		48/64 "	$3\frac{1}{4}$*d.*	13.
14.	New Zealand Lambs	Canterbury		28/36 "	$5\frac{1}{8}$*d.*	14.
15.	Do.	Do.		36/42 "	$4\frac{7}{8}$*d.*	15.
16.	Do.	Do.		42/50 "	$4\frac{5}{8}$*d.*	16.
*17.	Do.	Southland		28/42 "	$4\frac{7}{8}$*d.*	17.
*18.	Do.	North Island—Selected Brands		28/42 "	$4\frac{7}{8}$*d.*	18.
*19.	Do.	North Island		28/42 "	$4\frac{13}{16}$*d.*	19.
20.	Australian Lambs	Best Brands		28/42 "	$4\frac{1}{2}$*d.*	20.
21.	Do.	Fair Quality		28/42 "	$4\frac{1}{4}$*d.*	21.
22.	Do.	Inferior Quality		..	$4\frac{1}{4}$*d.*	22.
23.	River Plate Lambs	First Quality		..	$4\frac{5}{8}$*d.*	23.
24.	Do.	Second Quality		..	$4\frac{3}{8}$*d.*	24.
25.	New Zealand Frozen Beef	Ox Fores		160/200 lbs.	$2\frac{3}{8}$*d.*	25.
26.	Do.	Ox Hinds			$3\frac{1}{2}$*d.*	26.
27.	Australian Frozen Beef	Ox Fores		160/200 "	$2\frac{5}{16}$*d.*	27.
28.	Do.	Ox Hinds			$3\frac{1}{4}$*d.*	28.
29.	River Plate Frozen Beef	Ox Fores		160/220 "	$2\frac{3}{8}$*d.*	29.
30.	Do.	Ox Hinds			$3\frac{3}{8}$*d.*	30.
31.	River Plate Chilled Beef	Ox Fores		160/220 "	$2\frac{3}{8}$*d.*	31.
32.	Do.	Ox Hinds			$3\frac{7}{8}$*d.*	32.

15 *and* 16 *West Smithfield,*
London, E.C., 7th July, 1911.

For The Incorporated Society of Meat Importers,
Louis H. Furniss, *Secretary.*

* The differences in value between the light and heavy grades of weight of goods so marked, is approximately the same as in the detailed cases of Canterbury quotations.

APPENDIX V

COLD STORES AT THE CHIEF PORTS OF GREAT BRITAIN

	Capacity in 56-lb. carcasses of sheep.	Totals.
London. See map, Appendix V.		3,032,000
Liverpool.		
Union Cold Storage Co., Ltd., Canada Dock	666,000	
Union Cold Storage Co., Ltd., Banestre Street		
Union Cold Storage Co., Ltd., Williamson Square		
Union Cold Storage Co., Ltd., Albert Dock	200,000	
Eastmans, Ltd., Derby Road	150,000	
Lancashire Cold Storage Co., Ltd. (Borthwick's), Brunswick Place	175,000	
Bootle Cold Storage and Ice Co., Ltd, Canada Dock (New)	134,000	
James Nelson & Sons, Ltd.	130,000	
Imperial Cold Stores (London Central Markets Cold Storage Co., Ltd.)	116,000	
North Western Co-operative Cold Storage Co., Ltd., Redfern Street, Bankhall.	85,000	
Central Cold Storage Co., Ltd. (Ruddin's), Hood Street	80,000	
Compañia Sansinena de Carnes Congeladas, Sandon Dock	70,000	
Bootle Cold Storage and Ice Co., Ltd. Miller's Bridge (Old)	76,000	
Trent Cold Storage Co., Ltd., Sandhill's Lane	80,000	
Canadian Pacific Railway Co., Sandon Dock (1,550 tons)	186,000	
Cumberland Cold Storage Co.	20,000	
Eastmans, Ltd., Daulby Street	8,000	
		2,176,000
Glasgow.		
William Milne, Ltd., Old Wynd	350,000	
Union Cold Storage Co., Ltd., George Street	150,000	
Eastmans, Ltd., Cheapside Street	35,000	
W. McLachlan & Co., Logan and Bilbao Streets	35,000	
Sawers, Ltd., Rutherford Lane	35,000	
		605,000
Southampton.		
International Cold Storage and Ice Co., Ltd., The Docks	550,000	
River Plate Fresh Meat Co., Ltd., High Street	12,000	
		562,000
Manchester.		
Union Cold Storage Co., Ltd., Weaste	175,000	
Manchester Corporation Cold Stores, Elm Street, and Smithfield Market	152,000	
Union Cold Storage Co., Ltd., Miller Street	80,000	
		407,000

	Capacity in 56-lb. carcasses of sheep.	Totals.
CARDIFF, BARRY, NEWPORT, AND SWANSEA.		
CARDIFF.		
Cardiff Pure Ice and Cold Storage Co., Ltd.	394,000	
Cardiff Railway Co.	60,000	
H. Woodley & Co., Roath Dock Stores	30,000	
H. Woodley & Co., Godfrey Street	13,000	
Cardiff Railway Co. (chill room accommodating 500 sides of beef).		
		497,000
BARRY.		
Cardiff Pure Ice and Cold Storage Co., Ltd.	100,000	
		100,000
NEWPORT.		
Cardiff Pure Ice and Cold Storage Co., Ltd.	41,482	
		41,482
SWANSEA.		
Cambria Cold Storage and Ice, Ltd.	25,000	
		25,000
HULL.		
Union Cold Storage Co., Ltd., Sir William Wright Dock	200,000	
Compañia Sansinena de Carnes Congeladas, Alexandra Dock	54,000	
Union Cold Storage Co., Ltd., Alexandra Dock	50,000	
Union Cold Storage Co., Ltd., Blackfriargate	40,000	
		344,000
NEWCASTLE.		
Northern Counties Ice Making and Cold Stores Co., Ltd., Quayside	140,000	
Eastmans, Ltd.	35,000	
Compañia Sansinena de Carnes Congeladas, The Close	22,000	
Thomas Brown (Newcastle), Ltd., Newgate Street	14,000	
		211,000
BRISTOL.		
Bristol Corporation Cold Stores (three stores), Avonmouth Docks	80,000	
Eastmans, Ltd., Barton Hill	40,000	
Benjamin Perry & Sons, Ltd., Temple Street	15,000	
Benjamin Perry & Sons, Ltd., Redcliff Street. (This store is mainly used in connection with ice making and storage of provisions.)		
William Burgess, Welsh Back	15,000	
River Plate Fresh Meat Co., Ltd., Wapping	6,000	
Messrs. Spear Brothers and Clark, Broad Plain, and Messrs. Pullin, Thomas and Slade, Temple Street, have small cooling plants for dairy produce.		
		156,000
Total cold storage capacity in 56-lb. carcasses of mutton		8,156,482

APPENDIX VI

COLD STORAGE MAP OF LONDON

(Map References.)

	Name and Address of Stores.	Capacity Carcasses.
1.	London Central Markets Cold Storage Co., Ltd., W. Smithfield, E.C.	125,000
2.	London Central Markets Cold Storage Co., Ltd., Charterhouse Street, E.C.	78,000
3.	Port of London Authority, St. John Street, Smithfield, E.C.	95,000
4.	Palmer's Cold Air Stores, Charterhouse Street, E.C.	30,000
5.	Times Cold Storage Co., Ltd., Long Lane, E.C.	20,000
6.	Compañia Sansinena de Carnes Congeladas (private stores), Long Lane, E.C.	60,000
7.	James Nelson & Sons, Ltd. (private stores), W. Smithfield, E.C.	40,000
8.	Colonial Consignment and Distributing Co., Ltd., Nelson's Wharf, Upper Ground Street, S.E.	200,000
9.	Union Cold Storage Co., Ltd., Upper Ground Street, S.E.	70,000
10.	Blackfriars Cold Storage Co., Ltd., Purfleet Wharf, Upper Thames Street, E.C.	130,000
11.	Towers & Co., Ltd. (private stores), St. Paul's Pier Wharf, Upper Thames Street, E.C.	14,000
12.	Towers & Co., Ltd. (private stores), Brook's Wharf, Upper Thames Street, E.C.	15,000
13.	Thames Cold Storage Co., Ltd., Kennet Wharf, Upper Thames Street, E.C.	150,000
14.	Crown Wharf Cold Storage Co., Park Street, Borough, S.E.	60,000
15.	Union Cold Storage Co., Ltd., Cannon Street Station, E.C.	220,000
16.	Union Cold Storage Co., Ltd., Redbull Wharf, Upper Thames Street, E.C.	135,000
17.	River Plate Fresh Meat Co., Ltd. (private stores), Commercial Wharf, Upper Thames Street, E.C.	60,000
18.	The Cold Store, Ltd., Leadenhall Market, E.C.	20,000
19.	The Proprietors of Hay's Wharf, Ltd., New Hibernia Wharf, London Bridge, S.E.	135,000
20.	The Proprietors Hay's Wharf, Ltd., Cotton's Wharf, Tooley Street, S.E.	220,000
21.	Tooley Street Cold Storage Co., Ltd., Tooley Street, S.E.	40,000
22.	Anglo American Cold Storage Co., Ltd., Greenbank and Stoney Lane, Tooley Street, S.E.	150,000
23.	J. & J. Lonsdale, Ltd. (private stores), St. Katharine's Dock, E.	70,000
24.	Johnson, Cole, Brier & Cordrey, Ltd., Bermondsey, S.E.	18,000
25.	Port of London Authority, West Warehouse, Surrey Commercial Docks, S.E.	75,000 (25 and 26)
26.	Port of London Authority, East Warehouse, Surrey Commercial Docks, S.E.	
27.	Port of London Authority, West India Dock, E.	100,000
28.	Port of London Authority, South West India Dock, E. (Dismantled)	——
29.	Deptford Foreign Cattle Market (Chill Rooms)	——
30.	London Central Markets Cold Storage Co., Ltd., Poplar, E.	150,000
31.	Port of London Authority, Royal Victoria Dock, E.	552,000
	Total cold storage capacity in 56-lb. carcasses of mutton	3,032,000

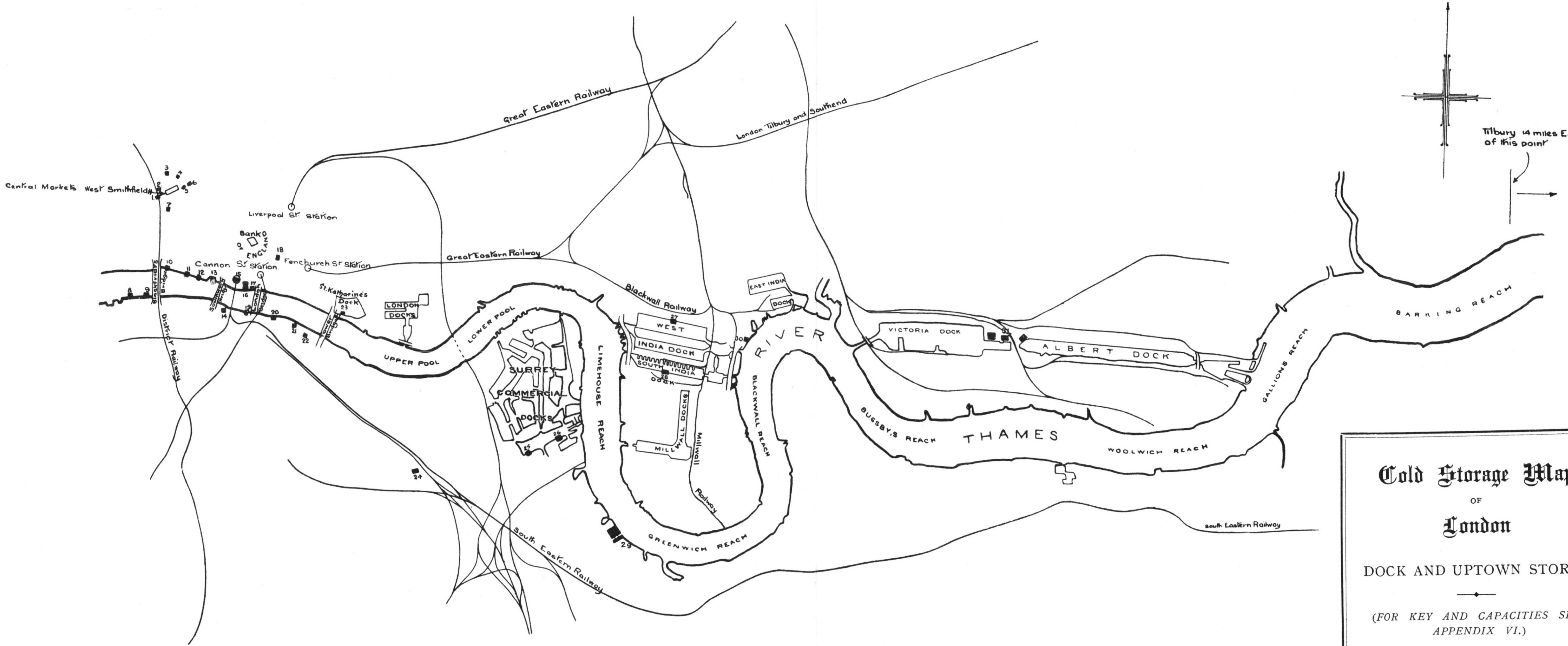

Cold Storage Map

OF

London

DOCK AND UPTOWN STORES

(*FOR KEY AND CAPACITIES SEE APPENDIX VI.*)

APPENDIX VIII

THE REFRIGERATED FLEET

THE following is a list of the shipping lines carrying chilled and frozen meat from Australia, New Zealand, and South America, to Europe and the East, with particulars of the number of vessels engaged in the trade, and their total refrigerated capacity in cubic feet. The figures have been supplied by the companies.

Name of Company	No. of ships.	Capacity in cubic feet.
Aberdeen Line (George Thompson and Co., Ltd.)	6	726,712
Argentine Cargo Line, Ltd.	3	1,107,000
Blue Star Line, Ltd.	3	612,234
British India Steam Navigation Co.	2	560,758
British and South American S.N. Co., Ltd. (R. P. Houston and Co., Managers)	7	438,119
Burns, Philp and Co., Ltd.	2	44,624
Canadian Australian Line	3	86,100
Chargeurs Réunis	4	480,300
China Navigation Co., Ltd.	2	66,734
Arch. Currie and Co.	1	9,000
Eastern and Australian Steamship Co., Ltd.	4	147,000
Federal Steam Navigation Co.	13	3,600,000
Furness, Withy and Co., Ltd.	3	222,201
German Australian Line	4	315,000
Glasgow Steam Shipping Co., Ltd.	2	178,925
Gulf Line, Ltd.	1	88,000
Houlder Brothers and Co., Ltd.	11	2,541,694
R. M. Hudson and Sons	1	97,680
Imataka Steamship Co., Ltd.	1	48,805
Indra Line, Ltd.	3	869,000
Italian Lloyd	1	5,755
Lamport and Holt, Ltd.	3	372,000
La Veloce Nav. Italiana	1	2,800
Manchester Liners, Ltd.	2	404,995
Nautilus Steam Shipping Co., Ltd.	2	160,540
Nelson's Steam Nav. Co., Ltd. } H. and W. Nelson, Ltd., Managers.	10	3,408,350
Nelson Line (Liverpool), Ltd. } H. and W. Nelson, Ltd., Managers.	10	2,346,609
New Zealand Shipping Co., Ltd.	18	4,826,000
Norddeutscher Lloyd	12	202,681
Ocean Steam Ship Co., Ltd. (Alfred Holt and Co.)	4	444,800
Orient Steam Navigation Co.	10	820,000
Peninsular and Oriental Steam Navigation Co. (including the Company's Branch Service)	24	2,419,240
Royal Mail Steam Packet Co.	16	2,400,976
Shaw, Savill and Albion Co., Ltd.	17	4,729,000
Shire Line	7	1,844,883
Société Générale Maritime	2	7,350
Star Line, Ltd.	7	1,524,942
Turner, Brightman and Co.	5	810,015
Tyser Line	8	1,931,000
Unione Austriaca de Navigazione	2	250,000
White Star Line	14	2,800,000
Totals	251	43,951,822

(Reckoning $2\frac{1}{2}$ to 3 cubic feet to one carcass of mutton, the carrying capacity of these 251 vessels may be given as 17,581,000 to 14,651,000 carcasses.)

APPENDIX IX

IMPORTS OF FROZEN MUTTON AND LAMB INTO THE UNITED KINGDOM FROM 1880 TO 1910

	Australia.		New Zealand.		South America.		Falkland Islands.	
	Cwts.	Carcasses.	Cwts.	Carcasses.	Cwts.	Carcasses.	Cwts.	Carcasses.
1880		400						
1881		17,275						
1882	31,469	57,256	5,814	8,839				
1883	31,747	63,733	71,942	120,893	3,571	17,165		
1884	63,511	111,745	240,613	412,349	40,230	108,823		
1885	53,574	95,051	284,013	492,269	112,223	190,571		
1886	37,711	66,960	346,565	655,888	190,409	434,699	17,411	30,000
1887	42,445	88,811	395,638	766,417	251,273	641,866	26,436	45,552
1888	44,489	112,214	498,628	939,231	345,392	924,003		
1889	42,100	86,547	568,499	1,068,286	395,303	1,009,936		
1890	109,826	207,984	787,322	1,533,393	435,084	1,196,531	5,901	10,168
1891	167,331	334,684	896,126	1,894,105	436,358	1,111,137	10,967	18,897
1892	211,726	504,738	765,668	1,539,605	471,128	1,247,861	10,341	17,818
1893	287,158	636,917	900,300	1,857,598	515,611	1,373,723	9,532	16,425
1894	468,430	939,360	971,072	1,958,259	585,729	1,414,815	6,776	11,675
1895	499,920	1,005,503	1,171,048	2,412,331	715,296	1,615,795	11,281	19,438
1896	774,020	1,653,243	1,079,109	2,211,895	801,733	1,792,280		
1897	706,752	1,407,417	1,302,333	2,703,845	908,623	2,121,481		
1898	619,489	1,248,653	1,314,619	2,784,101	1,106,201	2,397,337		
1899	525,733	1,204,601	1,475,719	3,250,100	1,141,208	2,414,718		
1900	446,049	943,924	1,487,197	3,157,060	1,114,795	2,332,837		
1901	518,639	1,226,558	1,488,217	3,234,119	1,271,654	2,634,105		
1902	279,134	724,297	1,635,037	3,668,061	1,352,501	2,827,496		
1903	181,269	478,037	2,035,434	4,583,760	1,485,770	3,118,437		
1904	163,014	450,323	1,626,893	3,885,231	1,422,397	2,937,701		
1905	505,413	1,368,438	1,524,981	3,704,566	1,462,537	3,203,210		
1906	616,870	1,732,328	1,748,188	4,148,288	1,433,097	2,919,276		
1907	858,226	2,337,931	2,005,078	4,797,367	1,402,302	3,015,229		
1908	636,034	1,831,246	1,737,606	4,234,158	1,556,746	3,586,014		
1909	943,753	2,678,838	1,987,023	5,035,103	1,646,264	3,353,211		
1910	1,525,399	4,219,012	2,104,173	5,407,470	1,631,185	3,353,726		
Totals ...	11,391,231	27,834,024	32,454,855	72,464,587	24,234,620	53,293,983	98,645	169,973

Grand Totals from all sources :
3,408,967 tons,
or
153,762,567 carcasses.

APPENDIX X

TABLE SHOWING THE YEARLY AVERAGES AT TOP PRICES AT SMITHFIELD MARKET, 1883—1910.

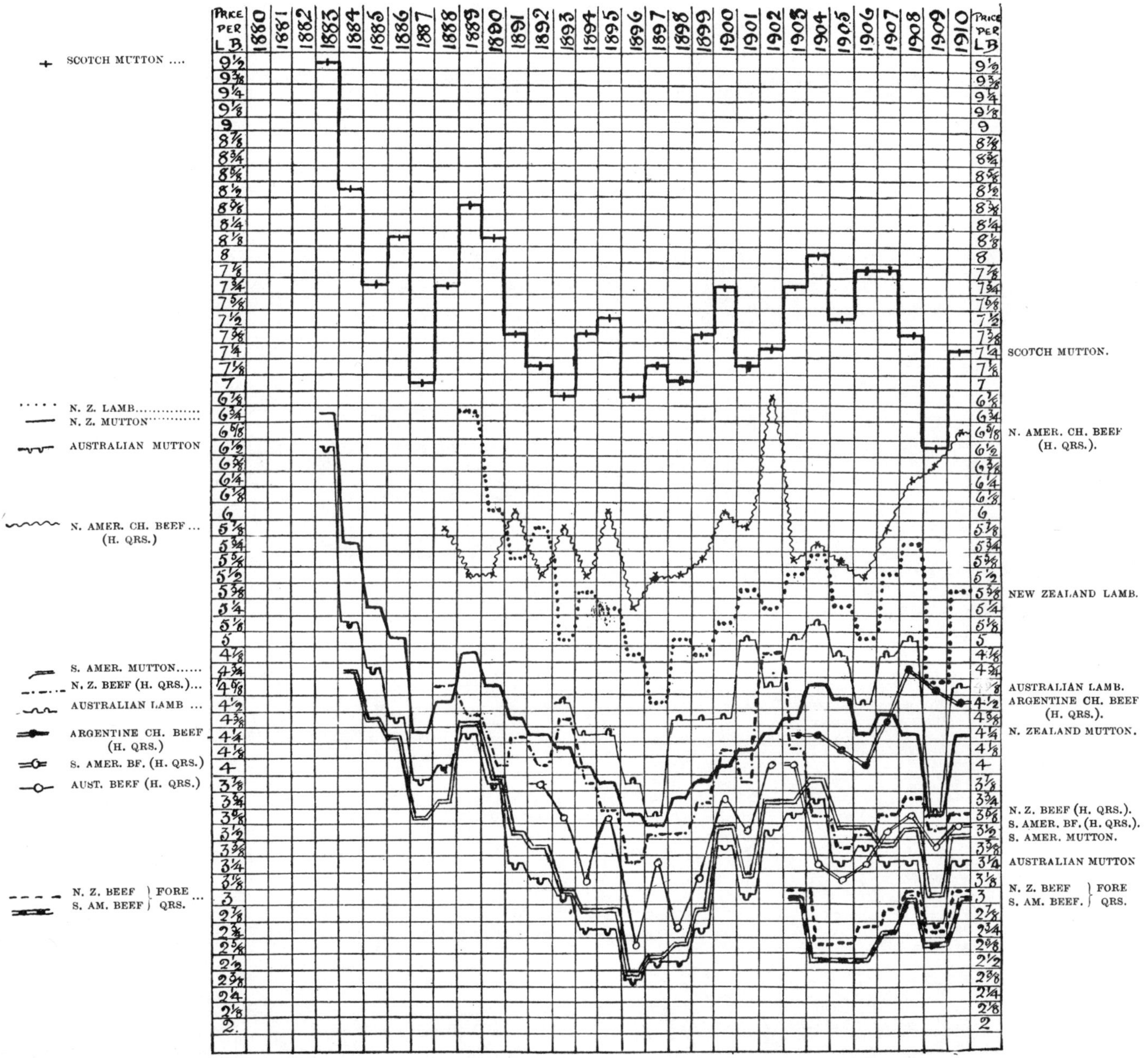

APPENDIX XI

IMPORTS OF FRESH BEEF (FROZEN AND CHILLED) INTO THE UNITED KINGDOM FROM 1874 TO 1910

	United States of America.	Australia.	New Zealand.	South America.	
	Cwts.	Cwts.	Cwts.	Cwts.	(Chilled beef). Cwts.
1874	1,095				
1875	3,098				
1876	144,336				
1877	443,042				
1878	483,012				
1879	559,730				
1880	724,272				
1881	747,785				
1882	446,778				
1883	730,966	1,522			
1884	811,644	2,754		500	
1885	854,845	7,778	8,844	930	
1886	762,226	10,195	10,962	6,736	
1887	643,320	18	8,398	270	
1888	784,429	864	40,490	3,678	
1889	1,275,948	24,865	75,131	8,665	
1890	1,693,148	21,426	88,495	8,933	
1891	1,747,578	41,615	107,433	14,485	
1892	1,951,887	56,568	62,065	8,309	
1893	1,489,949	210,983	14,686	35,383	
1894	1,775,538	301,896	2,617	5,279	
1895	1,649,473	485,851	16,317	23,384	
1896	2,074,644	494,975	28,803	50,095	
1897	2,242,063	560,829	73,426	84,673	
1898	2,301,956	531,651	92,756	108,288	
1899	2,756,458	609,216	134,427	150,368	
1900	2,867,238	413,991	310,667	412,262	2,114
1901	3,180,291	243,348	228,126	771,929	40,000
1902	2,290,465	65,860	237,257	923,748	152,000
1903	2,693,920	77,656	159,830	1,152,211	229,000
1904	2,395,836	76,345	175,012	1,675,271	347,000
1905	2,232,206	19,025	145,338	2,580,152	653,560
1906	2,426,644	34,457	236,587	2,795,913	750,000
1907	2,417,604	126,030	391,299	2,691,554	705,000
1908	1,432,142	112,583	347,872	3,706,245	1,267,400
1909	856,805	411,577	454,368	4,340,653	1,826,612
1910	477,147	880,695	532,830	5,058,453	2,710,747
Totals	52,369,518	5,824,573	3,984,036	26,618,367	8,683,433

Grand total from all sources, 4,873,996 tons.

All the fresh beef from the United States of America arrived in a chilled condition (29° F. to 30° F.), excepting some small quantities hard frozen by ice and salt mixtures at the start.

All the fresh beef from Australia and New Zealand arrived in a frozen condition, say 10° F. to 15° F. (frequently much lower), excepting five experimental shipments in 1909—1911 in the s.s. *Marathon* under the Linley process. Several early experiments in bringing beef from Australasia in a chilled state are alluded to in the earlier part of this book.

All the fresh beef from Argentina prior to 1900 was brought frozen. Column 6 in the table shows the progress of the South American chilled beef trade from 1900 onwards.

APPENDIX XII

MECHANICAL REFRIGERATION PATENTS

THE following is a chronological list of the British patents on mechanical refrigeration recorded between the years 1819 and 1876. In this period of fifty-eight years the most interesting efforts in the realm of refrigerating invention were recorded; the field of research was virgin and boundless.

The particulars below contain only the briefest mention of the patents to which they refer. A glance will show the periods most fruitful in invention, and will also reveal some names famous in the annals of the industry. The various directions in which research was made by some of the pioneers is interesting, *e.g.*, the records of 1876 show F. P. E. Carré developing the ammonia absorption machine, while his brother, Edmond Carré, was patenting vacuum freezing apparatus. No mention is made in the list of the many schemes devised for refrigeration by ice and salt or other freezing mixtures.

1819.

Jan. 15 (No. 4,331). SALMON, R., and WARRELL, W.—Production of cold and condensation or congelation by causing a blast of air to act on the surface of water so as to vaporize it, and then re-collecting the water and repeating the process.

1824.

Jan. 1 and Aug. 28 (Nos. 4,884 and 5,001). VALLANCE, J.—Vacuum process of freezing water, employing sulphuric acid.

1834.

Aug. 9 (No. 6,655). WRIGHT, L. W.—Ice manufacture by air compression process.

Aug. 14 (No. 6,662). PERKINS, JACOB.—Production of cold by expansion of volatile fluids, such as ether.

1845.

May 3 (No. 10,652).—Ice manufacture.

1850.

July 3 (No. 13,167). Kingsford, J.—Refrigeration by different processes, vacuum and also compression, using air or any gas.

Aug. 22 (No. 13,234). Newton, W. E.—Refrigeration by air compression process.

1852.

Oct. 6 (No. 270). Grimes, J.—Refrigeration by air compression process.

Dec. 24 (No. 1,166). Nesmond, P. C.—Refrigeration by air compression process.

1853.

Jan. 15 (No. 106). Vion, H. C.—Refrigeration by evaporation of liquids or liquefied gas.

Jan. 20 (No. 147). Williams, W.—Refrigeration by air compression process.

1855.

Nov. 14 (No. 2,559). Tolhausen, A.—Ice manufacture by evaporation of water by air currents.

1856.

March 28 (No. 747). Harrison, James.—Production of cold by evaporation of volatile liquids in vacuo.

June 28 (No. 1,522). Sloper, B. G.—Refrigeration by expansion of air.

Oct. 13 (No. 2,397). Piatti, G. B.—Refrigeration by expansion of compressed air, ether, or carbonic acid gas.

1857.

July 29 (No. 2,064). Siemens, C. W.—Refrigeration by expansion of air or elastic fluids.

Sept. 10 (No. 2,362). Harrison, James.—Ice making by evaporation of volatile liquids; an apparatus requiring "a motive power of ten horses," and containing one hundred ice moulds.

1860.

Oct. 15 (No. 2,503). Davies, G. (communication from F. P. E. Carré).—Refrigeration by absorption of liquefied gases or condensed vapours.

1861.

July 4 (No. 1,705). Mennons, M. A. F. (communication from E. Blée).—Refrigeration by expansion and compression of volatile liquid.

Nov. 28 (No. 3,005). De Labaume, J. d'A. (communication from E. Blée).—Refrigeration by ether, evaporation, and compression.

1862.

March 21 (No. 782). Siebe, D. E.—Refrigeration by ether, evaporation, and compression.

April 25 (No. 1,218). Kirk, A. C.—Air compression refrigerating machine.

June 17 (No. 1,786). CRESTADORO, A.—Refrigeration by cooling effect of air current induced by furnace draught.

Oct. 15 (No. 2,788). BROOMAN, R. A. (communication from D. J. Kennelly).—Refrigeration by evaporation of volatile liquid.

1864.

Feb. 15 (No. 387). FONTAINEMOREAU, P. A. (communication from C. Tellier).—Manufacture of methylic ether and its application to production of artificial ice.

Oct. 28 (No. 2,666). LAIDLAW, D., and ROBERTSON, J.—Refrigeration by air compression process.

Dec. 8 (No. 3,062). BROOMAN, R. A. (communication from Marcar Beylih'g).—Refrigeration by vaporization of volatile fluid.

1865.

Sept. 6 (No. 2,292). PARKES, A. W.—Arrangement of ice houses, skating rinks, etc., where ice is produced by evaporation of ether, carbonic acid, etc.

Sept. 28 (No. 2,483). REECE, R.—Refrigeration by the evaporation of liquid sulphurous acid (absorption system).

1866.

Jan. 27 (No. 267). MENNONS, M. A. F. (communication from N. de Telescheff).—Air compression refrigerating machine.

Feb. 21 (No. 540). RICHARDSON, B. W.—Portable apparatus for freezing small quantities of liquid by direct action of pulverized spray of volatile liquid.

1867.

March 30 (No. 952). NEWTON, W. E. (communication from T. S. C. Lowe).—Refrigeration or ice making by expansion of carbonic acid gas compressed into liquid state. Latent heat evolved during compression removed by water cooling.

May 31 (No. 1,621). REECE, R.—Refrigeration by the evaporation of liquid sulphurous acid.

Aug. 10 (No. 2,303). CLARK, A. M. (communication from J. B. Toselli).—Refrigeration or ice making by evaporation of liquid or gas (absorption system).

Sept. 9 (No. 2,544). WELCH, E. J. C.—Ice making by evaporation of ether or some other hydro-carbon, and its subsequent condensation.

Nov. 23 (No. 3,323). MORT, WILLIAM (communication from Thomas Sutcliffe Mort).—Refrigeration by evaporation of ammonia (absorption system). The meat receivers are made with " a double casing to form a compartment having its walls perfectly tight to contain the liquefied gas supplied from the liquefied gas receiver " ; each is surrounded with non-conducting substance enclosed in a painted or varnished wooden covering.

1868.

Jan. 4 (No. 32). SPENCE, P., and SMITH, W. A.—Meat storage for transport, cooled " by compression or by the ammonia or other processes."

March 25 (No. 1,006). LITTLE, R. (communication from D. Little).—Meat refrigeration on board ship by dry air compression process.

July 29 (No. 2,384). JEFFREYS, J.—Refrigeration by ether expansion.

Sept. 3 (No. 2,719). KIRK, A. C. (communication from J. Kyle).—Ice block formation by hydraulic or other pressure in moulds.

Oct. 26 (No. 3,278). MORT, W. (communication from T. S. Mort and E. D. Nicolle).—Refrigeration by expansion of air or other permanent gases, also the manufacture of ice.

1869.

Jan. 20 (No. 178). SIDDELEY, J., and MACKAY, F. N.—Refrigeration and ice making by evaporation of volatile liquid in vacuo.

Feb. 6 (No. 368). DUFRENÉ, H. A. (communication from C. Tellier).—Ice manufacture by methylic ether or ammonia compression.

March 5 (No. 669). WINDHAUSEN, F.—Refrigeration and ice making by air compression.

March 27 (No. 935). HUCH, E. H.—Meat refrigeration on board ship by adaptation of Windhausen's air compression machine.

July 21 (No. 2,211). KIRK, A. C.—Refrigeration by compression and expansion of air with two cylinders.

July 24 (No. 2,249). PICTET, R.—Special construction of pumps for refrigerating machine.

1870.

Jan. 29 (No. 267). WEST, H. J.—Refrigeration by evaporation of ether or other volatile liquid.

May 30 (No. 1,579). MIGNOT, L.—Air compression refrigerating machine.

July 1 (No. 1,866). CLARK, A. M.—Chloride of ethyl refrigerating machine.

Aug. 29 (No. 2,363). HUGHES, E. T. (communication from C. Parker).—Vacuum freezing apparatus.

Nov. 2 (No. 2,891). REECE, R.—Ammonia absorption refrigerating machine.

Dec. 6 (No. 3,210). MORT, W. (communication from E. D. Nicolle and T. S. Mort).—Ammonia absorption refrigerating machine.

Dec. 16 (No. 3,296). GAMGEE, J.—Vacuum refrigerating machine.

1871.

Feb. 25 (No. 511). PAGET, F. A., and ASHER, J. W.—Air compression refrigerating machine.

April 14 (No. 992). MURDOCH, H. H. (communication from E. Roettger).—Air compression refrigerating machine.

April 20 (No. 1,042). NORMANDY, A. L.—Air compression refrigerating machine.

May 3 (No. 1,189).—Refrigerating machine employing certain specified volatile agents.

June 5 (No. 1,490). MARCHANT, R. M.—Multiple stage air compression refrigerating machine.

July 8 (No. 1,794). KAEUFFER, P. E.—Air compression refrigerating machine.

Aug. 18 (No. 2,173). NORMAN, J.—Air compression refrigerating machine.

Aug. 25 (No. 2,228). PAGET, F. A., and ASHER, J. W.—Air compression refrigerating machine.

Oct. 11 (No. 2,701). MURDOCH, H. E.—Air compression refrigerating machine.

Dec. 22 (No. 3,474). HUGHES, E. T. (communication from A. C. Twining).—Compression refrigerating machine.

1872.

Jan. 24 (No. 228). Dufrené H. A. (communication from C. Tellier).—Refrigerating apparatus employing volatile agents.

Jan. 29 (No. 272). Walker, W. G.—Compression refrigerating machine.

June 14 (No. 1,791). Johnson, J. H. (communication from A. Lemaire and J. Sonnois).—Compression refrigerating machine employing volatile agents.

June 19 (No. 1,853). Abate, E.—Method of preserving meat aboard ship by refrigerating process.

June 26 (No. 1,935). Purkis, R. A.—Ether refrigerating machine.

July 24 (No. 2,210). Lake, W. R. (communication from S. B. Martin and J. M. Beath).—Compression refrigerating apparatus.

Sept. 18 (No. 2,761). Hunt, B. (communication from R. Riley).—Water agitation in ice-making apparatus.

Nov. 16 (No. 3,422). Johnson, J. H. (communication from J. B. J. Mignon and S. H. Rouart).—Ammonia absorption refrigerating machine.

1873.

Jan. 24 (No. 290). Lake, W. R. (communication from S. B. Martin and J. M. Beath).—Plate ice-making apparatus.

Feb. (No. 627). Smith, F. J. (communication from P. Giffard, A. Sublet, and J. A. Jeune).—Air compression refrigerating machine.

Feb. 26 (No. 709).—Hydrogen compression refrigerating machine.

March 5 (No. 793). Morgan-Brown, W. (communication from F. Littmann).—Ammonia absorption refrigerating apparatus.

March 28 (No. 1,158). West, H. J.—Ether refrigerating machine.

April 14 and 21 (Nos. 1,346 and 1,443). Barlow, H. B. (communication from N. J. Galland).—Vacuum refrigerating apparatus.

July 15 (No. 2,434). Warren, F. P.—Railway and other vehicle refrigeration by ether and brine.

July 23 (No. 2,523). Fleury, F. G.—Domestic ice manufacture by co-operation of ammonia.

Sept. 26 (No. 3,142). Werth, F. (communication from F. Windhausen).—Air compression refrigerating machine.

Nov. 19 (No. 3,760). Harrison, J.—Construction of refrigerating chambers.

Dec. 18 (No. 4,161). Weatherby, C. P. N.—Combined air and volatile gas compression refrigerating machine.

1874.

Jan. 17 (No. 383). Harrison, J.—Improvements on Patents No. 747, 1856, and No. 2,362, 1857.

Feb. 4 (No. 451). West, H. J.—Ice-making apparatus.

March 14 (No. 936). Martin, D. B.—Ice-making and refrigerating apparatus.

April 30 (No. 1,523). Joyce, T. L. (communication from J. Selten).—Carbonic acid refrigerating machine.

May 4 (No. 1,573). Mort, W. (communication from E. D. Nicolle and T. S. Mort).—Refrigeration of produce in air-tight vessels.

May 29 (No. 1,873). Nehrlich, H.—Air compression refrigerating machine.

June 4 (No. 1,946). Marchant, R. M.—Air compression refrigerating machine.

June 17 (No. 2,115). WEATHERBY, C. P. N.—Refrigerating machine employing compression of air and also a volatile agent.

June 22 (No. 2,160). HUNT, B. (communication from A. F. C. Reynoso).—Refrigerated preserving processes, including freezing articles in water.

July 4 (No. 2,336). WEST, H. J.—Ice manufacture.

July 14 (No. 2,460). BENSON, M. (communication from J. M. G. Beath).—Ice manufacture on the ammonia compression system.

July 27 (No. 2,621). GAMGEE, J., and PURKIS, R. A.—Ether refrigerating machine.

Aug. 10 (No. 2,763). MORT, W. (communication from E. D. Nicolle and T. S. Mort).—Ice manufacture.

Sept. 22 (No. 3,241). KYLE, J.—Ice manufacture.

Oct. 7 (No. 3,425). SIDDELEY, J., and MACKAY, F. N.—Ether ice-making machine.

Oct. 17 (No. 3,571). NEILD, H. W.—Ammonia absorption refrigerating machine.

Oct. 22 (No. 3,638). LOW, R.—Apparatus for condensing steam and refrigerating.

Dec. 3 (No. 4,152). BENSON, M.—Ammonia absorption refrigerating machine.

Dec. 16 (No. 4,329). KIRK, A. C., and BEILBY, G. T.—Ammonia absorption refrigerating machine.

1875.

Jan. 29 (No. 351). WEATHERBY, C. P. N.—Air compression refrigerating machine.

March 24 (No. 1,073). MACKAY, F. N., and RAE, D.—Artificial ice rink.

May 10 (No. 1,726). SIDDELEY, J., and MACKAY, F. N.—Ice manufacture.

June 4 (No. 2,064). CLARK, A. M. (communication from P. Giffard).—Double-acting air compression refrigerating machine.

June 18 (No. 2,239). WEST, H. J., and DE JACOBI DU VALLON, G. C. J.—Artificial ice rinks.

July 3 (No. 2,413). COUGHLIN, D.—Ammonia absorption refrigerating machine.

Aug. 3 (No. 2,727). PICTET, R. P.—Sulphurous acid refrigerating machine.

Oct. 23 (No. 3,682). GAMGEE, J.—Clear ice manufacture.

Nov. 10 (No. 3,907). STANLEY, H. F.—Ammonia absorption refrigerating machine.

Nov. 22 (No. 4,054). WHEELER, E. G. (communication from D. Coughlin).—Ammonia absorption ice-making machine.

Dec. 20 (No. 4,412). GAMGEE, J.—Artificial ice rink.

Dec. 29 (No. 4,519). GAMGEE, J.—Ether ice-making machine.

1876.

Feb. 2 (No. 411). GAMGEE, J.—Artificial ice rink.

Feb. 3 (No. 432). STOKES, G. E.—Artificial ice rink.

Feb. 16 (No. 625). ROSE, T.—Artificial ice rink.

Feb. 24 (No. 789). MACKAY, F. N.—Artificial ice rink.

Feb. 26 (No. 813). NISHIGAWA, T. M.—Ammonia absorption refrigerating machine.

March 4 (No. 937). CLAMOND, C.—Artificial ice rink.

March 21 (No. 1,208). MORT, W. (communication from E. D. Nicolle and T. S. Mort).—Improvements in refrigerating, especially applicable for meat cargoes.

March 29 (No. 1,337). MACKAY, F. N.—Artificial ice rink.

April 5 (No. 1,458). LINDE, CARL VON.—Ice-making machine.

May 26 (No. 2,219). MARCHANT, R. M.—Refrigerating compressors.

June 23 (No. 2,601). WIRTH, F. (communication from F. Wicker).—Ammonia absorption ice-making machine.

Aug. 31 (No. 3,427). NISHIGAWA, T. M., and HILL, F. B.—Ammonia absorption refrigerating machine.

Sept. 13 (No. 3,593). HUNT, B. (communication from A. F. C. Reynoso).—Refrigerating of alimentary matters.

Sept. 13 (No. 3,594). CARRÉ, F. P. EDOUARD, and JULLIEN, E.—Ammonia absorption refrigerating machine.

Oct. 4 (No. 3,837). ADAMS, A. W.—Ice manufacture.

Oct. 26 (No. 4,164). CLARK, A. M. (communication from Edmond Carré).—Vacuum refrigerating machine.

Oct. 28 (No. 4,176). GAMGEE, J.—Artificial ice rink.

Nov. 17 (No. 4,446). KNOTT, K.—Refrigeration of animal and vegetable substances.

Nov. 21 (No. 4,504). MARCHANT, R. M.—Refrigerating compressors.

Dec. 1 (No. 4,659). TONGUE, J. G. (communication from C. L. Riker).—Air compression refrigerating machine.

Dec. 8 (No. 4,762). WINDHAUSEN, F.—Air compression refrigerating machine.

Dec. 12 (No. 4,803). SKENE, R.—Refrigerating and ice-making apparatus.

Dec. 15 (No. 4,859). DU VALLON, G., and CSETE, J.—Ice manufacture.

INDEX